Advances in
Oceanography

Advances in Oceanography

Edited by
Henry Charnock
The University
Southampton, England

and
Sir George Deacon
Institute of Oceanographic Sciences
Surrey, England

PLENUM PRESS · NEW YORK AND LONDON

Library of Congress Cataloging in Publication Data

Main entry under title:

Advances in oceanography.

   Papers presented in general symposia at the Joint Oceanographic Assembly held in Edinburg, Scotland, Sept. 13–24, 1976.
   1. Oceanography—Congresses. 2. Marine biology—Congresses. I. Charnock, H. II. Deacon, George Edward Raven, Sir, 1906-   III. Joint Oceanographic Assembly.
GC2.A38   551.4'6   78-17970
ISBN 0-306-40019-7

A collection of papers presented in general symposia at the Joint Oceanographic Assembly held in Edinburgh, Scotland, September 13–24, 1976

© 1978 Plenum Press, New York
A Division of Plenum Publishing Corporation
227 West 17th Street, New York, N.Y. 10011

All rights reserved

No part of this book may be reproduced, stored in a retrieval system, or transmitted, in any form or by any means, electronic, mechanical, photocopying, microfilming, recording, or otherwise, without written permission from the Publisher

Printed in the United States of America

PREFACE

The papers at the General Symposia of the Joint Oceanographic Assembly were presented by authors who had been invited to deal with topics of broad interdisciplinary interest. Together they gave a valuable account of the present state of marine research, its problems and potential.

There seems merit in publishing them in one volume and we are grateful to the authors who kindly agreed to prepare their papers for publication. This has taken some time (and not all authors found it possible) but we hope the resulting volume is still indicative of trends in modern oceanography. Particularly noticeable is the way in which the applied aspects are beginning to play a more important part in spite of worries about the effects of the Law of the Sea on our freedom to make observations.

We are especially grateful to the four authors who agreed to give lectures summarising the work in their own field which was reported at the Assembly. We also wish to thank the other authors, as well as all concerned with the Assembly and with the production of this volume.

H. Charnock
G.E.R. Deacon

CONTENTS

A Schematic Model of the Evolution of the South Atlantic . . .   1
    X. Le Pichon, M. Melguen, and J. C. Sibuet

Climate and the Composition of Surface Ocean Waters 49
    Edward D. Goldberg

Ocean Circulation and Marine Life 65
    Joseph L. Reid, Edward Brinton, Abraham Fleminger,
    Elizabeth L. Venrick, and John A. McGowan

The Baltic - A Systems Analysis of a Semi-enclosed Sea 131
    B.-O. Jansson

Tactics of Fish Movement in Relation to Migration Strategy and
      Water Circulation 185
    F. R. Harden Jones, M. Greer Walker, and G. P. Arnold

Variable Ocean Structure 209
    Walter H. Munk

Ocean Variability - The Influence of Atmospheric Processes . . 213
    R. W. Stewart

Variability of Ecosystems 221
    N. M. Voronina

Man's Use of Coastal Lagoon Resources 245
    Robert R. Lankford

Non-renewable Resources of the Sea 255
    F. Neuweiler

Salinity Power, Potential and Processes, Especially Membrane
      Processes 267
    Sidney Loeb, M. Rudolf Bloch, and John D. Isaacs

Physical Oceanography 289
    J. D. Woods

Biological Oceanography 297
    Martin V. Angel

The Marine Geosciences 307
    John G. Sclater

Applied Aspects of Oceanography 339
    T. D. Patten

Index 349

A SCHEMATIC MODEL OF THE EVOLUTION OF THE SOUTH ATLANTIC[*]

X. LE PICHON,[**] M. MELGUEN,[***] J.C. SIBUET[***]

[*]CONTRIBUTION NO. 518 DU DÉPARTEMENT SCIENTIFIQUE DU
CENTRE OCÉANOLOGIQUE DE BRETAGNE
[**]CENTRE NATIONAL POUR L'EXPLOITATION DES OCÉANS, 39,
AVENUE D'IÉNA, 75016, PARIS, FRANCE.
[***]CENTRE OCÉANOLOGIQUE DE BRETAGNE, B.P.337, 29273,
BREST CÉDEX FRANCE.

ABSTRACT

A schematic model of the evolution of the South Atlantic ocean is proposed to demonstrate the possibilities offered by the Deep Sea Drilling Project within the framework of Plate Tectonics to reconstruct a logical evolution of the history of the ocean basins and their paleoenvironment. The emphasis is put on the methodology available to reconstruct the ocean crust morphology and its sedimentary cover at all stages throughout the opening of the ocean. Paleobathymetric, deep paleo-water-circulation and paleo-sedimentary - facies maps at five different stages of the opening are presented and discussed. The last type of maps are in great part based on a new paleo - Carbonate Compensation Depth Curve for the South Atlantic which is tentatively proposed here.

INTRODUCTION

It has long been known that the sedimentary record should be much more complete in the ocean basins than on the continents, because erosion is a more widespread phenomenon on land than it is on the deep sea floor. As it is the sedimentary record which contains most of the information available to reconstruct the paleoenvironment at the time the sediment was deposited, any paleoenvironment reconstruction relies heavily on its study. Accordingly, the deep holes made by the Drilling Vessel <u>Glomar Challenger</u> for the Deep Sea Drilling Project in the world ocean have provided a completely new basis to approach a paleoenvironmental study of our planet. This is specially so because Plate

Tectonics has given a coherent framework within which the various data collected from the ocean floor can be integrated to reconstruct a logical evolution of the history of the ocean basins and their paleoenvironment.

Paleoenvironmental oceanography is a new science which is progressing very rapidly. Any synthesis in this domain would be difficult to do and beyond the scope of this paper. We have attempted here to give a specific example of what can be done to reconstruct the evolution of a given ocean, limited ourselves to the ocean crust morphology and its sedimentary cover. We have chosen the South Atlantic Ocean because the age of its crust is known everywhere, through magnetic anomaly identification, and because the age distribution shows that it has been steadily widening since its early opening 125 to 130 M.y (millions years) ago. There are and there have been no deep sea trenches along its borders. As a consequence, there is a marked contrast in the sedimentary provinces, the deep basins being easily accessible to the products of erosion of the continents whereas the axial mid-Atlantic ridge only receives pelagic sedimentation.

Although it is theoretically possible to reconstruct, at least in a schematic way, the complete evolution of an ocean, the data we have are still quite limited and, for this reason, many of the conclusions are not yet firm and may even be erroneous. The emphasis is on the demonstration of a new methodology available to us to make such a reconstruction. 24 successful holes (figure 1) have been realized by the Glomar Challenger in the South Atlantic during different legs (3, 36, 39 40). Two of the Initial Reports are published (3, Maxwell et al., 1970; 36, Barker et al., 1974). Legs 39 and 40 Initial Reports (Perch-Nielsen et al., 1975; Bolli et al., 1975) are still in press but should be published soon. Only 4 of these drillings have been made in deep basins (Argentine, Brazil and Cape basins). It is consequently often risky to interpolate between holes over such a wide distance, although some reasonable guesses can be made.

In the first part, we will discuss some of the main features of the present South Atlantic environment, as it gives us one of the keys to interpret the sedimentary record. In the second part, we will present the methodology used to reconstruct the evolution of the ocean. In the last part, we will briefly comment paleogeographic and paleosedimentary maps of the South Atlantic ocean, from its birth to its present stage.

I. PRESENT ENVIRONMENT OF THE SOUTH ATLANTIC OCEAN

Physiography

In a geological sense, the South Atlantic Ocean can be

considered to have been created by the drifting apart of the South
American and the African continents. It should consequently
extend northwards from the Falkland Plateau in the south. Thus
its northern limit corresponds to the very large equatorial
fracture zones, of which the Romanche fracture zone is the largest.
Its southern limit is the Falkland - Agulhas fracture zone system
(fig. 1). These fracture zones were flow lines of the relative
continent motion during the opening (e.g. LE PICHON and HAYES,
1971a).

It is now known that the depth of the ocean is governed by
the slow cooling of the plates as they move away from the ridge
crest where they were produced by sea-floor spreading (LANGSETH
et al., 1966; McKENZIE and SCLATER, 1969). At the accreting
plate boundary, whose surface expression is the ridge crest, the
plate, which has just been formed by rapid intrusion of hot
material from below, is extremely hot throughout its thickness.
From then on, its evolution will be essentially due to loss of
heat through the sea-floor which results in thermal contraction.
As isostatic equilibrium prevails, the sea-floor will deepen by
roughly 30 % more than the amount corresponding to the contraction
of the plate, to compensate for the increasing water load. Thus,
the depth of the ocean only depends on its age, as is convincingly
demonstrated by the empirical curve first compiled by SCLATER et
al. (1971)(see fig. 6). The present 0 M.y. isochron follows the
crest of the mid-Atlantic ridge and roughly corresponds to the
2600 m isobath. On each side, the sea-floor progressively deepens
to reach depths larger than 5000 m under the basins.

However, two main volcanic rises, which run approximately
east-west near 30°S, the Walvis Ridge and the Rio Grande Rise,
are superposed on this general regular bathymetric trend (fig. 1).
Together with the mid-Atlantic ridge, they divide the South
Atlantic ocean into four main basins, the Brazil and Argentine
basins to the west and the Angola and Cape basins to the east.
In addition, as noted previously, the major fracture zones, which
also run at cross-trend with the mid-Atlantic ridge, are large
topographic features which provide east-west gaps through the
ridge but prevent deep north-south flow.

Age of ocean floor

LADD (1976) has presented a synthesis of the VINE and
MATTHEWS (1963) magnetic anomaly identifications in the South
Atlantic ocean, from the Recent Gauss anomaly at the crest of the
ridge to anomaly M 13, at the foot of the continental margin, in
the Cape basin. Each of these anomalies corresponds to an
isochron whose age is approximately known through correlation with
D.S.D.P. results. We have used the time scales proposed by
THIERSTEIN (1977) for the Cretaceous and TARLING and MITCHELL

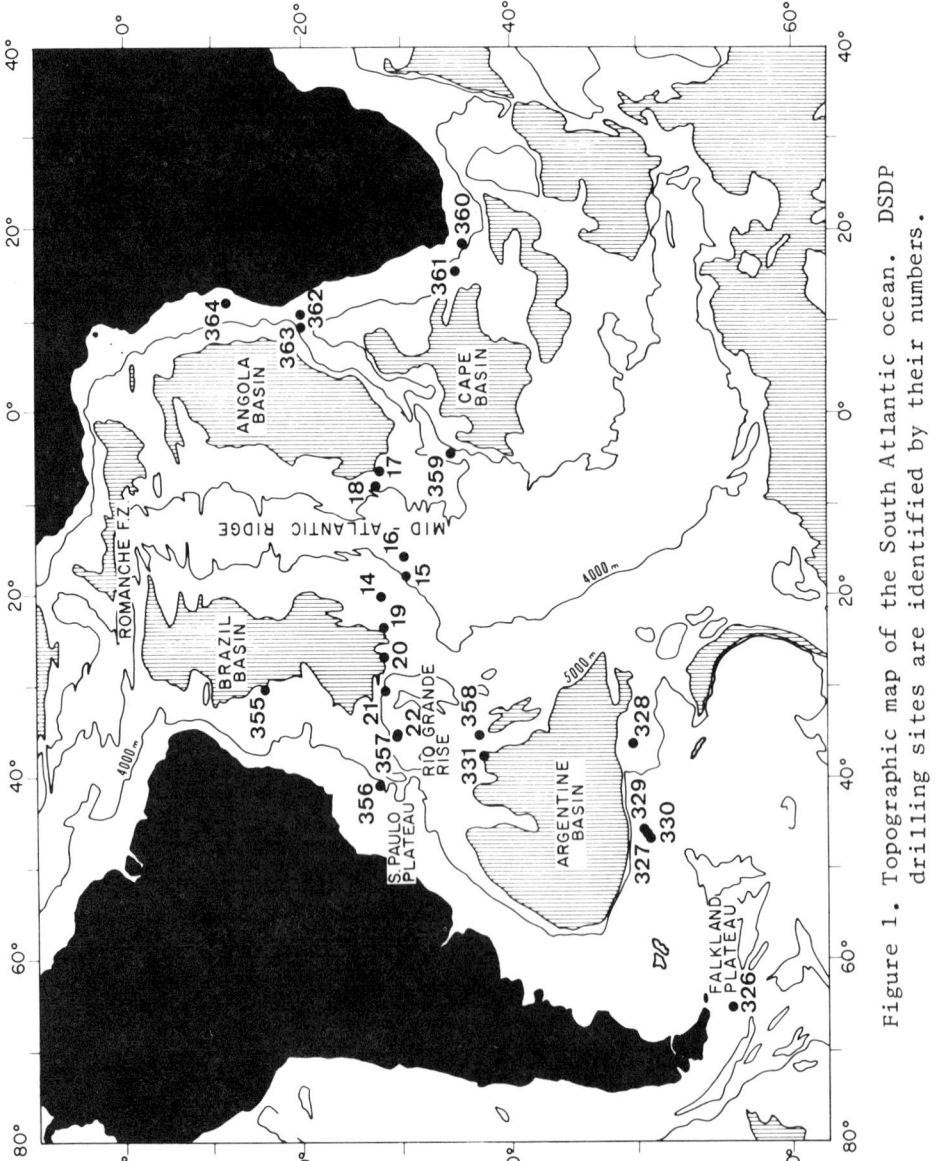

Figure 1. Topographic map of the South Atlantic ocean. DSDP drilling sites are identified by their numbers.

(1976) for the Cenozoic. The earliest anomaly identified corresponds to Valanginian, roughly 125 M.y. ago and it can be inferred that creation of oceanic crust started immediately before, sometime between 130 and 125 M.y. It is of course entirely possible that, prior to that time, a continental rifting stage was present for an extended period of time.

Sedimentary cover

It was previously mentioned that the depth of the sea-floor is directly related to its age. However this observation applies to the igneous oceanic crust as it has been produced at the ridge crest. As the plate moves away from the ridge crest, it is progressively covered by a layer of sediments which tends to increase with age. The thickness of the layer may reach several kilometers at the foot of the continental margin. It loads the crust which readjusts isostatically by sinking. If there were no isostatic adjustment, the water depth after sediment loading would be smaller than the ocean crust depth without sediment by a distance equal to the layer 1 thickness. As there is isostatic adjustment, the actual depth is decreased by 1/2 (using densities of 2.2 g/cm^3 for the sediments and 3.4 g/cm^3 for the mantle).

Figure 2 shows the isopach map of the sediment thickness modified after EWING et al. (1973) using some additional data from EMERY et al. (1975) off South Africa. It is obvious on this map that the thickness of sediments is negligible near the ridge crest, in the zone where the oceanic crust is very young. It is specially small between 30° and 10°S. It exceeds 1000 meters over most of the Argentine basin and part of the Brazil basin and over the continental margin area. Thus, except in the latter areas, the sea-floor depth is not significantly modified by the presence of sediments.

Water circulation

Since WUST (1939), it is known that a significant factor in the deep sea environment is the deep water circulation. As the deep basins of the South Atlantic ocean are open both to the north and south, they are swept by a vigorous deep water circulation which originates in high northern and southern latitudes. Figure 3 schematically shows the deep water circulation pattern. The bottom water is dominated by the spreading of the Antarctic Bottom Water (AABW) which is controlled by powerful western boundary currents. The resulting bottom water temperature is very low, less than 1.6°C in the three basins in which the Antarctic Bottom Current (AABC) enters, that is the Argentine and Brazil basins to the west and the Cape basin to the east. In contrast, the temperature is somewhat higher in the Angola basin (2.4°C).

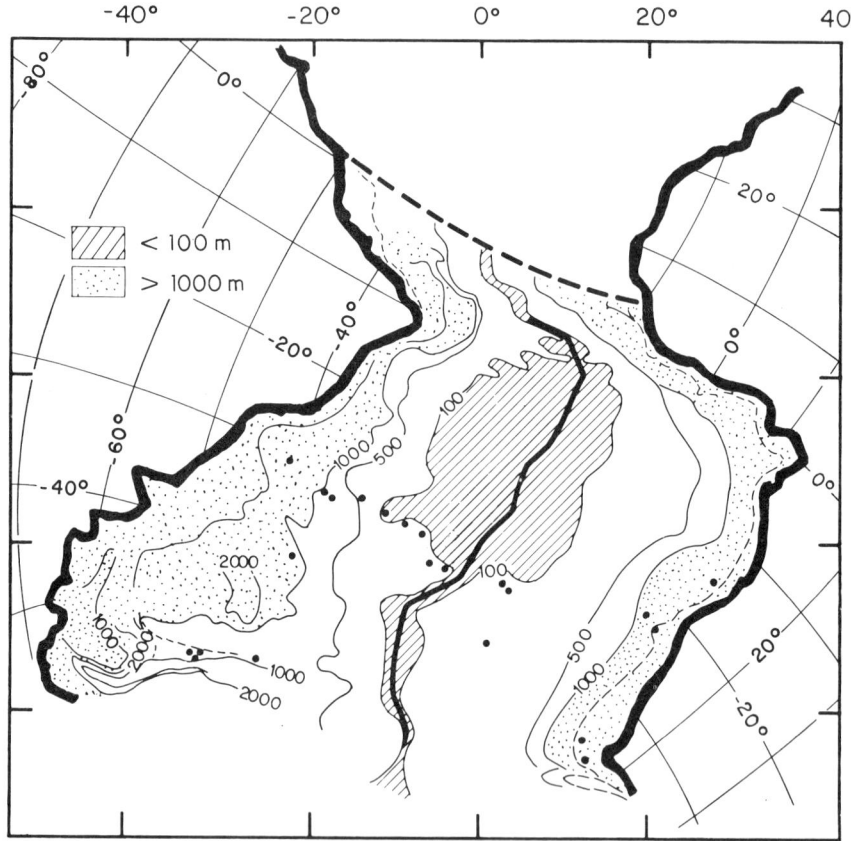

Figure 2. Oblique Mercator projection map of the South Atlantic showing the distribution of the thickness of the unconsolidated or semi-consolidated sediments after EWING et al. (1973). Isopachs are in meters. The pole of projection is at 30°N, 60°E. The equator of projection is vertical and is identified on the top of the figure to help in estimating the distortion. The same projection is used throughout this paper.

As the upper limit of the AABW is situated between 3500 and 4200 meters (LE PICHON et al., 1971 b; JOHNSON et al., 1975), the sea floor morphology controls its spreading. In particular, the passage of the AABW from one basin to the other often occurs through narrow passages or channels. It enters the Argentine basin from the south through a gap in the Falkland F.Z. (LE PICHON et al., 1971 a), follows the Argentine continental rise, enters the Brazil basin through a gap in the Rio Grande rise, called the Vema Channel (LE PICHON et al., 1971 b). It then continues to the north although a small branch enters the Angola basin through the Romanche F.Z. However, the sill depth in the Romanche F.Z. is only 3750 m (METCALF et al., 1964), so that the core of the AABW is prevented from entering the Angola basin.

Another branch of the AABC crosses the mid-Atlantic ridge and enters the Cape and Agulhas basins. Most of the water is prevented from entering the Angola basin by the Walvis ridge. There is only limited access near 6-7°W and 35-36°S (SHANNON and RIJSWIJCK, 1969; CONNARY, 1972; CONNARY and EWING, 1974).

Ewing et al.(1973) have shown that the AABC has controlled the distribution of sediments in the whole Argentine basin since the deposition of a major seismic reflector, called reflector A, which is now known to be quite probably Upper Oligocene ($\pm$ 30 M.y.) in age (Perch-Nielsen et al., 1975). This control includes non deposition or even erosion where the current flows and transport of large quantities of fine clay in suspension in what has been called the nepheloid layer. In addition, as we will see later, the deep cold water is highly corrosive and dissolves the carbonates. This has been clearly demonstrated by MELGUEN and THIEDE (1974, 1975) and CHAMLEY (1975) on the flanks of the Vema Channel and the adjacent Rio Grande Rise. Yet, it is quite clear that small changes in the morphology of the basins may deeply affect the circulation of the AABC. One should note further that, above the AABW, the North Atlantic Deep Water (NADW) flows from north to south. Its upper limit is near 1000 m.

Figure 3 also shows in a schematic fashion the main surface current pattern. This pattern controls the distribution of primary productivity in surface waters. It is specially high along the equatorial zone, off Angola and to the south.

Lysocline and CCD

A study of the preservation of the calcareous microfossils in recent sediments demonstrates that they suffer increasing dissolution with increasing depth, as has long been known. It is useful to characterize the degree of dissolution by two different levels, which have been called the lysocline and the Carbonate Compensation Depth (CCD).

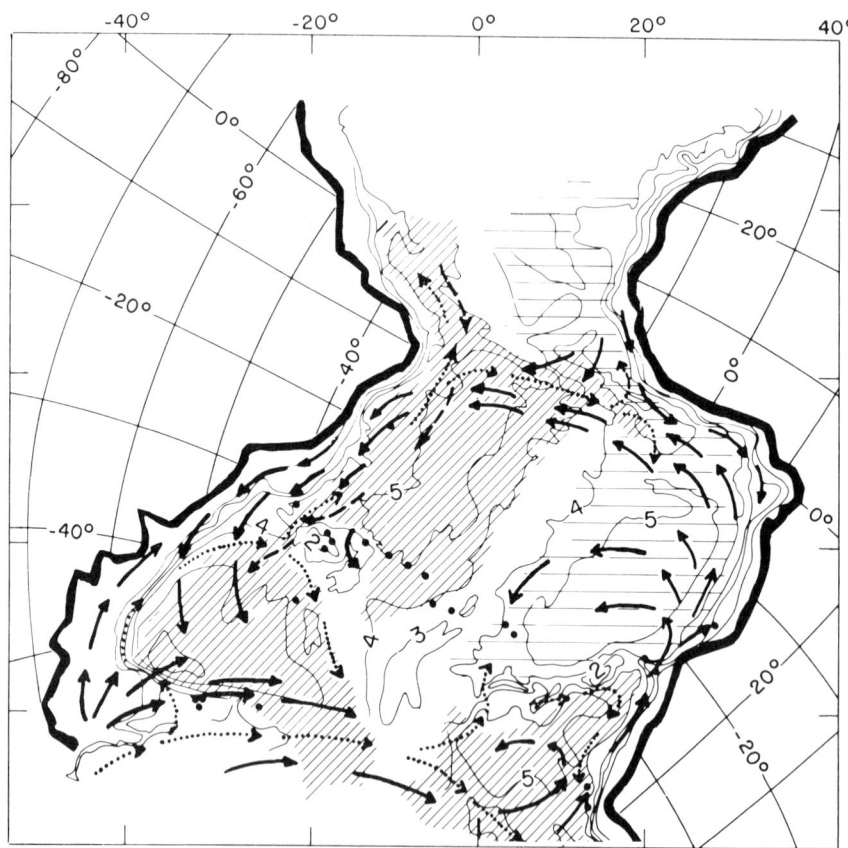

Figure 3. Bathymetric and paleocurrents map. Heavy arrows : surface currents; dashed arrows, NADC; dotted arrows, AABW. Horizontal hachures : bottom temperature greater than 1.6°C; oblique hachures, less than 1.6°C.

The lysocline has been defined by BERGER (1970 a) as the level at which the degree of dissolution of the calcareous microfossils shows a rapid increase. For example, in the case of planktonic foraminifera, the percentage of fragmentation of the tests of foraminifera passes at this level from 20 or less to more than 50% (MELGUEN and THIEDE, 1974). The lysocline is at different depths for different types of microfossils. It is for example deeper for the coccoliths than for the foraminifera (SCHNEIDERMANN, 1973; BERGER, 1973; BERGER and ROTH, 1975). Throughout this paper, the lysocline used will be the foraminiferal lysocline.

The lysocline is specially well defined in the Western South Atlantic where it coincides with the upper boundary of the AABW. In the Vema Channel and on the Rio Grande Rise, a detailed study by MELGUEN and THIEDE (1974) has shown that the lysocline sharply separates well preserved from poorly preserved calcareous assemblages. BERGER (1968) has shown that the lysocline is situated, as an average, 500 m. above the CCD and this has been verified in the South West Atlantic (MELGUEN and THIEDE, 1974).

The CCD was introduced by ARRHENIUS in 1952 to define the level at which the rate of supply of calcium carbonate is equal to the rate of dissolution, so that no more carbonate sediment is deposited (BRAMLETTE, 1961). Although the depth of the CCD is mainly controlled by the temperature of the water and the pressure, the rate of dissolution tending to increase as the temperature decreases and the pressure increases, many other factors are significant, such as the primary productivity (BRAMLETTE, 1965; TAPPAN, 1968; BROECKER, 1971; BERGER and ROTH, 1975), the carbonate supply from the continent and the distance to the continent (BERGER, 1970 a; BERGER and WINTERER, 1974), the succession of transgressions and regressions and the climate (SEIBOLD, 1970; BERGER and WINTERER, 1974) etc ... Therefore the CCD cannot be defined for the South Atlantic Ocean as a whole but it varies from basin to basin (fig. 4) and even within a basin.

According to ELLIS and MOORE (1973), the CCD varies from 5000 m. in the Argentine basin to 5200 m. in the Cape basin and to more than 5500 m. in the Angola Basin. For Berger (1968), the CCD varies from 5200 to 5800 m. from north to south in the Angola basin. BISCAYE et al. (1976) indicate a depth larger than 6000 m. in the Angola basin. Thus, there is still some disagreement among the different authors. We assume, for this paper, that the CCD in the Cape, Brazil and Angola basins is respectively 200, 500 and 1000 m. deeper than in the Argentine basin. These differences may be mainly explained by the AABC circulation previously discussed. In the Angola basin, where the input of AABC is minimum, the degree of saturation of the water in carbonate is higher at comparable depths than in the western basins, which results in a lower dissolving (TAKAHASHI, 1975). On the other

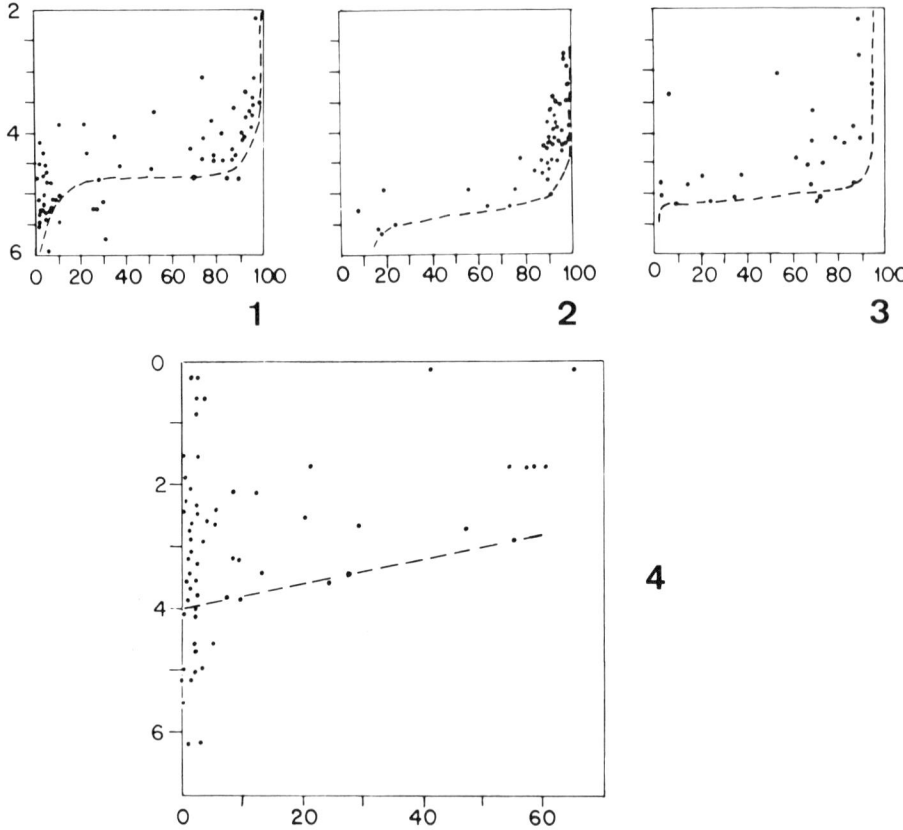

Figure 4. Calcium carbonate content in surface sediments of the South Atlantic Ocean in percent versus water depth in kilometers. 1,2,3, Argentine and Brazil basins, Angola basin and Cape basin respectively after ELLIS and MOORE (1973); 4, Falkland rise and 50°S after BISCAYE et al. (1976).

hand, the degree of under saturation (acidity) of the bottom water in the western basins is strongly affected by the AABC. This is demonstrated by the fact, mentioned previously, that the upper limit of the AABW coincides with the lysocline in the Vema Channel.

In the South Atlantic, the fertility of the surface waters is a significant factor in the control of the CCD. In areas of high productivity (e.g. equatorial zone and upwelling areas in general, STEEMAN-NEILSEN and JENSEN, 1957), two opposite processes affect the CCD. The increased supply of calcium carbonate tends to lower it whereas the additions of organic material tends to increase the acidity at the sediment-water interface and in the sediment itself, and thus to raise the CCD. This is why the CCD is shallower in the Falkland plateau area and along the Angola margin than in the adjacent basins.

Finally, the concentration of calcium carbonate in sediments is obviously affected by the input of non calcareous terrigenous material which dilutes it. Supply of terrigenous sediments is important along the continental margins, especially off large rivers, such as the Orange, the Congo, the Niger, the Amazon and the Parana. The dilution is maximum off the Congo and along northern part of the Argentine continental margin. In the Argentine basin, however, the main part of the terrigenous material comes from the south in suspension in the AABC (BISCAYE and DASCH, 1968; HOLLISTER and ELDER, 1969; EWING et al., 1973). Along the African margin, the terrigenous material derived from the Congo is transported by surface currents in the central part of the Angola basin and to the south of the Guinea basin along the coast of Gabon (Bornhold, 1973). South of the Walvis ridge, the terrigenous material derived from the Orange and Kunene rivers is transported by the Benguela current along the continental margin toward the southern part of the Angola basin. The Olifants and Berg rivers also supply terrigenous sediments to the Cape basin (SIESSER et al., 1974).

Surface sediment distribution

We have just seen that the surface sediment distribution on the sea-bottom is controlled by a multiplicity of complex factors among which the ocean basin morphology, the biogenic productivity, the terrigenous supply, the oceanic circulation and obviously the CCD play a major role. Figure 5 shows this distribution in a simplified manner, the major sedimentary facies being the only ones shown. This map is a compilation based on the numerous studies which have been made in the South Atlantic since the early work of MURRAY and RENARD (1891) during the Challenger expedition (1891). PRATJE, in 1939, taking into account the work of WUST (1936), pointed out the importance of the AABC in the distribution of the sedimentary facies. TUREKIAN (1964), GOLDBERG and GRIFFIN (1964), BISCAYE (1965), EWING (1965), LISITZIN (1971), EWING et al

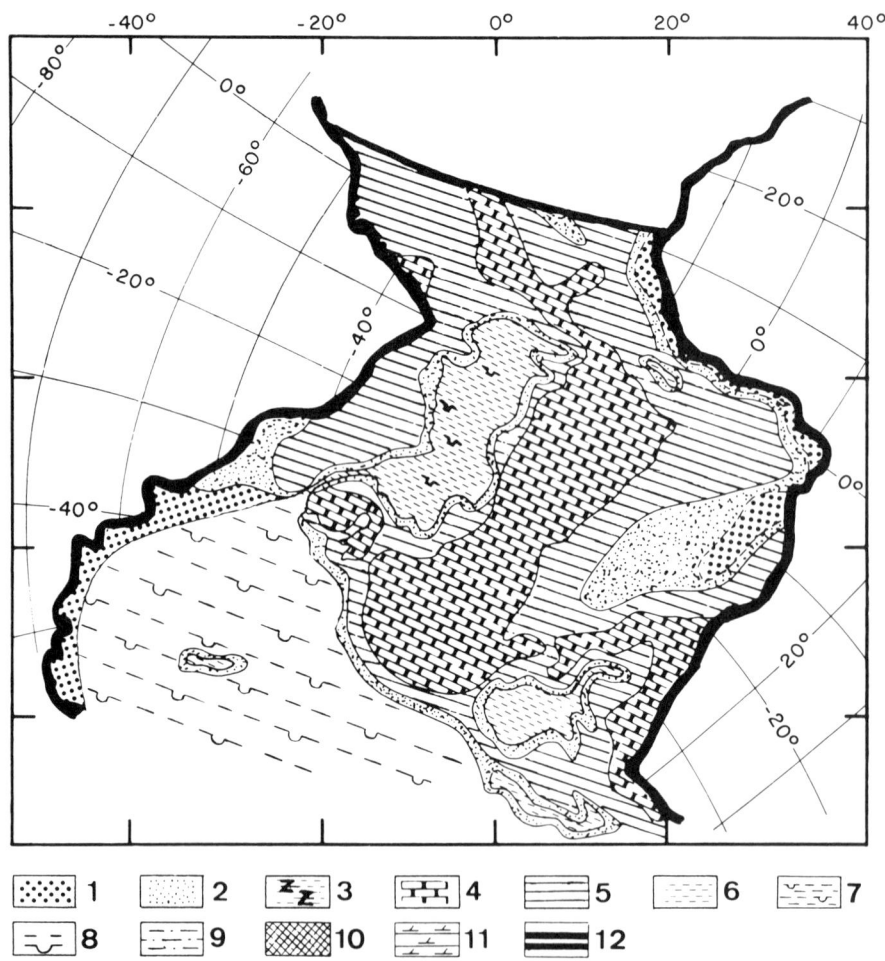

Figure 5. Map of present-day sedimentary facies distribution. 1, mud; 2, coccoliths and foraminifer-bearing mud; 3, zeolitic mud; 4, calcareous ooze or chalk; 5, marl; 6, pelagic clay; 7, diatom-bearing pelagic clay; 8, radiolarian and diatom-bearing mud; 9, shale and sandstone; 10, evaporites; 1, dolomitic limestone; 12, sapropel. The same symbols are used throughout this paper.

(1973), ELLIS and MOORE (1973), CALVERT and PRICE (1971), RAMSAY (1974), SIESSER et al. (1974), MELGUEN and THIEDE (1974), GARDNER (1975), BISCAYE et al. (1976), BURKE and STANTON (1976) have studied some of the major aspects of the sedimentation in the South Atlantic.

To define the major pelagic facies and their succession as a function of increasing water depth, we have used the common DSDP classification with modifications coming from the study of MELGUEN and THIEDE (1974) in the Vema Channel and on the Rio Grande rise. The continental margin sediment distribution is however more difficult to schematize in this way. We have been forced to adopt a highly simplified facies distribution which is in part based on our own observations in the Sao Paulo basin (CNEXO cruise, unpublished report), on the Niger Delta (THIDE et al., 1974), and on the Angola margin (MELGUEN et al., 1975).

The sedimentary facies adopted for this study are the following:

- mud consists essentially of silty clay with less than 10 % of calcareous or siliceous biogenic components. It is characteristic of parts of continental margins with large terrigenous input.

- coccolith and foraminifer bearing mud differs from the preceding by a greater abundance (10 - 30 %) of calcareous micro- and nanno-fossils as a result of a greater biogenic productivity or of a smaller supply of terrigenous material. It is characteristic of the base of the continental margin, the continental rise and the lower flanks of the mid-Atlantic ridge just above the CCD. In the latter case it depends not so much on the degree of dilution as on the degree of carbonate dissolution.

- radiolarian and diatomaceous bearing mud is similar to the previous one, but with siliceous instead of calcareous microfossils. It is, in general, characteristic of high productivity areas with high terrigenous material supply.

- marl contains 30 to 60 % of calcium carbonate, consisting essentially of microfossils and nannofossils. On the continental margin, however, marls may include terrigenous components.

- calcareous ooze or chalk contains more than 60 % of calcium carbonate (nannofossils and foraminifers). This facies, characteristic of oceanic ridges is also found along continental margins with little supply of terrigenous material.

- pelagic clay contains less than 2 % of calcium carbonate. It is found in deep basins, at or below the CCD and generally represents the residue of sediments heavily dissolved. Pelagic clays are often associated with zeolites, especially in areas influenced by volcanism.

MELGUEN and THIDE (1974) have determined the depth range of most of the pelagic facies with respect to the CCD in the area off Brazil. From bottom upward, pelagic clay is found below or at the CCD; coccolith and foraminifers bearing mud, from the CCD to 200 m. above it; then marl which extends up to approximately 1000 m above the CCD; and finally calcareous ooze and chalk above this 1000 m level. These considerations do not apply to the continental margin area where the major factor is not so much depth as the supply of terrigenous material. Thus, for example, the continental margin of Argentina is mostly covered by mud whereas the Cape basin continental margin is highly calcareous (fig. 5). In the same way, the distribution of biogenic siliceous mud is primarily related to the surface water productivity, and to the oceanic circulation. In the Argentine basin, for example, the distribution of displaced Antarctic diatoms, follows the flow of the Antarctic Bottom Water (BURCKLE and STANTON, 1976).

To conclude, we may point out the main characteristics of the map in figure 5. These are the wide area of calcareous facies over the mid-ocean ridge; the restriction of the terrigenous sedimentation to part of the Argentine and Angola basins; the restriction of siliceous mud to southwest, around the Falkland plateau and over the Argentine basin (small areas rich in siliceous debris, such as Walvis Bay are not represented); the variable extent of the pelagic clay area in the deep basins, reflecting changes in the CCD level. These changes are, as previously mentioned, strongly related to the flow of the AABC, which increases the carbonate dissolution and supplies the deep basins with suspended clay.

II. METHODS OF RECONSTRUCTION

Paleobathymetry

The preliminary basis for any reconstruction of the paleo-environment of an ocean is a paleobathymetric map. SCLATER and McKENZIE (1973) previously proposed a reconstruction of the paleobathymetry of the South Atlantic ocean. However, there are now much better magnetic data which allow us to obtain a more precise reconstruction.

The fit of the South American and African continents, as obtained by BULLARD et al. (1965), demonstrates that there has been no significant internal deformation as they moved apart and, consequently, that they have behaved as rigid plates throughout the opening. Consequently, we can obtain the past relative positions of the two continents at different stages during the opening by fitting together corresponding isochrons of the ocean crust. The linear magnetic anomalies associated with spreading of the sea-floor (Vine and Matthews lineations) were created at the accreting plate boundary (the ridge crest) and subsequently carried away on either side of the boundary. The finite rotation which will fit them together will also restore the plates to their proper positions with respect to the ridge crest, at the time the anomalies were created. It will also restore, of course, the older isochrons to their proper relative positions at the time, thus obtaining the corresponding distribution of sea-floor ages by subtracting from the rotated isochrons the age of the reconstruction. The process is equivalent to eliminating the portion of sea-floor younger than the age of the reconstruction.

LADD (1976) has obtained such fits by trial and error method for some of the most clearly identified anomalies. We have used his parameters and those of Sibuet and Mascle (in preparation) to reconstruct maps of the distribution of ocean crust isochrons at different geological stages, of which four are shown in this paper: the Albian (anomaly MO, 100 M.y.), the Coniacian-Santonian (anomaly 34, 86 M.y.), the Maastrichtian (anomaly 31, 68 M.y.) and the Lower Oligocene (anomaly 13, 34 M.y.). These maps are plotted on an oblique Mercator projection which avoids distortion. The corresponding parameters are given in table 1.

If, as discussed previously, a piece of ocean floor of a given age is associated with a given depth it is then trivial to convert these ocean floor paleo-isochrons maps into paleobathymetric maps. The empirical curve which relates depth to age is shown in figure 6. An empirical formula was obtained by least squares by LE PICHON et al., (1973) on the basis of data compiled by SCLATER et al. (1971). The curve may be 200 m too low between 80 and 120 M.y. on the basis of recently compiled North Atlantic basin data (TREHU et al., in press). However, this kind of error is probably well within the accuracy of this method.

We discussed earlier the effect of the sediment cover on the depth of the sea-floor. Given a sediment thickness of 1, the sea-floor depth will be smaller than the theoretical depth of 1/2. This correction will thus be significant only where the sediment thickness is larger than 500 meters, giving a correction larger than 250 m. In addition, it will get smaller with increasing age. As can be seen in figure 2, the correction will thus only be significant in the ocean basins. The estimate of the sediment thickness at the time of the reconstruction was made on the basis of figure 2 and of the different DSDP holes. We then obtained paleoisopach maps by successively peeling off the part of the sediment layer younger than the age of the reconstruction. Although the paleoisopach maps obtained (not shown for lack of space) are not very accurate, they are sufficient for correction purposes except over the continental margins where, anyway, the empirical law relating depth to age breaks down because of strong

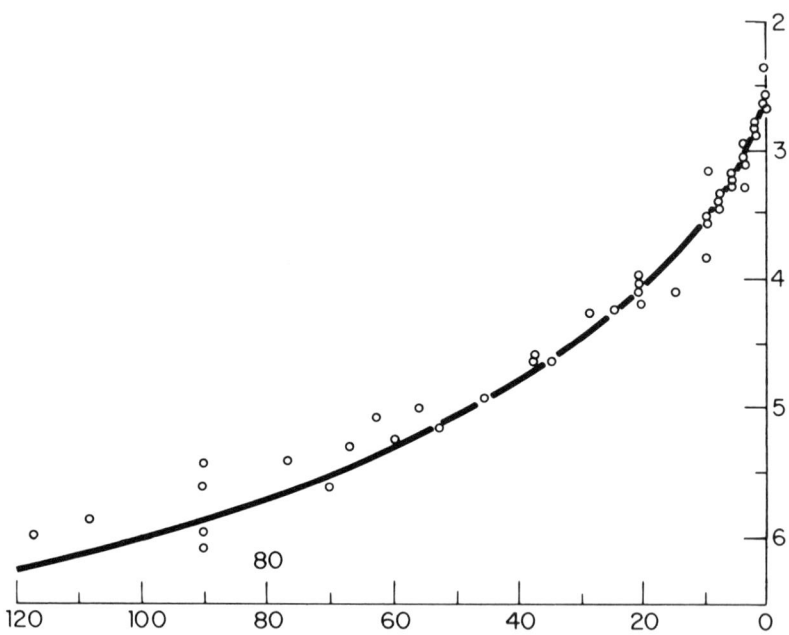

Figure 6. Empirical depth-age curve for the ocean bottom, for ages greater than 80 m.y. after LE PICHON et al. (1976) based on data of SCLATER et al. (1971) and TREHU et al. (in press). Continuous curve, D = 7 100 - 3 904 exp (A/78 - 606 exp (A/5.98) where D is the depth in meters and A is the age in m.y.

coupling between the continental lithosphere and the oceanic lithosphere.

However, even if the depth is anomalous with respect to the theoretical depth, available data suggest that the subsidence curve is not. It is then still possible to obtain an estimate of the paleodepth by proceeding backward from the present depth and removing the amount corresponding to the theoretical subsidence between the present and the age for which the reconstruction is made. This method can be applied not only to the lower continental margin area but also to other anomalous areas such as fracture zones and the Walvis and Rio Grande rises. It is thus implicitly assumed that these volcanic ridges have always maintained the same depth difference between their summit and the adjacent sea-floor. This hypothesis, which was already used by SCLATER and McKENZIE (1973) seems reasonable to us and leads to paleobathymetric reconstructions which seem to be in good agreement with drilling results.

To summarize, after applying the adequate sediment thickness correction, the depth of the transverse volcanic ridges and plateaus was obtained by assuming a constant depth difference between their summits and the adjacent sea-floor. The paleodepths are not correct in the continental margins areas due to insufficient data on sediment thickness.

This last restriction could be a very significant problem for the interpretation of the sedimentary record of the JOIDES drill holes, because we need an accurate reconstruction of the paleobathymetric evolution for this purpose. It can be seen in figure 1 that many of these holes are in so called "anomalous areas" where the use of the empirical curve of figure 6 may lead to errors larger than 1000 meters in the estimate of the present depth. Fortunately, for these holes, we have detailed information on the sediment thickness which enabled us to obtain with good accuracy the paleobathymetric evolution by proceeding backward from the present depth using the empirical curve as a subsidence curve and not as an absolute depth curve.

A final remark of technical nature should be made here. The empirical age-depth formula used here is based on the magnetic reversal time scale proposed by HEIRTZLER et al. (1968) which is slightly different from the TARLING and MITCHELL (1976) time scale used in this paper. This is not a major problem, however, as the empirical curve of SCLATER et al. (1971) relates depth to magnetic anomaly number. Thus, provided one uses the same time scale throughout the correction process, it is then possible to apply the new time scale to the corrected sea-floor paleobathymetry without error.

Paleo CCD

The discussion of the surface sediment distribution has demonstrated that the variation of level of the CCD is a crucial factor in the facies distribution. It is thus necessary to know its evolution through time as well as space in order to reconstruct maps of paleodistribution of sediments. Unfortunately, this is not an easy problem as was shown by the considerable differences in the estimates, made by recent papers, of the present CCD in the different basins of the South Atlantic. This is partly due to the fact that the CCD varies not only from basin to basin but also within the same basin, and partly to insufficient data.

With the present distribution of data, it is of course impossible to approach adequately the paleovariation of the CCD in space. It is necessary to make some simplifying assumptions in order to extrapolate data, from one to a few points at most, to the whole surface of the South Atlantic sea-floor. In this paper, we have tried to establish an average curve for the deep Argentine and Cape basins considering, by reference to present observations discussed earlier, that it is shallower than the CCD in the Brazil and Angola basins but deeper than the CCD over the high productivity areas of the Falkland plateau and the Angola continental margin. Thus our average curve provides a reference level from which we estimate by difference the CCD in the other locations.

This hypothesis seems to be confirmed by data available from the various drilled sites. If, for example, we compare the curves of evolution of paleodepth of the pelagic clay facies on the Falkland plateau to those in the Argentine and Cape basins, the former are generally situated 2 000 m shallower. A similar observation is made when we compare the paleodepths of the marls deposited at the lysocline level in the Cape basin (fig. 10) and in the area of high productivity of the easternmost Walvis ridge (sites 361 and 363). The lysocline, during middle Eocene and lower Oligocene, seems to have been from 1 000 to 2 000 m shallower in the latter area than in the Cape basin.

We base our estimate of this paleo CCD curve shown in figure 12 on an analysis of the facies of all JOIDES sediment cores recovered in the South Atlantic. For the periods where these data did not give us any reliable estimate of the paleo-CCD, we have used values previously proposed by other authors (RAMSAY, 1974); BERGER and ROTH, 1975; VAN ANDEL, 1975) based on data coming from other parts of the ocean.

The type of information coming from a description of the sedimentary facies at the different drilling sites is vividly illustrated by the 5 sites drilled on a transect of the mid-Atlantic ridge during leg 3 (fig. 7). Sites 16 to 19 are situated at increasing

MODEL OF THE EVOLUTION OF THE SOUTH ATLANTIC 19

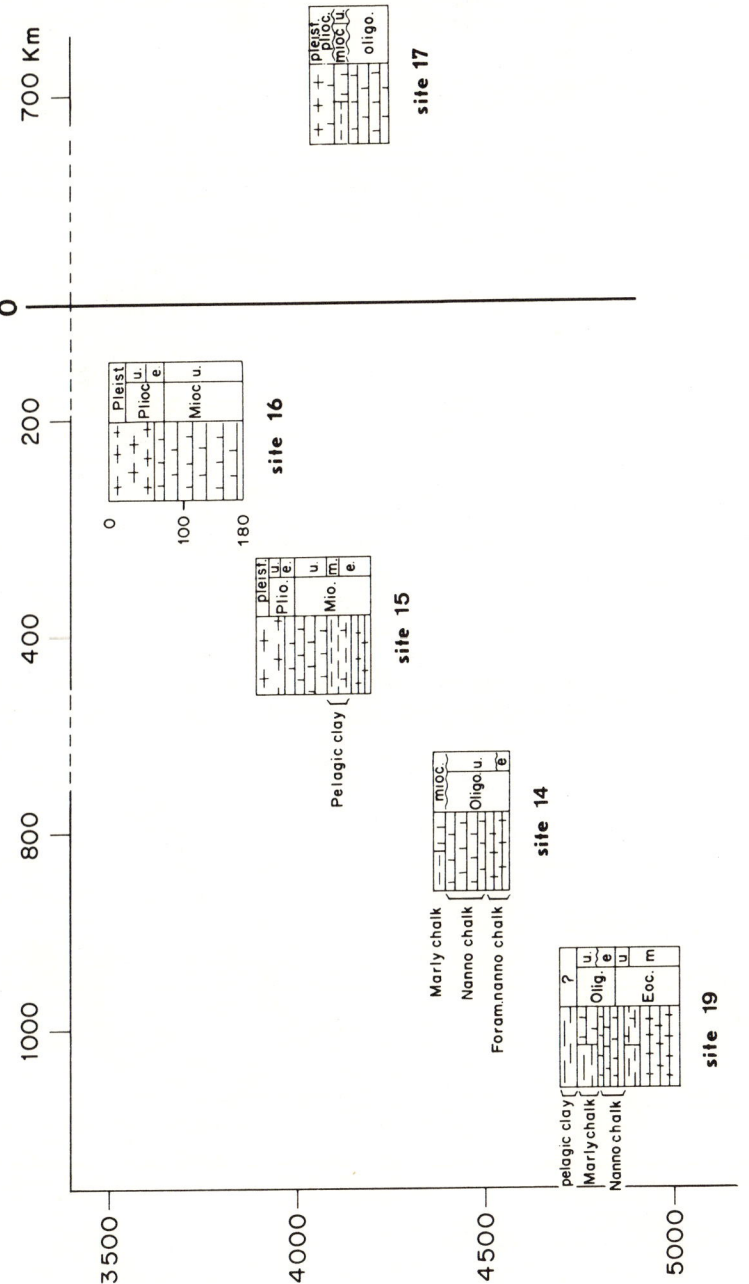

Figure 7. Sedimentary facies evolution as a function of the distance from the Mid-Atlantic Ridge axis. Sites 15 and 19, drilled by Glomar Challenger, leg 3 (MAXWELL et al., 1970) show a rise of CCD level in Eocene and Middle Miocene. Water depth in meters.

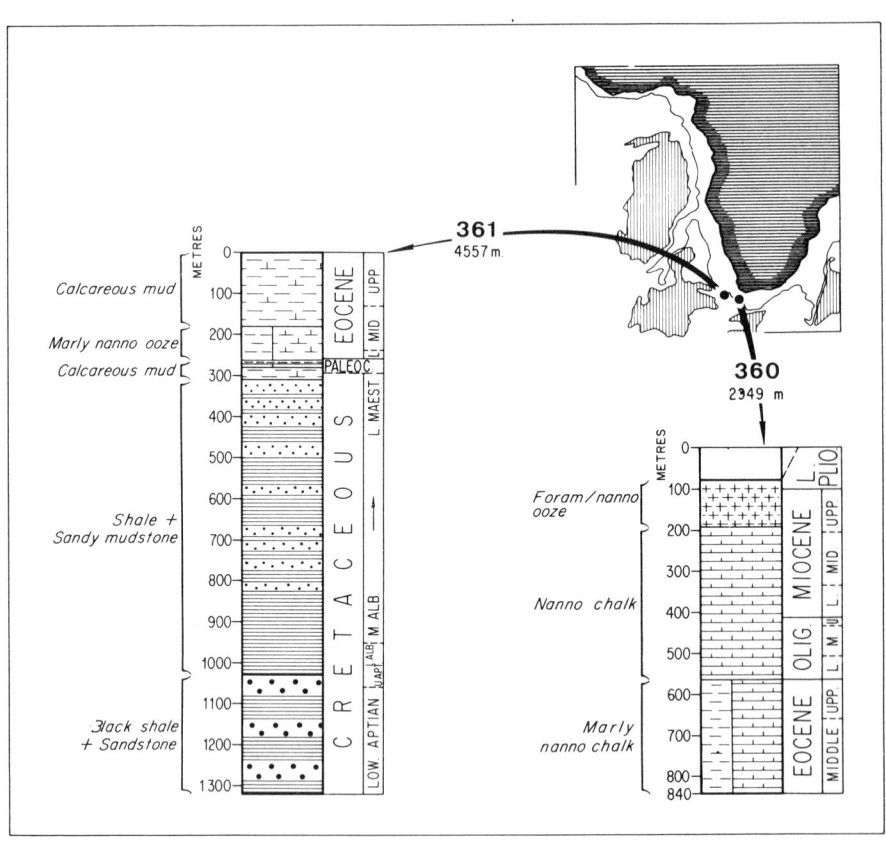

Figure 8. Evolution of sedimentary facies at sites 360 and 361 drilled on the Cape continental margin and in the Cape basin (DSDP leg 40; BOLLI et al., 1975).

distances from the ridge crest, and consequently increasing depths and ages. First, we note that the surface sediments pass from chalk to marly chalk to pelagic clay with increasing depth, indicating that the present CCD is situated between 4 300 and 4 700 meters here. Second, we know that, as the oceanic crust operates like a conveyor belt, progressively moving down the flanks of the ridge, the calcareous mud from the crest area will be transgressively covered by sediments more affected by dissolution. Thus, at site 19, the earliest Eocene sediments are Eocene chalk which changes to marly chalk in upper Oligocene and finally to pelagic clay after a hiatus. Third, the vertical succession of facies does not only reflect the deepening of the ocean crust but also the variations of the CCD. This is shown in site 15 where pelagic clay is present in middle Miocene between two chalk intervals, thus indicating a shallowing of the CCD at the time. Similarly, in site 19, marly chalk appears in Upper Eocene between two chalk intervals indicating that the lysocline reached site 19 in upper Eocene due to a shallowing of the CCD.

It is obviously not possible in this paper to describe in detail the information given by the 24 drill sites, and which permits us to estimate the paleo-CCD. We just give as an example in figures 8 and 9 the analysis of site 360 drilled in Cape basin in 2 949 m water depth. The sediments at this site have clearly recovered CCD variations in their facies (fig. 8), in the proportion of carbonate (fig. 9), in the sand fraction, its relative abundance, its composition and the preservation of the foraminifera tests that it contains (fig. 9).

The nature of the sediment facies was determined on smear slides. It is a more or less marly nanno chalk. The Eocene sediments are richer in clay and less calcareous than the Oligo-Pliocene sediments. This could result either from increased dilution by terrigenous material or by increased dissolution. It is the purpose of the study of the sand fraction (63 to 2 000 microns) (MELGUEN and THIEDE, 1974) to discriminate between the two possibilities. We know for example that marls deposited at the lysocline level are characterized by a degree of fragmentation of planktonic foraminifera close to or greater than 50 %. Figure 9 illustrates that the fragmentation was greater than 50 % in middle Eocene. Note also the relative abundance of dissolution resistant components such as benthic foraminifera and fish debris. Thus we can conclude that this site was close to the lysocline during the Eocene. Note also in figure 9 the indications of another level close to the lysocline during middle Miocene.

In practice, we have considered three types of facies as being specially significant in estimating the position of the CCD: pelagic clay deposited at or below the level of the CCD, marls

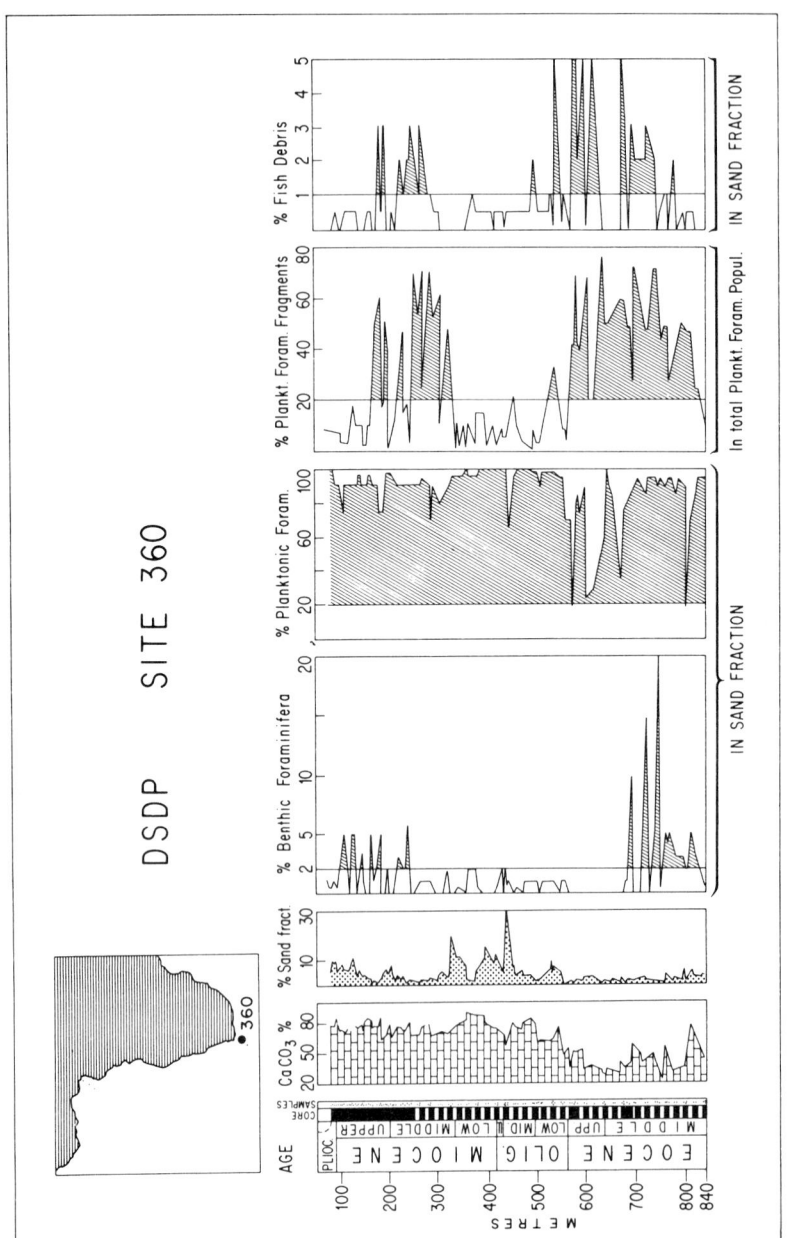

Figure 9. Carbonate content, coarse fraction abundance (63 - 2 000 μ) and composition in sediment.

deposited at the lysocline level, roughly 500 m above the CCD and chalk deposited at least 1 000 m above the CCD. We have then summarized the paleofacies information for each of the drill sites on a curve which gives the evolution of the paleodepth with time. Figures 10 and 11 give the two most important sets of such curves. Figure 12 shows the paleo-CCD curve that we have adopted. For comparison purposes, we have also shown this paleo-CCD curve on figures 10 and 11.

Below, we will discuss this paleo-CCD curve that has to be still considered as quite hypothetical and is only considered as an average curve for the deep Argentine and Cape basins.

- Aptian-Albian (see fig. 12 for corresponding age in M.y.)

The value of 1 500 m chosen, which is very shallow, is related to the period of deposition of black shales with intensive carbonate dissolution (MELGUEN, in preparation). Although part of the dissolution occurred after sedimentation, as shown by the existence of numerous moulds of coccoliths now dissolved (NOEL and MELGUEN, in press), and although there was abundant terrigenous supply at the time, we still consider the lack of calcium carbonate as reflecting a very high CCD (site 361), see fig. 10 for paleodepth). Black shales were also deposited during Albian on the Falkland plateau at a paleodepth close to 1 500 m.

- Coniacian - Santonian

    4 000 m after RAMSAY (1974)

- Santonian - Campanian

The value of 3 200 m is based on site 361 in the Cape basin which contains very rare and highly dissolved calcareous nanno-fossils. The paleodepth is 3 000 m and we have chosen 3 200 m for the paleo CCD to agree with the value proposed by VAN ANDEL (1975).

- Campanian - Maastrichtian

The value of 4 500 m chosen is the one proposed by RAMSAY (1974) as we have only indirect evidence, that is the presence of chalk during the Maastrichtian and the Campanian-Maastrichtian respectively in the Argentine (site 358) and Brazil (site 355) basins. The paleodepths of 2 700 and 3 100 m respectively suggest a CCD deeper than 3 700 and 4 100 m.

- Maastrichtian - Paleocene

The estimate of 3 000 m is based on the presence of pelagic

clay at a paleodepth of 3 000 m at site 328 in the Argentine basin. This value is confirmed by the presence of pelagic clay at sites 361 and 355 at the Maastrichtian/Paleocene boundary, at paleodepths of 3 200 - 3 400 m. It agrees with the estimate of RAMSAY (1974).

- Upper Paleocene

The estimate of 3 600 m is made on the basis of the presence of pelagic clays at a paleodepth of 3 600 m in the Brazil basin (site 355). This is in agreement with the 3 500 m proposed by VAN ANDEL (1975) and close to the 4 000 m proposed by RAMSAY (1974). However, there is chalk at a paleodepth of 3 600 m in the Cape basin (site 361) at that time which suggests a much deeper CCD. The chalk deposition is only a short event as the Paleocene/Eocene boundary is already marked by the presence of marls deposited close

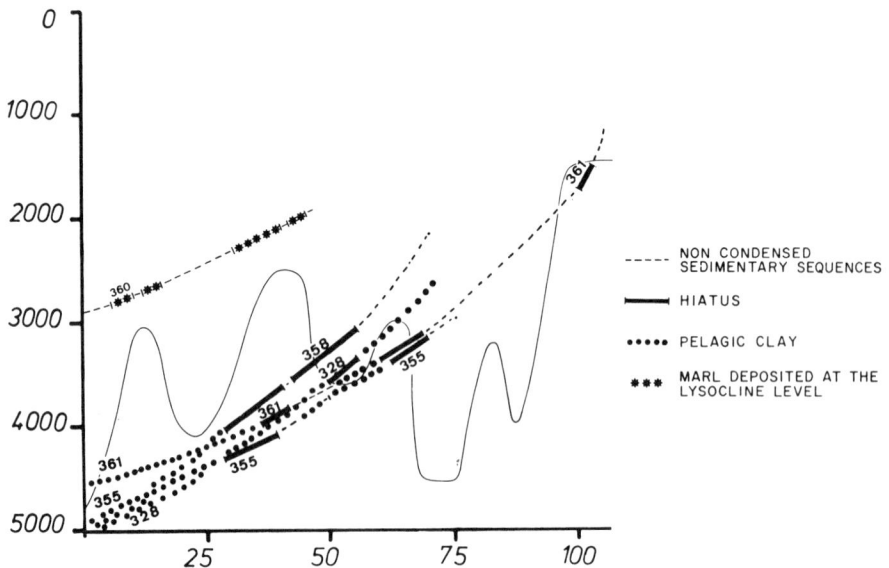

Figure 10. Paleobathymetric evolution of sites 355 (Brazil basin; leg 39), 360 and 361 (Cape basin, leg 40), 328 and 358 (Argentine basin, legs 36 and 39). Hiatuses indicate changes in oceanic paleocirculation. The CCD curve of figure 12 has been drawn in accordance with the paleobathymetry of representative facies, such as pelagic clay, marl deposited at the lysocline level and chalk (average values from Argentine and Cape basins). Water depths in meters, age in m.y.

to the lysocline. Unless the chalk is not in place, this suggests a very large difference in CCD between the Cape and Brazil basin which seems unlikely to us.

- Middle Eocene

The estimate of 2 500 m is based on good data from site 360 in the Cape basin, where marls were deposited at a paleodepth of 2 000 m close to the lysocline level (fig. 10). This is confirmed by the apparition of pelagic clay at a paleodepth of 3 700 m at site 361. BERGER and ROTH (1975) have proposed a paleo-CCD of 3 000 m which is in reasonable agreement considering that its level is generally deeper in the Cape and Argentine basins than to the north.

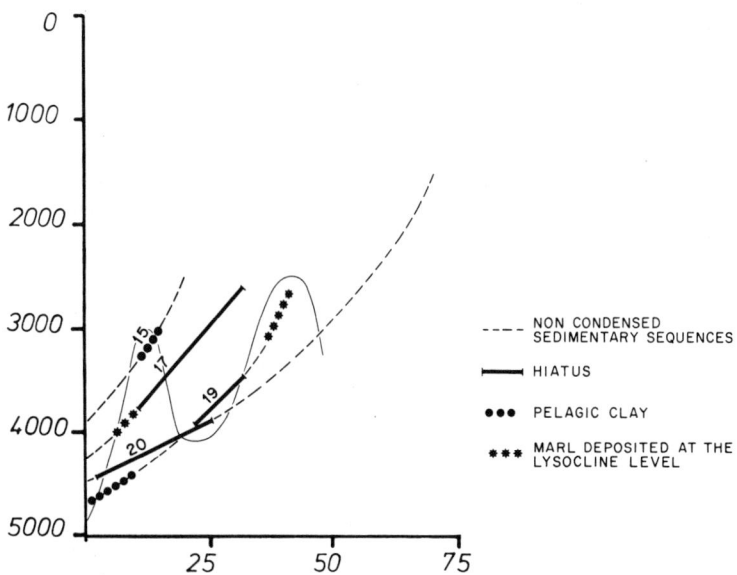

Figure 11. Paleobathymetric evolution of sites 15, 19, 20, drilled on the flanks of the Mid-Atlantic Ridge (DSDP, leg 3, MAXWELL et al., 1970). The CCD curve is shallower than the lysocline level of sites 17 and 19. Both sites are situated in the Angola basin, where the CCD is much deeper than in the Argentine and Cape basins. Water depths in meters, age in m.y.

- Lower Oligocene

We have adopted a value of 3 000 m close to the 3 200 m of BERGER and ROTH (1975). The only available data is a slightly higher degree of fragmentation of planktonic foraminifera (30 %) at site 360 at a paleodepth of 2 250 m (fig. 10). This suggests a level slightly above the lysocline, hence our choice.

- Middle Miocene

The estimate of 3 000 m is based on the presence of marls deposited at the lysocline level at a paleodepth of 2 500 m at site 360 in the Cape basin and on the presence of pelagic clays at a paleodepth of 3 000 m at site 15 on the Brazil basin flank of the mid-Atlantic ridge (fig. 7 and 11). Our estimate is 500 m deeper than the one proposed by BERGER and ROTH (1975).

From the Middle Miocene to present time, the CCD has dropped to 4 500 - 4 800 m and more depending on the different basins (fig. 4).

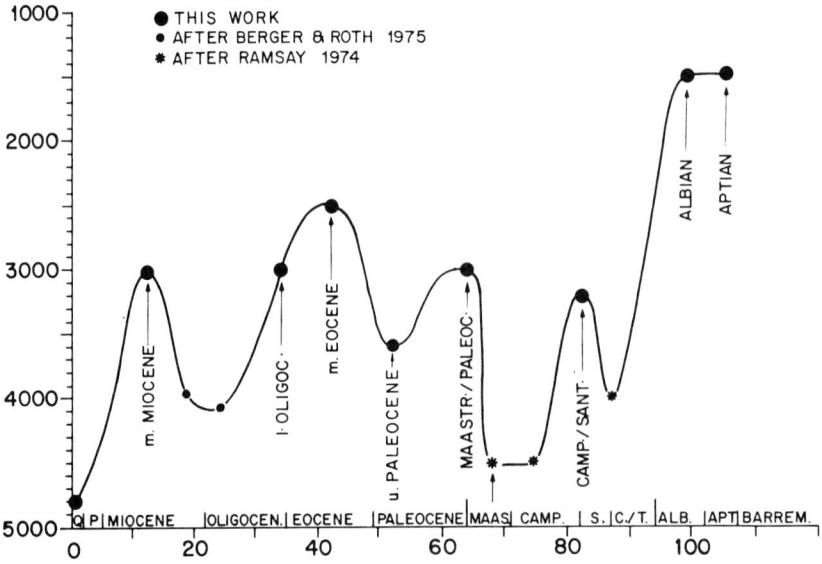

Figure 12. CCD fluctuations since Aptian time in the South Atlantic ocean (average values from Argentine and Cape basins). Water depths in meters, age in m.y.

The paleo-CCD curve of figure 12, which tries to give a rough approximation of the general evolution of the CCD in the Argentine and Cape basins, will thus be the basis of our attempt at reconstructing the paleofacies maps. Although this curve will undoubtedly be modified, it will allow us to demonstrate the basic methodology of reconstruction.

Deep water circulation

We have seen the importance of the water circulation in the present deep sea floor of the South Atlantic ocean, and especially of the AABC which plays a major role in the processes of transport of sediment, erosion and sedimentation control and dissolution. It is consequently essential to be able to reconstruct, at least in a general way, the paleocirculation of the deep sea floor. A first approach, which was followed by EWING et al. (1971), is to use the information given by seismic reflection data on the relationship of seismic reflector characters to the AABW circulation. These authors showed that all over the Argentine basin a new type of sedimentation clearly related to the AABC was installed at the age of a major seismic reflector, called reflector A, which is now known to date from the Upper Oligocene (PERCH-NIELSEN and SUPKO, 1975). A second approach is related to the fact that bottom currents result in dissolution and erosion which appear as hiatuses in the sedimentary column. Thus a study of the distribution of hiatuses on the ocean gives some basic information for a reconstruction of paleocurrents.

Hiatuses become widespread during the Maastrichtian and most specially at the limit Upper Cretaceous/Cenozoic over the Falkland plateau, the Rio Grande Rise and the Sao Paulo plateau. They suggest the presence of strong bottom water circulation at depths ranging between 1 000 and 3 000 m. Over the Walvis Ridge, hiatuses are frequent at the Coniacian-Santonian limit, but not in Maastrichtian (site 363).

In Early Cenozoic, hiatuses appear in the Argentine and Brazil basins, during the Paleocene/Eocene at sites 358 and 328 and especially during the Eocene/Oligocene at sites 329, 327, 358, 22 and 355. They suggest the installation of deep water circulation at that time. It seems that the AABC first appears in the Cape basin in Eocene but it probably only becomes a strong well-defined current in Oligocene.

These observations agree with the conclusions of MARGOLIS and KENNETT (1970), KENNETT et al. (1974), SHACKLETON and KENNETT (1975 a and b) that the AABW circulation begins in Eocene/Oligocene and that the AABC becomes well established in Upper Oligocene. Major fluctuations of the AABC occurred in Miocene, Pliocene and Pleistocene during peaks of the Antarctic glaciation (BERGER, 1973;

1974; WATKINS and KENNETT, 1971).

We have no direct information either on the first passage of the AABC through the Romanche Fracture Zone or on the installation of the NADC. SCLATER and McKENZIE (1973) have proposed an Oligocene age for the former (35 M.y.) whereas BERGGREN and HOLLISTER (1974) proposed a later age between Oligocene and Upper Miocene. The NADC seems to be well established in the North Atlantic by lower/middle Eocene (BERGGREN and HOLLISTER, 1974).

The major point is that in the first stage of opening of the ocean, in Lower Cretaceous, there is no evidence for any significant deep or intermediate water circulation in the South Atlantic.

Facies distribution map

Knowing the paleobathymetry and the approximate paleo-CCD, it is possible to reconstruct a paleofacies distribution map, on the basis of the present distribution of facies with respect to CCD and of the probable relative abundance of the terrigenous sedimentation, especially on the Argentine and Angola continental margins.

It is necessary, of course, to try to account for the spatial variation of the CCD at any given time. This is mostly related to the influence of the AABC after its establishment as discussed for the present situation. Prior to its establishment, we have supposed that the CCD is at the same depth in the Brazil and Angola basins, roughly 500 m deeper than in the Cape and Argentine basins to take into account the easier access to cold water there.

There is however an exception which concerns the early periods of stagnation during which the CCD was very shallow. This was for example the case during the Coniacian north of the Rio Grande-Walvis topographic barrier. The CCD, there, seems to have been intermittently close to the photic zone whereas it was at a paleodepth of 3 200 m in the Cape and Argentine basins.

III. SCHEMATIC MODEL OF EVOLUTION

Figures 13 to 22 present a series of paleobathymetric, paleocurrent and paleofacies maps constructed according to the methodology previously discussed and which schematize the evolution of the south Atlantic ocean, as it widens, deepens and as a vigourous deep thermohaline circulation progressively gets established. Three main periods can be recognized: from Valanginian to Santonian, this is the early opening stage which goes from the continental rift to the narrow confined basin stage; from Campanian to early Oligocene, the ocean progressively opens to the north and to the south as the topographic barriers formed by the large fracture zones and the Walvis-Rio Grande ridges break; the deep water circulation is initiated; from upper Oligocene to the present, the modern pattern of deep-sea circulation and sedimentation is established and is mostly modulated by the climatic fluctuations associated with the Antarctic glaciation, which control, in a great part, the CCD variations and the AABC circulation.

Early opening: Valanginian-Santonian (127 - 82 M.y.)

Fig. 13, 14, 15, 16.

Figures 13 and 15 illustrate that we are dealing with two narrow basins, closed to the north and south, but fairly deep as they are already deeper than 3 km 100 M.y. ago and they are deeper than 4 km 80 M.y. ago when the first breaks in the topographic barriers appear, first to the south, then to the north 'Equatorial Fracture Zone) and in the center (Walvis - Rio Grande).

Figure 14 shows the distribution of sediment 100 M.y. ago; figure 16 shows it just prior to the break (roughly 85 M.y. ago) whereas figure 17 shows it immediately after the break (80 M.y. ago). The South Atlantic consists now of two narrow confined basins, characterized by a frequent stagnant environment where black shales and sapropels (black deposits very rich in organic matter) are sedimented. To the South of the Walvis-Rio Grande barrier, these black shales are deposited during Aptian Albian at depths greater than 3 000 m (fig. 13, 14) over a thickness of several hundred meters. They are very fine, very homogeneous and very poor in microfossils. Intermittent layers of coarse sandstones, often rich in plant debris, have been deposited by turbidity currents.

To the north, evaporites (products of evaporation of salt water which include anhydrite and halite, associated with dolomite) are deposited within salt basins whose limits are outlined in fig. 14 (LEYDEN et al., 1976). Seismic reflection data suggest that the evaporite thickness exceeds several hundred meters. However, we have no direct information on these evaporites. The evaporites seem to be overlain by dolomitic marls which have been reached in site 364 on the Angola margin (BOLLI et al., 1975).

To the south, the sapropelitic sedimentation ends during Albian time, which confirms the paleobathymetric information of an earlier opening across the Falkland-Agulhas barrier than across the Equatorial Fracture Zone barrier. Thus the oceanic sedimentation regime progressively gets established to the south and is reflected on the mid-Atlantic flanks by the deposition of pelagic calcareous sediments and on the southern border where siliceous mud is deposited under a high productivity zone. To the north, the sapropelitic sedimentation intermittently continues and the corresponding deposits have been cored on the Angola margin, on the northern flank of the Walvis Ridge and on the Sao Paulo plateau (BOLLI et al., 1975; PERCH-NIELSEN et al., 1975). Marls and

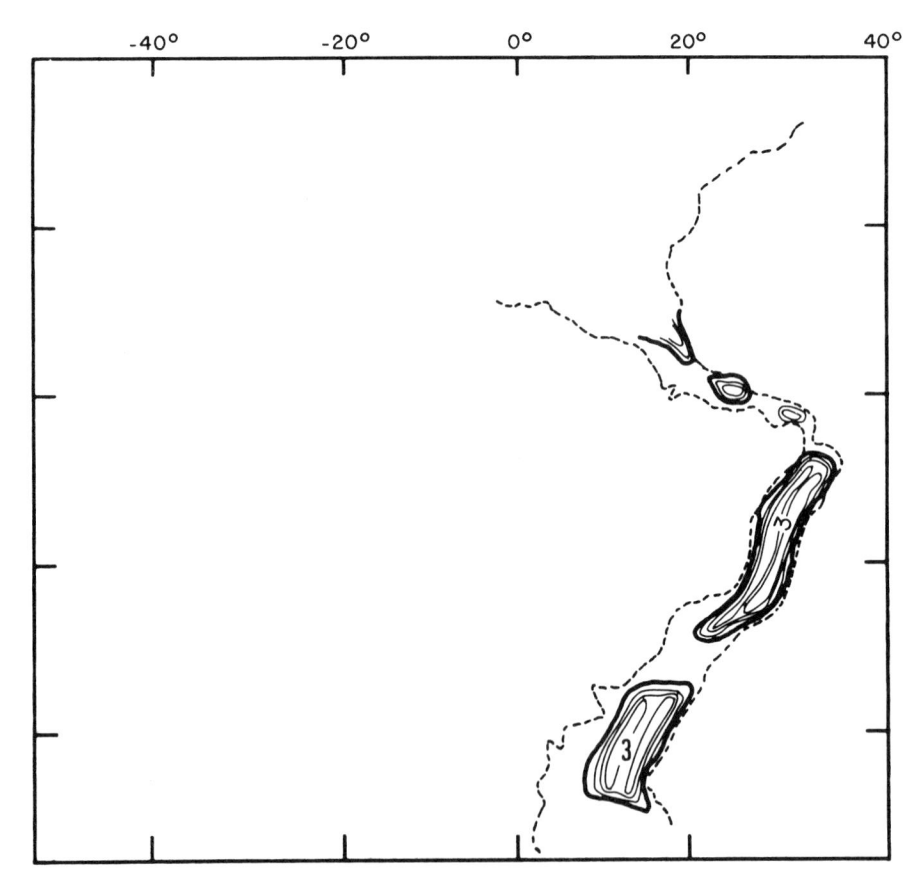

Figure 13. Paleobathymetry of the South Atlantic Ocean at Albian time (anomaly MO = 100 m.y.). Two major basins are well differentiated north and south of the Rio Grande Rise and Walvis Ridge. Water depths in km.

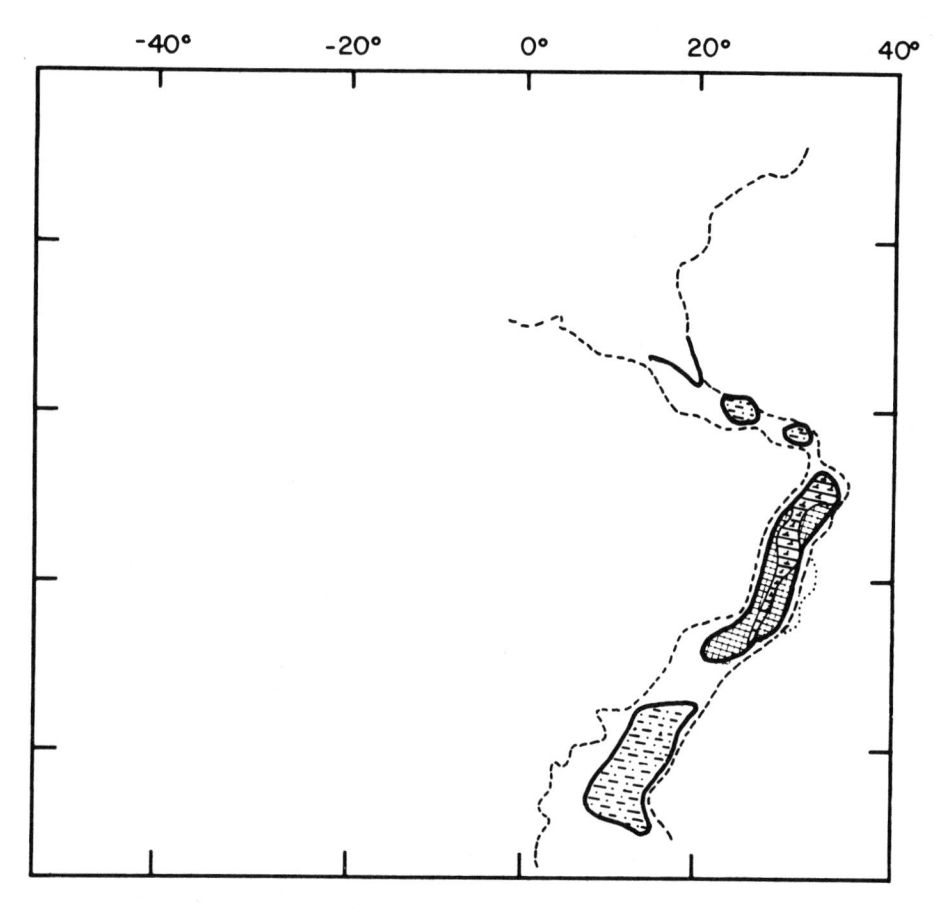

Figure 14. Albian (anomaly m O = 100 m.y.) sedimentary facies distribution based essentially on DSDP legs 36 and 40 (BARKER, DALZIEL et al., 1974; BOLLI, RYAN et al., in press). Limits of the salt basins are from LEYDEN et al. (1976). The CCD in the southern basin is around 1 500 m.

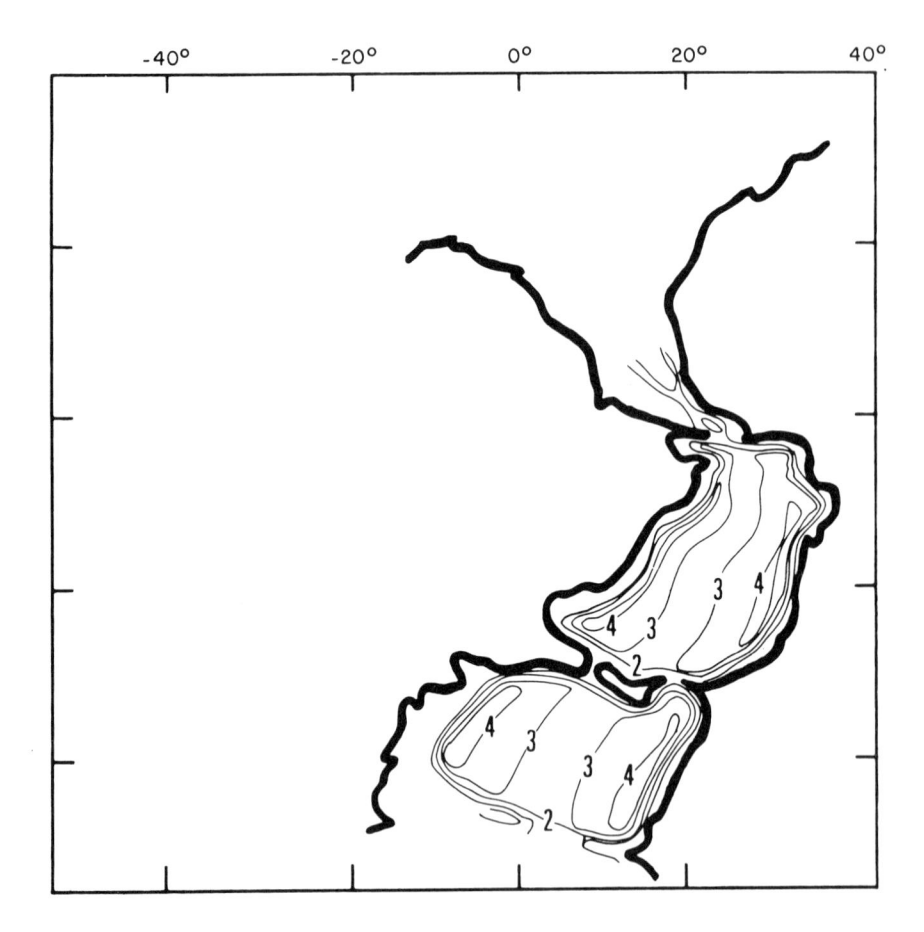

Figure 15. Paleobathymetry of the South Atlantic Ocean during Coniacian/Santonian time (anomaly 34 : 86 m.y.) Water depth in km.

MODEL OF THE EVOLUTION OF THE SOUTH ATLANTIC

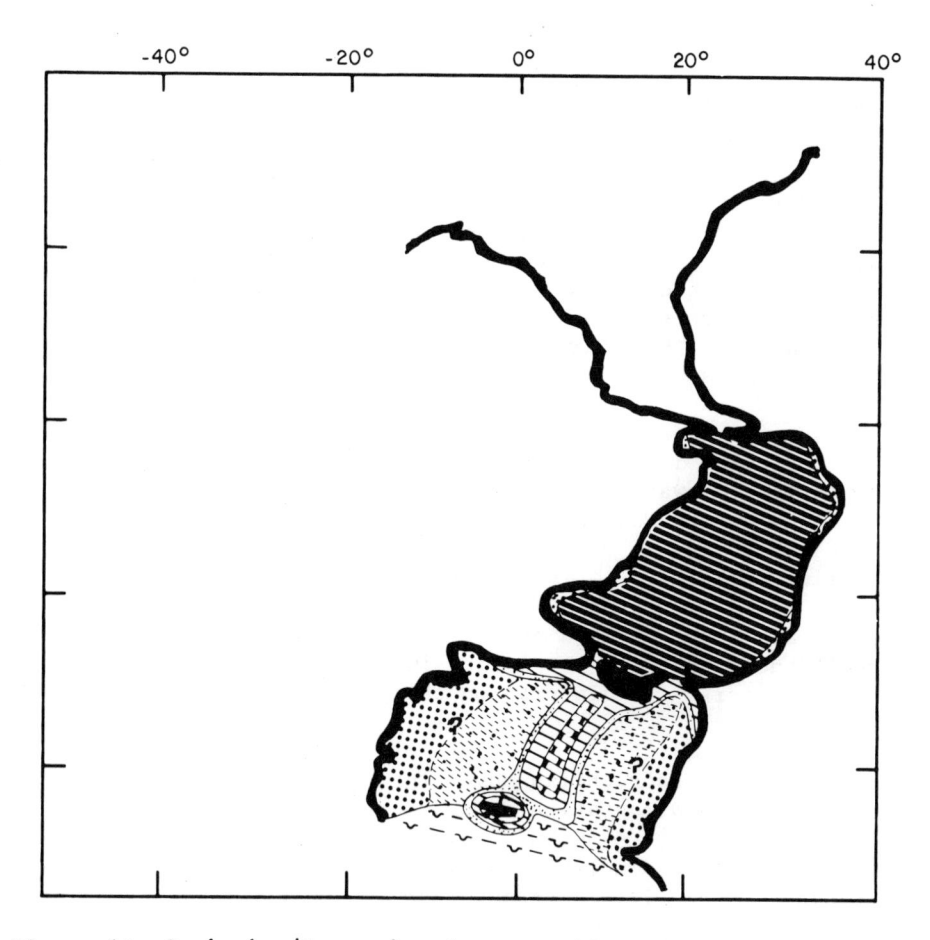

Figure 16. Coniacian/Santonian (anomaly 34 : 86 m.y.) sedimentary facies distribution based on data from DSDP legs 36, 39, 40 (BARKER, DALZIEL et al., 1974; PERCH-NIELSEN, SUPKO et al., in press; BOLLI, RYAN et al., in press). CCD level around 3 200 m in the Cape and Argentine basins, and around 2 000 m in the Brazil and Angola basins. For facies symbols see figure 5.

marly limestones alternate with the sapropels. They contain microfaunas which are quite similar to those living in the Tethys (the ocean situated between Eurasia and Africa) at the time (BOLLI et al., 1975).

Progressive establishment of oceanic circulation:

Campanian-Early Oligocene (82 - 35 M.y.) Fig. 17, 18, 19, 20.

Figure 18 shows that this is the stage of complete disruption of the different east-west topographic barriers. At anomaly 34 time (82 M.y., fig. 15), there is already a 200 km gap between the well defined marginal fracture zones of the equatorial fracture zones. Similarly, to the south, the long Falkland-Agulhas fracture zone has already been broken and the opening enters into a new stage where it is not constrained any more by this strong Africa-South America coupling (LE PICHON and HAYES, 1971). The Walvis-Rio Grande barrier is still nearly continuous but the first gaps appear.

This opening of the Brazil and Angola basins to the north is clearly reflected in the sedimentary records from the Santonian/Campanian limit upward, as marls have been drilled, for example, on the flank of the Walvis ridge (site 363), on the Angola margin (site 364), on the Sao Paulo plateau (site 356) and on the Rio Grande rise (site 357). These marls are overlain by chalks (BOLLI et al., 1975; PERCH-NIELSEN et al., 1975). Campanian chalk has also been drilled in the Brazil basin (site 355). Figure 17 shows that, to the south, in the Cape and Argentine basins, there is no striking sedimentation change from the Santonian to the Campanian.

During the Maastrichtian, over the whole South Atlantic, there is an extensive sedimentation of chalk, which is related to a deepening of the CCD (4 500 m, fig. 12). These chalk deposits are especially widespread north of Walvis-Rio Grande, where the terrigenous supply seems less abundant than to the south.

It is probably during Maastrichtian, as previously mentioned, than an intermediate depth circulation (1 000 - 3 000m) progressively gets established, as numerous hiatuses appear at this paleo-depth on the Rio Grande rise and Falkland plateau. This paleo-circulation is tentatively sketched on fig. 18, although the arrows are highly hypothetical.

At the beginning of Paleocene (fig. 20). the CCD gets much shallower (3 000 m in the Cape and Argentine basins and this is reflected in a regression of the chalk facies and an extension of the pelagic clay facies. This shallowing of the CCD seems to be part of a worldwide phenomenon described by WORSLEY (1974). Thus, the Paleocene is characterized by highly condensed sedimentary series, strongly affected by dissolution.

MODEL OF THE EVOLUTION OF THE SOUTH ATLANTIC 35

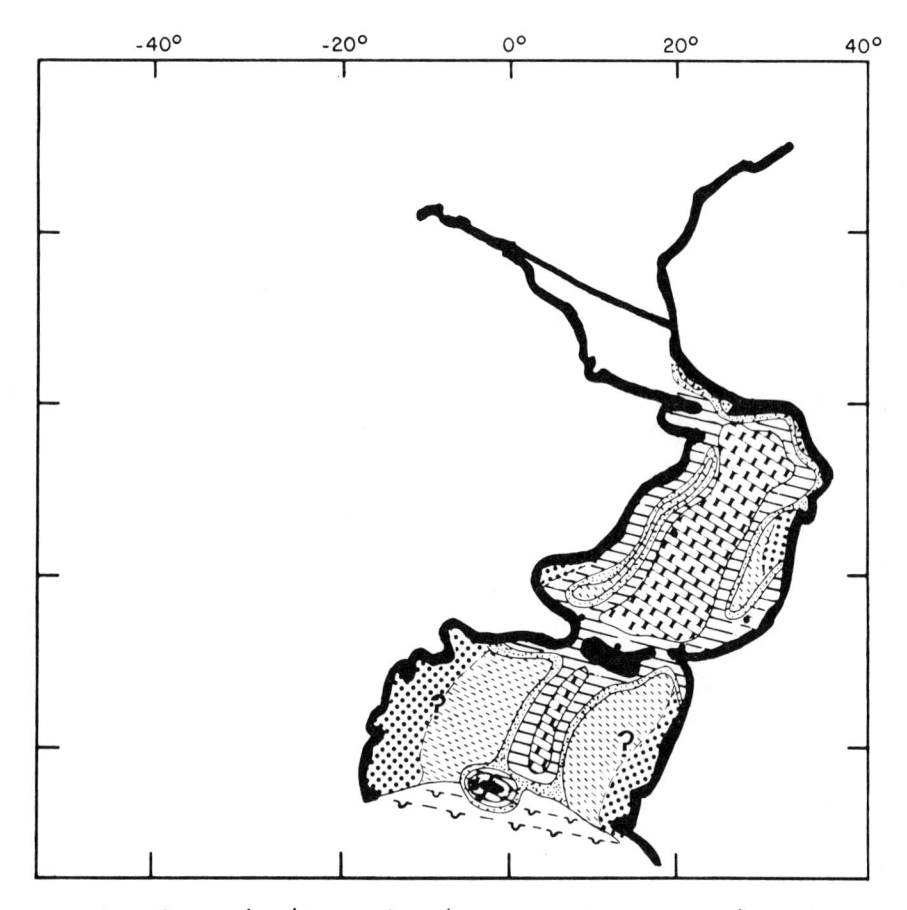

Figure 17. Santonian/Campanian (anomaly 34 : 82 m.y.) sedimentary facies distribution based on DSDP legs 36, 39, 40 (BARKER, DALZIEL et al., 1974; PERCH-NIELSEN, SUPKO et al., in press; BOLLI, RYAN et al., in press). CCD level around 3 200 m in the Cape and Argentine basins, and 3 700 m in the Brazil and Angola basins.

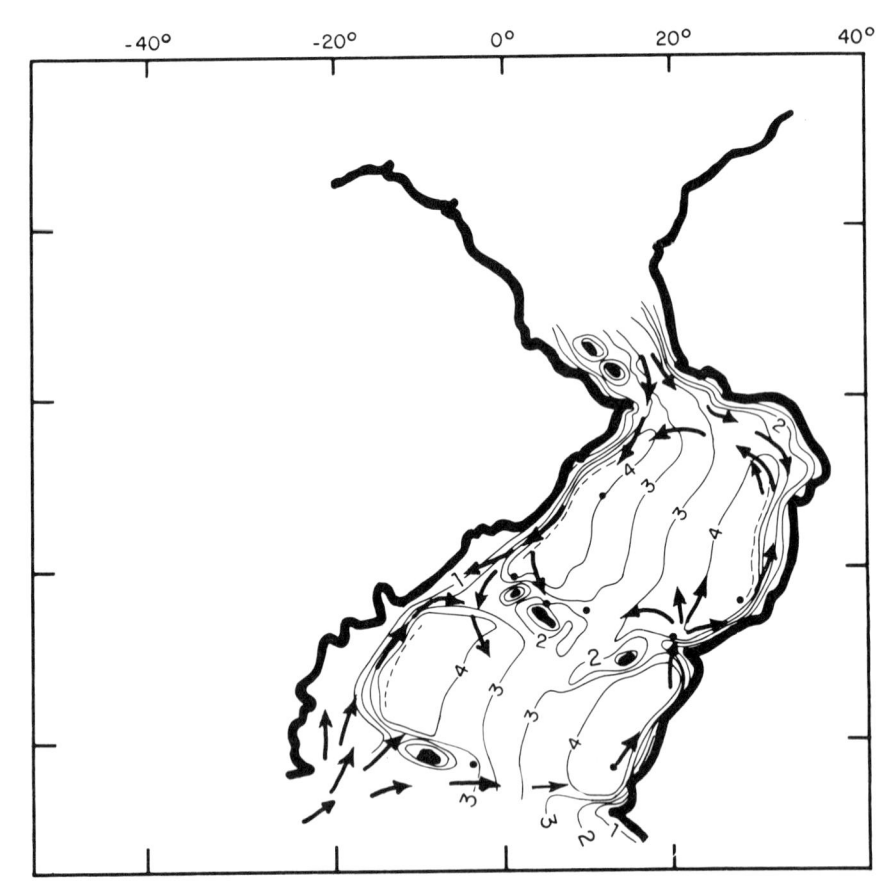

Figure 18. Paleobathymetry and paleocirculation of the South Atlantic Ocean during Maastrichtian time (anomaly 31 : 68 m.y.). Water depths in kilometers.

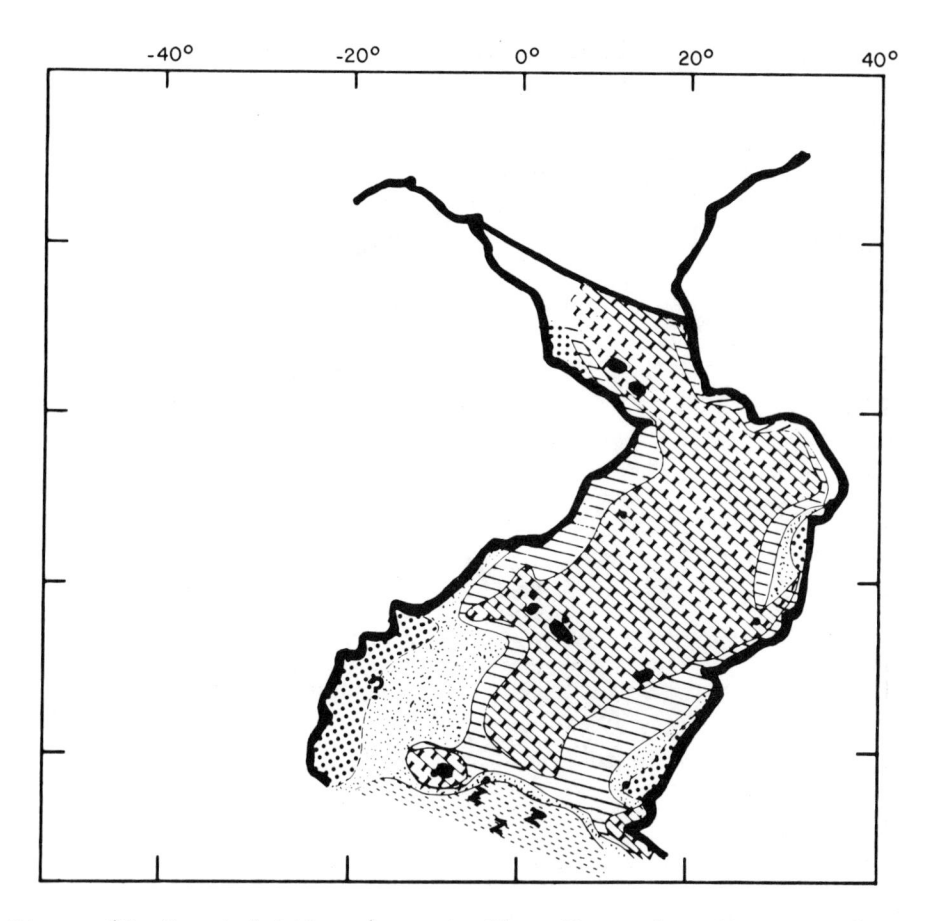

Figure 19. Maastrichtian (anomaly 31 : 68 m.y.) sedimentary facies
distribution based on DSDP legs 36, 39, 40 (BARKER,
DALZIEL et al., 1974; PERCH-NIELSEN, SUPKO et al., in
press; BOLLI, RYAN et al., in press). CCD level
around 4 500 m according to RAMSAY (1974). For facies
symbols see figure 5.

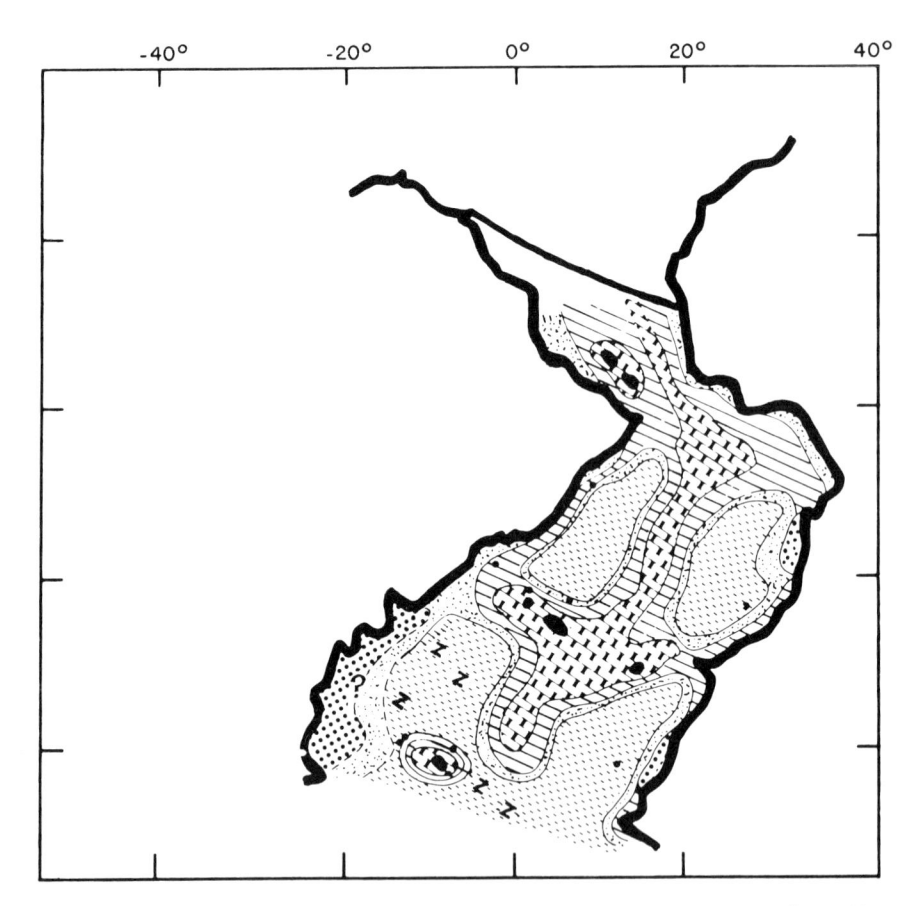

Figure 20. Maastrichtian-Paleocene (anomaly 29 : 64 m.y.) sedimentary facies distribution based on DSDP legs 36, 39, 40 data (BARKER, DALZIEL et al., 1974; PERCH-NIELSEN, SUPKO et al., in press; BOLLI, RYAN et al., in press). We have used the same reconstruction as in figure 19. CCD level around 3 000 m in the Cape-Argentine basins, and 3 500 m in the Angola and Brazil basins. For symbols see figure 5.

MODEL OF THE EVOLUTION OF THE SOUTH ATLANTIC 39

Present-day pattern: Upper Oligocene-Present (30-0 M.y.) Fig. 21, 22.

The modern deep-sea circulation was firmly established in Oligocene. This is the time at which the narrow passages, such as the Falkland and Vema channels, reached a depth **greater than** 4 000 m. It is also the time at which the new current-related sediment deposition pattern gets established in the Argentine basin (reflector A of EWING et al., 1973). Hiatuses are fairly widespread all over the western side of the South Atlantic ocean. The deep AABW circulation might have been the cause of the shallow CCD (3 000 m in the Cape and Argentine basins). In any case, the pelagic clay facies covers a large surface.

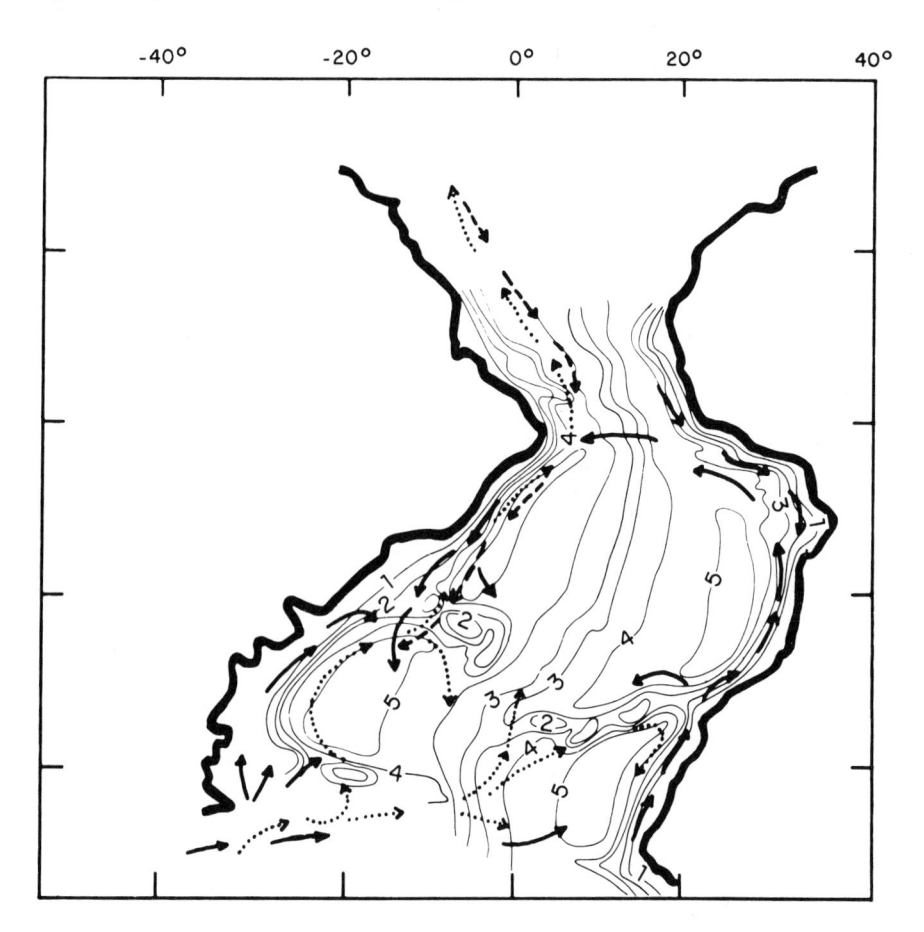

Figure 21. Paleobathymetry and paleocirculation of the South Atlantic Ocean during lower Oligocene time (anomaly 13 : 34 m.y.). Heavy arrows : intermediate/surface currents; dashed arrows : NADC; dotted arrows : AABW. Water depths in km.

It is clear that, by lower Oligocene, the ocean basins have reached a size, which is comparable to their present size; their depths exceed 5 000 m and the deep water circulation is well established. Thus, from then on, the facies distribution can be considered as representative of the present oceanic environment. Its evolution is going to reflect, primarily, the progressive deterioration of the climate which leads to progressively colder surface temperatures, variations in the intensity of the AABW circulation and in the continent erosion pattern.

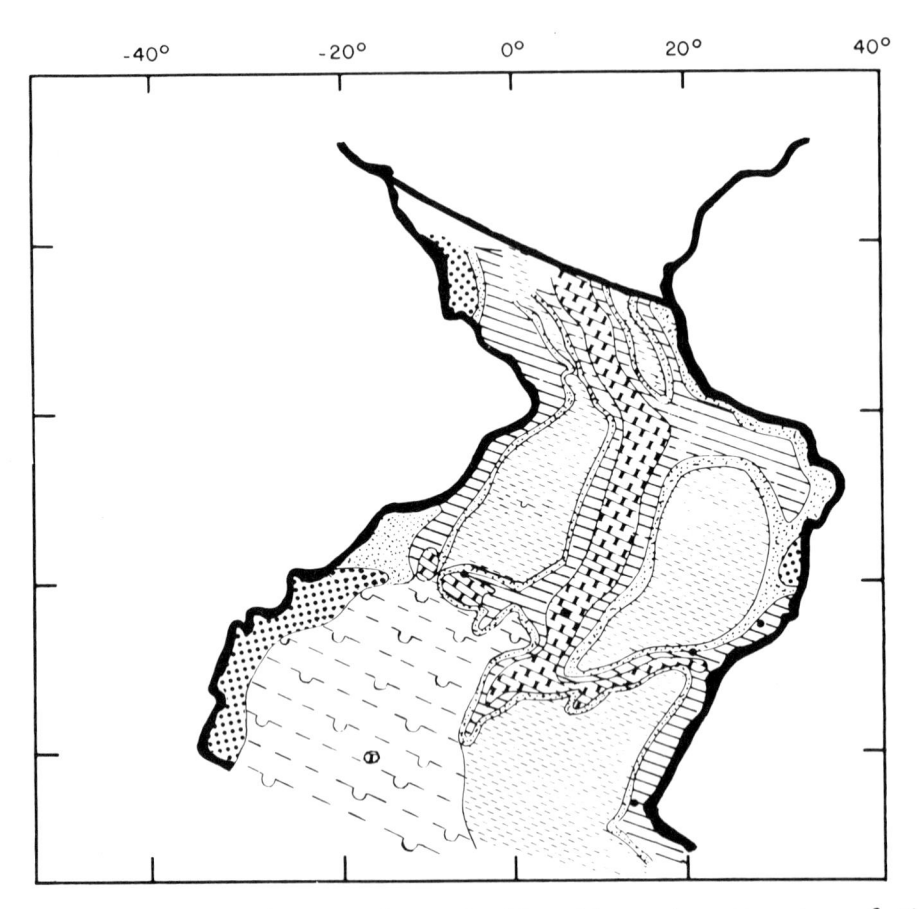

Figure 22. Lower Oligocene (anomaly 13 : 34 m.y.) sedimentary facies distribution based on DSDP leg 3, 36, 39, 40 data (MAXWELL et al., 1970; BARKER et al., 1974; PERCH-NIELSEN et al., in press; BOLLI et al., in press). CCD level around 3 000 m in the Cape and Argentine basins, 3 500 m in the Brazil basin, 4 000 m in the Angola basin. The CCD differentiation from basin to basin is due to the establishment of the AABW circulation. For facies symbols see figure 5.

TABLE 1: Parameters of reconstruction of positions of South America with respect to Africa

EPOCH	MAGNETIC ANOMALIES	Age in M.Y.	LATITUDE	LONGTITUDE	ANGLE OF ROTATION
Albian[1]	MO	100	46.64 N	31.09 W	52.39
Conacian-Santonian[2]	34	86	63. N	36. W	33.8
Maastrichtian[2]	31	68	63. N	36. W	25.8
Lower Oligocene[2]	13	34	58. N	35. W	13.5

1. after SIBUET and MASCLE (in preparation)
2. after LADD (in press)

CONCLUSION

As mentioned in the introduction, paleoenvironmental oceanography is a new science which is progressing very rapidly. This study is not a synthesis but tries to offer a working model of the evolution of the sea-floor of the South Atlantic ocean, based on plate tectonics considerations and integrating recent deep sea drilling results. The emphasis is put on the methodology now available to reconstruct the ocean crust morphology and its sedimentary cover. Three main stages are recognized in the opening of the South Atlantic, a confined basin stage, from 130 to 80 M.y., an embryonic deep circulation stage from 80 to 35 M.y. and a modern stage from 35 m.y. to present. Paleobathymetric, deep paleo-water circulation and paleo-sedimentary facies maps at five different ages illustrate this schematic reconstruction.

ACKNOWLEDGEMENTS

We are grateful to Dr. Eric SIMPSON who invited us to present this paper at the International Oceanographic Congress in Edinburgh. The data on which a great part of this paper is based come from the Deep Sea Drilling Project and we gratefully acknowledge the invaluable information obtained from the Initial Reports descriptions, some of which are still in press (PERCH-NIELSEN et al.). One of us (M.M.) thanks her colleagues of leg 40 of the Glomar Challenger for letting her use data prior to publication of the Initial Report.

BIBLIOGRAPHY

ARRHENIUS, G.O., 1952: Sediment cores from the East Pacific. Rep. Swed. deep-sea Exped. (1947-1948), parts 1-4, 5, p. 1-288.

BARKER, P.F., DALZIEL, I.W.D. et al., 1974: Southwestern Atlantic, leg. 36 Geotimes, v. 19, n° 11, p. 16-18.

BERGER, W.H., 1968: Planktonic Foraminifera - selective solution and paleo-climatic interpretation - Deep-Sea Research, 15, 31-43.

BERGER, W.H., 1970 a: Planktonic foraminifera - selective solution and the lysocline; Marine Geology, v. 8, p. 111-138.

BERGER, W.H., 1970 b: Biogenous deep-sea sediments: fractionation by deep-sea circulation. Bull. Geol. Soc. Am., 81, p. 1385-1402.

BERGER, W.H., 1973, Deep-Sea Carbonates: evidence for a coccolith lysocline, Deep-Sea Research, vol.20, p. 917-921.

BERGER, W.H., 1974: Sedimentation of deep-sea carbonates - maps and models of variations and fluctuations, in "Marine Planktonic sediments", edit. by W.R. Riedel and T. Saito, Micropaleontology Press, New-York, (in press).

BERGER, W.H., and E.L. WINTERER, 1974: Plate stratigraphy and the fluctuating carbonate line, in "Pelagic sediments on Land and under the Sea", edit. by K.J. Hsu and H. Jenkyns, spec. Publ. Internat. Assoc. Sedim. vol. 1, p. 11-48.

BERGER, W.H., and P.H. ROTH, 1975: Oceanic Micropaleontology - Progress and Prospect; Reviews of Geophysics and space physics, vol. 13, n° 3, p.561-636.

BERGGREN W.A. and C.D. HOLLISTER, 1974: Paleogeography, paleobiogeography and the history of circulation in the Atlantic Ocean - in Studies in Paleo-oceanography, W.W. Hay, edit., Societ. of Econ. Paleontol. and mineralogists. Spec. public. n° 20 p. 126-186.

BISCAYE, P.E., 1965, Mineralogy and sedimentation of recent deep-sea clay in the Atlantic Ocean and adjacent seas and oceans. Bull. Geol. Soc. Amer., 76 - p. 803-832.

BISCAYE, P.E., and E.F. DASCH, 1968: Source of Argentine Basin Sediment, Southwestern South Atlantic Ocean. Geol. Soc. Amer., Abstracts, p. 28.

BISCAYE, P.E., V. KOLLA and K.K. TUREKIAN, 1976: Distribution of calcium carbonate in surface sediments of the Atlantic ocean. Geophys. Research., vol. 81 n° 15, p. 2595-2603.

BOLLI, H.M., W.B.F. RYAN et al., 1975: Basins and margins of the eastern South Atlantic; Geotimes, v. 20, n° 6.

BOLLI, H.M., W.B.F. RYAN et al. Initial Reports of the Deep-Sea Drilling Project, vol. 40, in press.

BORNHOLD, B.D., 1973: Late Quaternary sedimentation in the eastern Angola Basin - Woods Hole Oceanographic Institution - Technical Report (unpublished manuscript).

BRAMLETTE, M.N., 1961: Pelagic sediments, in : Oceanography, Lectures at the International Oceanographic Congress, pp. 345-366. Mary Sears, Editor. Amer. Ass. Advan. Sci. Publ. 67, New-York, 635 p.

BRAMLETTE, M.N., 1965: Massive extinctions in biota at the end of Mesozoic time, Science, 148, p. 1696-1699.

BROECKER, W.S., 1971: A kinematic model for the chemical composition of sea water; Quaternary Res., 1, p. 188-207.

BULLARD, E.C., EVERETT, J.E. and SMITH, A.G., 1965. The fit of the continents around the Atlantic. In : P.M.S. Blackett, E. Bullard and S.K. Runcorn (Editors), A symposium on Continental Drift. Phil. Trans. Roy. Soc. London, A, 1088 : 41-51.

BURCKLE, L.H. and D. STANTON, 1976: Distribution of displaced Antarctic diatoms in the Argentine Basin. Third symposium on Recent and Fossil Marine Diatoms, Kiel, September 9-13, 1974, edit. by Reimer Simonsen, 1975. Royal Octavo. V. III, p. 283-292.

CALVERT, S.E., and N.B. PRICE, 1971: Recent sediments of the south West African Shelf. In : The Geology of the East Atlantic Continental Margin (Ed. by F.M. Delany), p. 173-185. Inst. Geol. Sci. Rept. n° 70/16.

CHAMLEY, H., 1975: Influence des courants profonds au large du Brésil sur la sédimentation argileuse récente. IXe Congres International de sédimentologie, Nice 1975, Thème 8, p. 13-17.

CONNARY, S.D., 1972: Investigations of the Walvis Ridge and environs - Ph. D. dissert, Columbia Univ., 228 p.

CONNARY, S.D., and EWING, M., 1974 : Penetration of Antarctic water from the Cape Basin into the Angola Basin. J. Geophys. Res., vol. 79, n° 3, p. 463-469.

ELLIS, D.B., and T.C. MOORE, Jr., 1973 : Calcium carbonate, opal and quartz in Holocene pelagic sediments and the calcite compensation level in the South Atlantic Ocean. Journ. Marine Research, vol. 31, n° 3, p. 210-227.

EMERY, K.O., UCHUPI, E., BOWIN, C., PHILLIPS, J., and E.S.W. SIMPSON, 1975 : Continental margin off Western Africa - Cape St Francis (South Africa) to Walvis Ridge (South-West Africa); A.A.P.G. Bull., v. 59, n° 1, p. 3-59.

EWING, M., 1965 : The sediments of the Argentine Basin - Quarterly journ. of the Roy. Astronom. Society, vol. 6, p. 10-27.

EWING, M., CARPENTER, G., WINDISCH C. and J. EWING, 1973 : Sediment distribution in the Oceans. The Atlantic Geol. Soc. of Amer. Bull., v. 84, p. 71-88.

GARDNER, J.V., 1975 : Late Pleistocene carbonate dissolution cycles in the eastern equatorial Atlantic - in "Dissolution of Deep Sea Carbonates", spec. public. n° 13, p. 129-141, Cushman foundation for foraminiferal research.

GOLDBERG, E.D. and GRIFFIN, J.J., 1964 : Sedimentation rates and mineralogy in the south Atlantic - J. Geophys. Res., v. 69, p. 4293-4309.

HEIRTZLER, J.R., DICKSON, G.O., HERRON, E.M., PITMAN III, W.C. and LE PICHON, X., 1968. Marine magnetic anomalies, geomagnetic field reversals, and motions of the ocean floor and continents. J. Geophys. Res., 73 : 2119-2136.

HOLLISTER, C.D., and R.B. ELDER, 1969: Contour currents in the Weddell Sea. Deep-Sea Research, 16 p. 99-101.

JOHNSON, D., McDOWELL S.E. and J.L. REID Jr., 1975 : Antarctic Bottom Water Transport through the Vema Channel (in press).

KENNETT, J.P., BURNS, R.E., et al., 1972 : Australian - Antarctic Continental Drift, Palaeocirculation changes and Oligocene Deep-Sea Erosion; Science, vol. 239, n° 91, p. 51-55.

KENNETT, J.P., HOUTZ, R.E., et al., 1974 : Development of the circum-Antarctic Current, Science, 186, p. 144-147.

LADD, J.W., 1976. Sea-Floor spreading in the South Atlantic. Geological Soc. of America. In press.

LANGSETH, M.G., LE PICHON, X. and EWING, M., 1966. Crustal structure of the mid-ocean ridges, 5. Heat flow through the Atlantic Ocean floor and convection currents. J. Geophys. Res., 71 : 5321-5355.

LARSON, R.L., and J.W. LADD, 1973 : Evidence for the opening of the south Atlantic in the Early Cretaceous - Nature, vol. 246, p. 209-212.

LE PICHON X., and D.H. HAYES, 1971 a : Marginal offsets, fracture zones and the early opening of the South Atlantic. J. Geeophys. Research., v. 76, p. 6283-6293.

LE PICHON, X., EWING, M., and TRUCHAN, M., 1971 b : Sediment transport and distribution in the Argentine Basin, 2. Antarctic bottom current passage into the Brazil Basin. In "Physics and Chemistry of the Earth, VIII, edit. by L.H. Ahrens, pp. 31-48.

LE PICHON, X., J. FRANCHETEAU and J. BONNIN, 1973. Plate Tectonics. Elsevier, 312 p.

LE PICHON, X., SIBUET, J.C. and FRANCHETEAU, J., 1976. The fit of the continents around the North Atlantic ocean. Tectonophysics, v.38, 169-209.

LEYDEN, R., ASMUS, H., ZEMBRUSKI, S., and G. BRYAN, 1976 : South Atlantic Diapiric Structures, A.A.P.G. Bull., v. 60, n° 2, p. 196-212.

LISITZIN, A.P., 1971 : Distribution of siliceous microfossils in suspension and in bottom sediments. In "The micropaleontology of Oceans", p. 173-195, Edit. by B.M. FUNNEL and W.R. RIEDEL, Cambridge University Press, Cambridge, 828 p.

LONARDI, A.G., and EWING, M., 1971 : Sediment transport and distribution in the Argentine Basin, 4. Bathymetry of the continental margin, Argentine Basin and other related provinces. Canyons and sources of sediments. In : L.H. AHRENS (Edit.), Physics and Chemistry of the Earth, VIII, pp. 73-121.

MARGOLIS, S.V. and J.P. KENNETT, 1970 : Antarctic glaciation during the Tertiary recorded in sub-Antarctic deep-sea cores : Science, v. 170, p. 1085-1087.

MAXWELL, A.E. et al., 1970. Initial Reports of the Deep Sea Drilling Project, vol. III, 806 p. U.S. Government Printing Office, Washington.

McKENZIE, D.P. and SCLATER, J.G., 1969. Heat flow in the eastern Pacific and sea-floor spreading. Bull. Volcanol. 33 (1) : 101-118.

MELGUEN, M., 1976. Cretaceous and Cenozoic facies evolution and carbonate dissolution cycles in sediments from the Eastern South Atlantic (DSDP Leg 40) : The correlation with major changes in the South Atlantic paleoenvironment since its opening. Initial Reports of the Deep Sea Drilling Project, vol. 40. In press.

MELGUEN, M., and THIEDE, J., 1974 : Facies distribution and dissolution depths of surface sediment components from the Vema Channel and the Rio Grande Rise (Southwest Atlantic Ocean). Marine Geology, v. 17, p. 341-353.

MELGUEN, M., and J. THIEDE, 1975 : Influence des courants profonds au large du Brésil sur la distribution des faciès sédimentaires récents. IXe Congrès International de sédimentologie, Nice 1975, Theme 8, p. 51-55.

MELGUEN, M., BOLLI, H.M., et al., 1975 : Facies evolution and carbonate dissolution cycles in sediments from basins and continental margins of the eastern South Atlantic since early Cretaceous. IXe Congrès International de sédimentologie, Nice, 1975. Thème 8; p. 43-50.

METCALF, W.G., HEEZEN, B.C., and M.C. STALCUP, 1964 : The sill depth of the Mid-Atlantic Ridge in equatorial regions - Deep Sea Research, v. 11, p. 1-10.

MURRAY, J., and RENARD, A.F., 1891 : Report on deep-sea deposits based on the specimens collected during the voyages of the H.M.S. Challenger in the years 1872 to 1976. In "Challenger Reports", 525 p., HMSO, Edinburgh.

NOEL, D., and M. MELGUEN, 1977 : Nannofacies of Cape Basin and Walvis Ridge sediments (Leg 40). Initial Reports of the Deep Sea Drilling Project, vol. 40 - in press.

PERCH-NIELSEN, K., SUPKO, R., et al., 1975 : Leg 39 examines facies changes in south Atlantic; Geotimes, v. 20, n° 3, p. 26-28.

PERCH-NIELSEN, K., SUPKO, R., et. al., 1977. Site Chapters. Initial reports of the Deep-Sea Drilling Project, in press.

PRATJE, O., 1939 : Sediments of South Atlantic. Bull. Amer. Assoc. Petrol. Geol., 23, p. 1666-1672.

RAMSAY, A.T.S., 1974 : The distribution of calcium carbonates in deep sea sediments; in "Studies in Paleo-oceanography", edit. by Hay, W.C. - Societ. of Econom. Paleontolog. and Mineralog., Spec. public. n° 20, p. 58-76.

SCHNEIDERMANN, N., 1973 : Deposition of coccoliths in the compensation zone of the Atlantic Ocean, in Smith, L.A., and HARDENBOL, J. Eds., Proc. Symposium calcareous nannofossils, Gulf Coast Sec., Soc. Econ. Paleontologists and mineralogists, Houston, Texas, p. 140-151.

SCLATER, J.G., ANDERSON R.N. and BELL, M.L., 1971. Elevation of ridges and evolution of the central eastern Pacific. J. Geophys. Res., 76 : 7888-7915.

SCLATER, J. and D.P. McKENZIE, 1973 : Paleobathymetry of the South Atlantic. Geol. Soc. Amer. Bull. v. 84, p. 3203-3215.

SEIBOLD, E., 1970. Nebenmeere in humiden und ariden klimabereich. Geol. Rundschau; vol. 60, p.73-105.

SHACKLETON, N.J. and J.P. KENNETT, 1975 a : Paleotemperature history of the Cenozoic and the initiation of Antarctic glaciation: oxygen and carbon isotope analyses in DSDP sites 277, 279 and 281; in KENNETT, J.P., HOUTZ, R.E. et al., Initial Reports of the Deep Sea Drilling Project. volume XXIX, Washington (U.S. Government Printing Office), p. 743-755.

SHACKLETON, N.J., and KENNET, J.P., 1975 b : Late Cenozoic oxygen and carbon isotopic changes at DSDP site 284 - implications for glacial history of the northern hemisphere and Antarctica - in KENNETT, J.P., HOUTZ, R.E. et al., Initial Reports of the Deep Sea Drilling Project, volume XXIX, Washington (U.S. Government Printing Office), p. 801-807.

SHANNON, L.V., and M. Van RIJSWIJCK, 1969 : Physical Oceanography of the Walvis Ridge region - Investl. Rept. Div. Sea Fisheries South Africa, n° 70.

SIBUET, J.C. and J. MASCLE. South Atlantic equatorial fracture zones and plate kinematics, in preparation.

SIESSER, W.G., SCRUTTON, R.A., and E.S.W. SIMPSON, 1974 : Atlantic and Indian ocean Margins of Southern Africa, in "The Geology of continental Margins", C.A. BURK, C.L. DRAKE, Edit., Springer-Verlag, New York, p. 641-654.

STEEMAN NIELSEN, E., and A.E. JENSEN, 1957 : Primary oceanic production - Galathea Reports, v. 1, p. 47-136.

TAKAHASHI, T., 1975 : Carbonate chemistry of seawater and the calcite compensation depth in the oceans. In Dissolution of Deep-Sea Carbonates, edit. by W. SLITER, A.W.H. BE and W. BERGER, Spec. Public., 13, p. 11-26, Cushman Foundation for Foraminiferal Research, Washington, D.C.

TAPPAN, H., 1968 : Primary production, isotopes, extinctions and the atmosphere. Paleogeogr. Paleoclimatol. Paleoecol., 4, p. 187-210.

TARLING, D.H., and J.G. MITCHELL, 1976 : Revised Cenozoic polarity time scale. Geology, March 1976, p. 133-136.

THIEDE, J., L. PASTOURET and M. MELGUEN, 1974 : Sédimentation profonde au large du delta du Niger (golfe de Guinée) - C.R. Acad. Sc. Paris, t. 278, Série D, p. 987-990.

THIERSTEIN, H.R., 1977 : Biostratigraphy of marine Mesozoic sediments by calcareous nannoplankton, Third Planktonic Conference Proceedings (in press).

TREHU, A., J. SCLATER and J. NABELEK, 1976. The depth and thickness of the ocean crust and its dependence upon age. Bulletin de la Société Géologique de France, V.18, 917-930.

TUREKIAN, K.K., 1964 : The geochemistry of the Atlantic Ocean basin, Trans. N.Y. Acad. Sci., Ser. 2, 26, p. 312-330.

VAN ANDEL, T.H., 1975 : Mesozoic Cenozoic calcite compensation depth and the global distribution of calcareous sediments. Earth. Planetary Science letters, 26, p. 187-194.

VINE, F.J. and D.H. MATTHEWS, 1963 : Magnetic anomalies over oceanic ridges. Nature, 199, p. 947-949.

WATKINS, N.D., and J.P. KENNETT, 1971 : Antarctic bottom water - a major change in velocity during the late Cenozoic between Australia and Antarctica, Science, v. 173, p. 813-813.

WORSLEY, T., 1974 : The Cretaceous - Tertiary Boundary event in the ocean In HAY, W.W. (edit.), studies in Paleo-oceanography, Societ. of Econom. Paleontolog. and Mineralog., Spec. public., p. 94-120.

WUST, G., 1936 : Das Bodenwasser und die gliederung der Atlantischen Tiefsee. Dtsch. Atlant. Exped. "Meteor", 1925-1927, 6, p. 1-106.

CLIMATE AND THE COMPOSITION OF SURFACE OCEAN WATERS

EDWARD D. GOLDBERG

SCRIPPS INSTITUTION OF OCEANOGRAPHY
LA JOLLA, CALIFORNIA 92093

ABSTRACT

The climate of the earth may govern the surface ocean water concentrations of its highly reactive components (i.e. those chemicals with short residence times). Geological and biological processes that result in the mobilization of materials from the continents to the oceans are dependent upon such characteristics of climate as crustal temperature, rainfall and the extent and types of vegetation. The atmospheric fluxes of some metals to an oceanic area may be of the same order of magnitude as their fluxes from the mixed layer to the deep ocean.

The high atmospheric concentrations of such metals as copper, zinc, manganese, lead and nickel have been related to their volatilization from the earth's crust. The relative importance of high temperature processes (volcanism) and low temperature processes such as the sublimation of metals from rocks and soils, has not as yet been established. Nevertheless, the subsequent dry or wet fallout of volatilized species may regulate their levels in surface ocean waters. This appears to be the case for mercury. In addition, volatilized metals can be sorbed onto atmospherically transported rock and soil debris. Consequently upon fallout of these particles into marine waters, the metals may be desorbed. In the ocean adjacent to arid land areas such a process can significantly influence the metal contents of surface waters.

Biological processes potentially can influence surface water concentrations. The volatilization of plant exudates such as terpenes to the atmosphere can result in a flux of such substances or their degradation products to the oceans. Recently, field

studies have indicated that heavy metals in the plants may be
complexed by these organic species and the complexes may be volatilized. Perhaps of even greater importance are the high
temperature processes such as forest and plant fires which release
annually amounts of materials to the atmosphere comparable to
other fluxes in the major sedimentary cycle. The distribution of
complex polycyclic aromatic compounds over a wide range of depositional environments has been attributed to such activities. The
importance of inorganic species so mobilized is yet to be assessed.

Clearly, climate affects the magnitude of material transfer
to the atmosphere. The flow of river-borne phases and the preceding weathering processes are governed by climate. The quantitative importance of all such processes, coupled to a description
of a past climate, can in principle lead to a formulation of surface
ocean water composition for the time period of concern.

INTRODUCTION

Changes in the composition of the oceans over geological time
have constituted a field of inquiry visited by many investigators.
Usually the major constituents of seawater have been considered.
They have long residence times, of the order of millions of years,
and alterations in their abundances have been sought over such
periods. The behavior of the major constitutents in the weathering
cycle guided these studies. On the other hand, very little attention has been paid to the elements in very small concentrations in
seawater, such as the metals at micromolar levels or less. Their
species may be involved in biological processes and some have been
postulated to influence the overall productivity of the marine
system.

There are several reasons for this lack of activity with
regard to variations of metal contents of ancient oceans. First
of all, the processes today that regulate the fluxes of metals to
the oceans and the reactions that result in their removal from
seawater are poorly known. These substances are delivered to the
oceans from winds, rivers, glaciers and volcanic activity.
Whereas the major seawater constituents enter the oceans primarily
as dissolved and particulate phases of the rivers, some of the
trace metals may have different pathways. For example, the primary
flux of mercury from the continents to the oceans appears to take
place in the atmosphere, with the river flow movement being significantly less important (Weiss et al., 1971). But the factors
that determine the river or atmospheric concentrations of a metal
are still poorly defined. Thus, the parameters that define present day oceanic concentrations of heavy metals are as enigmatic
as those for the past.

Also, in addition to the movement of substances about the
surface of the earth as a consequence of weathering reactions,

there is the mobilization of metals by man. In some instances the
anthropogenic fluxes from one geological reservoir to another far
exceed the natural ones. Such is the case with lead going from
the continents to the atmosphere. The man-mobilized lead, as a
consequence of its use in internal combustion engines as an anti-
knock agent, usually exceeds and sometimes masks the natural lead
in nearly every situation, whether the measurements are made in
glaciers, the atmosphere, or seawater. Man's activity as a
geologic agent can complicate the identification and quantification
of natural processes that result in the seawater concentrations of
some metals.

The goal of this paper will be a consideration of a small part
of this general problem -- what processes result in the flow of
heavy metals from the continents to the marine environment through
the atmosphere. This initial entry to a consideration of metal
concentrations in ancient oceans on the basis of understanding
modern levels may provide a pattern to investigate in detail heavy
metal paths to the oceans occurring today and in the past.

Natural Processes (Non-biological)

The high concentrations of some metals in the atmosphere have
led Duce et al. (1975) and Goldberg (1976) to propose that
vaporization plays an important role in their environmental
chemistries. To gain an insight into the sources of atmospheric
trace metals, Duce et al. (op. cit) define an enrichment factor
EF for an element X in atmospheric particulates relative to the
earth's crust:

$$EF_{crust} = \frac{(X/Al)_{air}}{(X/Al)_{crust}}$$

where $(X/Al)_{air}$ and $(X/Al)_{crust}$ refer, respectively, to the ratio
of the concentration of X to that of Al in the atmosphere and in
the average crustal material. Values near unity are indicative
of a crustal origin for the atmospheric contents of the element.
Values significantly higher or lower than one suggest sources other
than crustal materials or fractionation processes relative to
aluminium for the element X entering the atmosphere. Enrichment
factors for a group of elements in the atmospheres of the North
Atlantic and of the South Pole are presented in Table 1.

The elements Cr, Zn, Cu, Pb, Sb and Se are markedly enriched
in the atmosphere (Table 1). The concentrations for some of these
elements vary over three orders of magnitude between these two
areas. Yet, the similarity of enrichment factors for the North
Atlantic and South Polar airs suggests a natural source for these
metals as opposed to an anthropogenic one. Duce et al. (op.cit.)
indicate that about 90% of the pollutants in the troposphere

Table 1: Mean enrichment factors for elements in the atmosphere of the North Atlantic and of the South Pole (Zoller et al., 1974; Duce et al., 1975).

Element	EF_{crust} for North Atlantic	EF_{crust} for South Pole
Al	1.0	1.0
Sc	0.8	0.8
Fe	1.4	2.1
Co	2.4	4.7
Mn	2.6	1.4
Cr	11.	6.9
Zn	110.	69.
Cu	120	93.
Pb	2200.	2500.
Sb	2300.	1300.
Se	10000.	18000.

originate in the Northern Hemisphere. Interhemispheric mixing takes place in times of the order of six months to a year. Yet the residence times of these substances in the atmosphere is of the order of weeks to a month. There clearly is inadequate time for the transport of these metals from the northern hemisphere to the south pole.

Such arguments support natural sources. Some of these elements with high enrichment factors have volatile compounds. Duce et al. (op.cit.) acknowledge both low and high temperature volatilization processes to transfer the metals to the atmosphere. They also indicate the possibility of biologically mediated vaporizations and of the fractionation of these metals during the formation of sea spray and suggest that these processes are worthy of assessment.

Goldberg (1976) noted that the abundance patterns of five elements in the atmosphere and in rains were similar to the order of volatility found in spectrographic studies with the dc arc. In airs and precipitation from both hemispheres and from several continents, from urban and rural areas and from all seasons of the year, the sequence of decreasing concentrations follows the ranking of Pb, Zn, Cu, Mn and Ni (Table 2). For the sulfides, oxides, sulfates, carbonates, silicates and phosphates the sequence of

Table 2: Trace metal concentrations in aerosols, rains and crustal rocks *

	Pb	Zn	Cu	Mn	Ni
Mean concentrations in South Pole atmosphere in units of 10^{-12} grams/m^3 (Zoller et al., 1974)	630	30	36	10	
English rain, Suffolk in units of 10^{-6} grams/liter (Peirson et al., 1974)	40	260	28	12	6.4
English atmosphere, Shetland Isle in units of 10^{-9} grams/kg of air (Peirson et al., 1974)	31	25	4	3.2	2.9
Atmosphere, Gas Platform, North, Sea, in units of 10^{-6} grams/kg of air (Peirson et al., 1974)	190	2000	82	60	23
Average concentrations in Tuscon Air in units of 10^{-6} g/m^3 (Ranweiler and Moyers, 1974)	0.47	0.15	0.41	0.013	0.021
Crustal abundance in ppm (Turekian and Wedepohl, 1961)	10-15	32-58	37-72	670-1000	37-72
Soils, in percent (Vinogradov, 1959)	1×10^{-3}	5×10^{-3}	2×10^{-3}	6.7×10^{-3}	4×10^{3}

*Concentrations are taken directly from references and have not been corrected to a simple set of units inasmuch as their order is of significance.

volatilization from the dc arc is similar. This concentration pattern does not resemble that found in crustal rocks or soils (Table 2) where, for example, manganese is up to an order of magnitude higher in concentration than the four other metals.

Goldberg (1976) proposed a low temperature volatilization to account for these observations and indicated that low vapor pressures of metallic species are sufficient to account for steady state fluxes to the atmosphere. On the basis of a model from kinetic theory, a relationship between vapor pressure p, the molecular weight of a substance M, its concentration in the atmosphere C and its concentration in crustal rocks, f, was derived:

$$p = 3.4 \times 10^{-2} (T/M)^{1/2} (C/f)$$

where T is the absolute temperature. The model assumes that vaporization takes place to a tropospheric height of 13 km. and that the vaporized materials are removed from the atmosphere every ten days by dry fallout or rain.

Computed values of p may be obtained with the data in Table 1. For Cu, an atmospheric concentration of 50×10^{-15} grams/cc and a crustal abundance of 50 parts per million yields vapor pressures for its species of 6.6×10^{-11} atmosphere at $300°K$ and 12×10^{-11} atmospheres at $1000°K$. Some insight into these numbers may be gained by noting that DDT has been dispersed about the earth, primarily in the vapor phases (Bidleman and Olney, 1974), and its vapor pressure is of the order of 2×10^{-10} atmospheres (Nisbet and Sarofim, 1972) at about $300°K$. Mean concentrations in the air over the Sargasso Sea were of the order of 0.42×10^{-16} g/cc (Bidleman and Olney, 1974).

Although there is a paucity of data on the high temperature volatilization of heavy metals from volcanoes and fumaroles, work from several investigations clearly suggests that such mobilizations do take place. Cadle et al. (1973) found substantial concentrations of heavy metals upon filters through which passed the fumes from Hawaiian volcanoes. After a 15 hour run with a gas flow of 15 liters per minute the following amounts of some elements were collected in nanogram quantities: Na, 500; Cr, 200; Pb, 125, Cu, 900; Mn, 250; Hg+Se, 80; As, 24; and Cd, 37.

But perhaps more indication of the mobilization of metals in these high temperatures processes comes from a consideration of the minerals produced (Stoiber and Rose, 1974). These workers examined the fumarole incrustations at 14 areas in Guatemala, El Salvador, Nicaragua and Costa Rica. Forty-seven minerals were identified. The most frequently found and abundant minerals are: sulfur, hematite, halite, sylvite, gypsum, ralstonite ($NaMgAl(F,OH) \cdot H_2O$), anhydrite, thenardite (Na_2SO_4) and langbeinite ($K_2Mg_2(SO_4)_3$). There were also five minerals containing copper and one containing lead, but none containing manganese. From a literature search there were reports of 51 other volcanic minerals not found in South America during this investigation. Five more copper minerals and 4 more lead minerals were reported. In addition, there were two minerals each containing manganese and arsenic and one mineral with either tin or zinc. One cannot escape the induction that the volatility of these elements and their compounds results in their ready movement about the earth's crust at high temperatures.

The relative contributions of high temperature and low temperature volatility processes to the heavy metal burden of the atmosphere have not as yet been made. The episodic nature of volcanic events is difficult to reconcile with the maintenance of constant levels of the heavy metals in the atmosphere. Still, the stratosphere could be a reservoir for heavy metals, where they spend a year or so residence, following volcanic injections. Weiss et al. (1975) explain the seasonal variations of concentration of copper in Greenland glacial strata by the fallout of stratospheric volcanic debris. The highest concentrations occurred in the fall of 1968

Table 3: The copper content in seasonal strata of Greenland glacier (Weiss et al., 1975)

Year	Season	Cu Content (ng/kg of water)
1971	Summer	162
1971	Spring	134
1970-1971	Winter	217
1970	Fall	1020
1970	Summer	625
1970	Spring	955
1969-1970	Winter	5120
1969	Fall	841
1969	Summer	415
1969	Spring	244
1968-1969	Winter	2500
1968	Fall	15700
1968	Summer	1920
1968	Spring	1480
1967-1968	Winter	1880
1967	Fall	165
1967	Summer	224
1967	Spring	523
1966-1967	Winter	1360
1966	Fall	653

and the winter of 1969-1970 with values of 15.7 and 5.1 ppb, respectively. There are also higher levels of copper in the fall of 1970 and in the winter of 1966 - 1967 over the samples taken in adjacent seasons (Table 3).

These data suggest that the copper peaks might be related to the periods of high stratospheric fallout that occur during the winter and fall at midlatitudes. It is reasonable to consider that copper was introduced into the northern hemispheric stratosphere by the Icelandic, Aleutian or other volcanic events.

Perhaps mercury is the best documented example of a metal whose atmospheric concentrations appear to be governed by a vaporization process. The mercury content of the atmosphere is of the order of 1 - 2 nannograms/m^3 (Chris Junge, personal communication). Assuming the lower value to be representative and a washout time of ten days, Weiss et al. (1971) calculate a flux to the atmosphere of about 1.5×10^{11} grams per year, as calculated from the mercury contents of rains and from the mercury content of glaciers.

This amount of mercury cannot be mobilized by man's activities. The total mercury production in 1968 was 8.8×10^9 grams and perhaps a third of this was lost to the environment in applications as pesticides and as losses from chemical plants. The loss from the burning of fossil fuels is around 1.6×10^9 grams per year and from cement production about 10^8 grams per year. The roasting of sulfide ores at most releases about 2×10^9 grams/year. Thus, it appears that the atmospheric background concentration of mercury arises from the degassing of the lithosphere. The view finds support in the observation that mercury is enriched in sediments as compared to igneous rocks.

Natural Processes (Biological)

The fluxes of organic materials from plants to the atmosphere, through both low and high temperature processes, are comparable in value to other fluxes in the major weathering cycle. Are there any inorganic species accompanying the organics in the volatilization processes? There are a few reports in the literature that suggest an affirmative answer.

A world production of plant volatiles was estimated by Rasmussen and Went (1965) to be 438 megatons per year on the basis of measurements of atmospheric organic species in presumably uncontaminated samples. They identified alpha and beta pinenes, myrcene and isoprene, well known plant terpenes. The vaporization is greater in summer than in winter and is especially evident with the dying of leaves and the mowing of meadows.

Pea plants grown in solutions containing radioactive zinc released the metal to the atmosphere (Beauford et al., 1975). The rate of zinc mobilization appears to be of the order of 1 µg/h/m^2 of leaf. Any extrapolation of these results to global movements clearly would be unwise. Still, the data do suggest there may be an appreciable loading of the atmosphere with zinc, and perhaps other heavy metals, by plant emissions.

Further supportive evidence comes from field studies of plant exudates (Curtin et al., 1974). Polyethylene bags were placed around pine and fir trees and the exudates, condensed in these bags over periods of five to ten days, were collected and analyzed. The ash of the residue of volatile exudates contained <u>lithium</u>, beryllium, boron, <u>sodium</u>, <u>magnesium</u>, titanium, vanadium, <u>chromium</u>, manganese, iron, cobalt, <u>nickel</u>, <u>copper</u>, <u>zinc</u>, gallium, arsenic, <u>strontium</u>, yttrium, zirconium, <u>molybdenum</u>, <u>silver</u>, lead, <u>bismuth</u>, <u>cadmium,</u> <u>tin</u>, antimony, barium and lanthanum. The italicized elements were most markedly enriched in the exudates compared to the ash of the associated vegetation. Curtin et al. (<u>op</u>.cit.) suggest the complexing of some of the metals with terpenes may be the mechanism for the volatilization and urge the formulation of an airborne sampling program to assess this possibility.

Special attention should be paid to biological methylation processes for the transference of metals from the earth's surface to the atmosphere. Attention was decisively drawn to methyl metal compounds in nature with the Minimata Bay incident where thousands of Japanese fishermen and their families fell victim to poisoning by methyl mercury introduced to the marine environment in discharges from a chemical plant. Subsequently it was found that the principal form of mercury in fish was in the dimethyl mercury form, most probably produced enzymatically.

The natural combustion of plants and trees appears to contribute substantial quantities of particulates to the atmosphere and may as well introduce large quantities of volatile phases. Robinson and Robbins (1971) have reviewed estimates of the production of forest fire particles and present a range of 27 megatons/year for the U.S. alone with an addition of 17 megatons per year from slash burning and forest litter control operations to a global value of 150 megatons per year. There is in general a latitudinal zonation of forest types about the earth's surface. Tropical forests span the near equatorial regions. The combustion of tropical forest organic matter is through low temperature, biologically mediated processes. The high temperature burning takes place primarily in the mid-latitudes of the northern hemisphere where there are extensive forested areas of conifers and other temperate climate trees.

The principal wind systems that move the debris of high temperature burning about the earth are zonal. Coupling the activities of the winds to the zonal distribution of forests results in a latitudinal transport of the forest fire products with maximum fluxes at mid-latitudes. This phenomenon is reflected in the content of carbon (soot) particles in deep sea sediments which goes through a maximum at mid latitudes (Smith et al., 1973). The fallout of carbon to the earth's surface is estimated to be about 0.1 microgram/cm^2/year.

Blumer and Youngblood (1975) have attributed the distribution of a complex polycyclic aromatic assemblage over a wide range of depositional environments to a formation of the organics in forest fires and a subsequent dispersion in wind systems. Large quantities of these organics (anthracenes, phenanthrenes, pyrenes, fluoranthenes, chrysenes, triphenylenes, benzanthracenes and others) are formed in forest and prairie fires. Blumer and Youngblood suggest that natural fires form and that air currents disperse a complex mixture of organics that eventually accumulates in soils and in sediments.

What inorganic species might accompany the polycyclic aromatics and the elemental carbon? Forest fires create temperatures of 400° C and upwards. Clearly, there must be a group of inorganic species that are mobilized to the atmosphere by such episodes.

Anthropogenic Fluxes

Several of man's activities appear capable of mobilizing amounts of heavy metals to the atmosphere that are comparable or perhaps even greater than those introduced during natural processes. It should be noted that the anthropogenic sources are localized in the mid-latitudes of the northern hemisphere where the measurable alterations of the environment should be greatest.

Estimates of the fossil fuel burning fluxes of heavy metals have been made by Bertine and Goldberg (1971). Their model utilized the consumption of fossil fuels in 1967: 1.75×10^{15} g. of coal; 1.04×10^{15} g. of lignite; 1.63×10^{15} g. of oils and 0.66×10^{15} g. of natural gas. The literature was surveyed for reasonable values of the elemental contents of these fuels and estimates of the amounts released to the atmosphere were based upon the following assumptions: the fly ash released to the atmosphere from the burning of coals and oils is about ten percent of the total ash; and 50% of the coal is used in the manufacture of coke. The results are given in Table 4. For such elements as barium and mercury, fossil fuel mobilization to the atmosphere appears to be within an order of magnitude of the river fluxes of these metals to the oceans.

Selective volatilization can introduce the more readily volatile materials into the atmosphere in higher concentrations than those that are indicated in Table 4. Bertine and Goldberg (op.cit.) suggest that those elements that will be mobilized more effectively through volatilization processes may be found in the emissions from the dc arc as observed by spectrographers. On such a basis they suggest a preferential transfer, greater than that indicated in Table 4, for arsenic, mercury, cadmium, tin, antimony, lead, zinc, thallium, silver and bismuth.

Smaller but similar fluxes of heavy metals to the atmosphere appear to be a consequence of cement production. In 1970 the world cement production was about 5.7×10^{14} grams per year (USDI, 1972). About 95% of this output is Portland cement whose chemical formulation can be considered as one-third shale and two-thirds limestone. Cement is produced by roasting such a mixture at temperatures between 1450° and 1600°C attained in the burning zones of the kilns. Particulate emissions from cement manufacturing have been estimated to range between 116 and 171 kg per ton in the U.S. (EPA, 1972). Applying a value of 150 kg per ton to world production, there are 860,000 tons of particulates emitted annually to the atmosphere from this activity.

An estimate of the volatile substances emitted from cement manufacture can be obtained in the following way. About 36% of the ignition loss is in the liberation of carbon dioxide from limestone. Hence, the initial amount of limestone needed to produce the cement is $5.7 \times 10^{14} \times 2/3 \times 100$ g $CaCO_3$ per 56 g CaO = 6.8×10^{14} grams per year. The initial amount of shale required is $1/3 \times 5.7 \times 10^{14}$ grams per year or 1.9×10^{14} grams per year.

The time of passage through a kiln is two to four hours. Those metals whose oxides have boiling points below 2000° C may be expected to enter the vapor phase in substantial amounts. Those below 1500° C may be expected to be totally volatilized. As an approximation, I assume that those metal oxides with the following boiling points would be volatilized to the associated percentage of the amount in the cement components:

<	1500	100%
	1500 - 1600	50%
	1600 - 1700	40%
	1700 - 1800	30%
	1800 - 1900	20%
	1900 - 2000	10%
>	2000	1%

The predicted mobilization of some heavy metals to the atmosphere as a result of cement production is given in Table 5. The vaporization by cement production appears to be greater than that by fossil fuel burning for such elements as arsenic, boron, lead, selenium and zinc. This may be true for other metals for which data on boiling points are not presently available. In some cases where the oxides decomposed upon heating below 2000° C, the volatility of the metal was used.

Table 4: Amounts of some elements mobilized in the atmosphere as a result of weathering processes and the combustion of fossil fuels (Bertine and Goldberg, 1971)

Element	Fossil Fuel Concentration (ppm) Coal	Oil	Fossil Fuel Mobilization 10^9 grams/year	Weathering Mobilization 10^9 grams/year* River Flow	Sedimentation
Ba	500	0.1	70	360	500
As	5	0.01	7	72	
B	75	0.002	10.5	360	
Cd		0.01	0.002		11
Cs					
Pb	25	0.3	1.2	2.5	1.0
Li	65			110	12
Hg	0.012	10	1.6	2.5	11
Rb	500		70	1800	600
Se	3	0.17	0.45	7.2	
Zn	50	0.25	7	720	80

*Two different techniques were employed to calculate the weathering fluxes of the metals. One was based upon marine sedimentation computations and the other upon river flow. The difference of the results between the two methods usually reflects the inadequacies of existing data.

Table 5: Emissions of volatile oxides from the production of cement

Element	B.P. of Oxide	ppm in Shales	ppm in limestones	Grams in 5.7×10^{14} grams cement	Emission in grams per year
Sb	1550	1.5	0.2	4.2×10^8	2.1×10^8
As	315	13.	1.	3.2×10^9	3.2×10^9
B	ca 1860	100.	20.	3.2×10^{10}	3.3×10^{10}
Cd	d. 900-1000	0.3	0.035	8.1×10^7	8.1×10^7
Cs	690 (metal)	6	6	5.2×10^9	5.2×10^9
Pb	1744 (metal)	20	9	9.9×10^{10}	3.0×10^{10}
Li	1200^{600} mm Hg	66	5	1.4×10^{10}	1.4×10^{10}
Hg	356 (metal)	0.4	0.04	1.0×10^8	1.0×10^8
Rb	d. 400	140	3	2.0×10^{10}	2.9×10^{10}
Se	subl. 350	0.6	0.88	7.1×10^8	7.1×10^8
Tl	1457 (metal)	1.4	0.0	2.7×10^8	2.7×10^8
Zn	907 (metal)	95.	20.0	3.2×10^{10}	3.2×10^{10}

Other activities of man such as smelting of ores, the production of chemicals or agricultural field burning (Shum and Loveland, 1974) may introduce substantial amounts of metals into the atmosphere and subsequently into the oceans. As yet, there have been no attempts to evaluate their quantitative importance in the major sedimentary cycle.

An Enigma

There may be a movement of heavy metals from the surface of the sea to the atmosphere by an effect as yet not elucidated. In both rains and air collected from a gas platform in the North Sea the heavy metal concentrations were an order of magnitude higher, or even greater, than those collected from stations in England (Table 2, Peirson et al., 1974). The investigators suggest that the excess of trace metals in the rains most plausibly enters the atmosphere with sea spray, driven by wind from the ocean surface. Further, they point out that other workers have noted enrichments of heavy metals in the surface layers of the oceans. Thus, for the time being, one must consider the possibility of the sea surface acting as a source of heavy metals for the atmosphere.

Overview

A number of processes which can transfer metals from the continents to the oceans through the atmosphere have been identified; some or all may provide fluxes for a given metal which are comparable to those for rivers. But most are influenced by the climatic conditions of the earth. The characteristics of climate that are relevant to this discussion are mean temperature, annual precipitation and the type and amount of vegetation coverage for a given area.

Crustal rock volatility is temperature controlled; clearly, the higher the temperature, the greater the amount of degassing. Low temperature vaporization of metals from plants depends upon the extent and types of plants in a given area, again a consequence of temperature and the amount of rain. The higher temperature volatilisation from plant and tree burning again reflects the type and extent of forest and brush coverage.

The intensity of the weathering of exposed crustal rocks and the resultant transport of metals from the continents to the oceans via the rivers is governed by the parameters of climate.

The recent arguments that some of the heavy metal contents of seawater are too high as a result of contamination, primarily in the sampling phase (for example, see Sclater et al., 1976) are reducing the residence times of concerned elements often by values up to a factor of ten. Lower values of metal residence times are observed in coastal waters, those zones most significantly affected by both atmospheric and fluvial entries.

Thus on the basis of shorter residence times, multiple transport paths from the continents, and of climatic control of fluxes from the continents there is every reason to suspect that there have been non-trivial variations in metal contents of ocean waters over geologic times.

REFERENCES

Beauford, W., J. Barber and A.R. Barringer. Heavy metal release from plants into the atmosphere. Nature 256, 35-37 (1975).

Bertine, K.K. and E.D. Goldberg. Fossil fuel combustion and the major sedimentary cycle. Science 173, 233-235 (1971).

Bidleman, T.F. and C.E. Olney. Chlorinated hydrocarbons in the Sargasso Sea atmosphere and surface water. Science 183, 516-518 (1974).

Blumer, M. and W.W. Youngblood. Polycycle aromatic hydrocarbons in soils and recent sediments. Science 188, 53-55 (1975).

Cadle, R.D., A.F. Wartburg, W.H. Pollock, B.W. Gandrud and J. Shelovsky. Trace constituents emitted to the atmosphere by Hawaiian volcanos. Chemosphere No.6, 231-234 (1975).

Curtin, G.C., H.D. King and E.L. Mosier. Movement of elements into the atmosphere from coniferous trees in subalpine forests of Colorado and Idaho. J. Geochem. Explor. 3, 245-263 (1974)

Duce, R.A., G.L. Hoffman and W.H. Zoller. Atmospheric trace metals at remote northern and southern hemisphere sites: pollution or natural? Science 187, 59-61 (1975).

EPA. Compilation of air pollutant emission factors. Office of Air Programs, Research Triangle Park, North Carolina, U.S.A. (1972).

Goldberg, E.D. Rock volatility and the composition of aerosols. Nature 260, 128-9 (1976).

Nisbet, I.C.T. and A.F. Sarofim. Rates and routes of transport of PCBs to the environment. Environmental Health Perspectices 1, 21-38 (1972).

Peirson, D.H., P.A. Cawse and R.S. Cambray. Chemical uniformity of airborne particulate material and a maritime effect. Nature 251, 675-679 (1974).

Ranweiler, L.E. and J.L. Moyers. Atomic absorption procedure for analysis of metals in atmospheric particulate matter. Environ. Sci. Technol. 8, 152-156 (1974).

Rasmussen, R.A. and F.W. Went. Volatile material of plant origin in the atmosphere. Proc. Nat. Acad. Sci. (U.S.) 53, 215-220 (1965).

Robinson, E. and R.C. Robbins. Final Report API SRI Project. SCC-8507 (Stanford Research Institute (1971).

Sclater, F.R., E. Boyle and J.M. Edmond. On the marine geochemistry of nickel. Earth Planet. Sci. Letters 31, 119-128 (1976).

Shum, Y.S. and W.O. Loveland. Atmospheric trace element concentrations associated with agriculture field burning in the Willamette Valley of Oregon. Atmos. Environ. 8, 645-655 (1974).

Smith, D.M., J.J. Griffin and E.D. Goldberg. Elemental carbon in marine sediments: a baseline for burning. Nature 241, 268-270 (1973).

Stoiber, R.E. and W.I. Rose, Jr. Fumarole incrustations at active Central American Volcanoes. Geochim. Cosmochim. Acta 38, 495-516 (1974)

Turekian, K.K. and K.H. Wedepohl. Distribution of the elements in some major units of the earth's crust. Bull. Geol. Soc Amer. 72, 175 (1961).

U.S.D.I. Minerals Yearbook, 1970, Volume 1, United States Department of the Interior, Washington D.C., 1235 pp (1972).

Vinogradov, A.P. The geochemistry of rare and dispersed chemical elements in soils. Consultants Bureau Inc., New York, 209 pp. (1959).

Weiss, H., K. Bertine, M. Koide and E.D. Goldberg. The chemical composition of a Greenland glacier. Geochim. Cosmochim. Acta 39, 1-10 (1975).

Weiss, H.V., M. Koide and E.D. Goldberg. Mercury in a Greenland Ice Sheet: evidence of recent input by man. Science 174, 692-694 (1971).

Wood, J.M. Biological cycles for toxic elements in the environment. Science 183, 1049-1052 (1974).

Zoller, W.H., E.S. Gladney and R.A. Duce. Atmospheric concentrations and sources of trace metals at the South Pole. Science 183, 198-200 (1974).

OCEAN CIRCULATION AND MARINE LIFE

JOSEPH L. REID,[*] EDWARD BRINTON,[*] ABRAHAM FLEMINGER,[*]
ELIZABETH L. VENRICK[*] and JOHN A. McGOWAN[*]

[*]SCRIPPS INSTITUTION OF OCEANOGRAPHY, UNIVERSITY OF
CALIFORNIA, SAN DIEGO, LA JOLLA, CALIFORNIA 92093, USA

ABSTRACT

The geostrophic nature of the gross patterns of ocean circulation, with the wind-driven convergences within the anticyclonic gyres and the divergences within the cyclonic gyres, along the equator and the eastern boundaries, provides a set of quite different biological provinces. Because of their several climates and differences in vertical circulation, the various gyres contain different sets of nutrient and temperature characteristics, and these provide separate oceanic habitats. The principal cyclonic gyres are in the subarctic and subantarctic latitudes and have equatorward extensions along the eastern boundaries. They are cold, high in nutrients, and undergo large seasonal changes: a relatively small number of species is indigenous to these gyres, but the biomass is relatively large.

The principal anticyclonic gyres are in the subtropical zones, are warm and low in nutrients, with less seasonal variations than at higher latitudes: a larger number of species is indigenous to these gyres, but the biomass is small. The subequatorial zone contains a series of alternate eastward and westward flows, with associated ridging in thermocline depth. It is the warmest of the zones at the surface but is colder beneath the upper layer than the anticyclonic gyres. It also contains a large number of species and a large biomass, but there is substantial east-west variation: some species are confined to the east. There are also species that inhabit the zones between the subtropical and subarctic gyres. It is not certain how these maintain their geographical position within predominantly eastward flow. Perhaps some extend farther equatorward at greater depths, where the westward flow extends into somewhat higher latitudes; perhaps a resident population along the western boundary can maintain these mid-ocean patterns.

While these general patterns are observed in all oceans, there are notable differences. The tropical gyre of the North Atlantic, which is the warmest of the oceans, extends into higher latitudes; some species can extend all the way from the central North Atlantic into the Barents Sea and as far as Novaya Zemblya (75°N), at least in summer.

In the eastern tropical oceans the waters just beneath the upper layer are cold and, farthest from their surface sources, are lowest in oxygen concentration. Many of the shallow-living zooplankters that inhabit the anticyclonic gyres extend into the eastern tropical Pacific without showing any effect from these subsurface oxygen minima, but some of the more deeply vertically-migrating forms, while present in the surrounding waters, are excluded from the areas of lowest subsurface oxygen concentration. Some species, however, occupy only the waters in and above these low-oxygen layers.

Within this system of circulation the number of plankton species appears to vary with the biological province. The fewest species occur at latitudes poleward of about 45° within the subarctic and subantarctic gyres. The numbers increase abruptly equatorward from there within the anticyclonic gyres and remain high almost to the equator where there may be a small decrease. Perhaps the highest numbers are found where the subtropical and subequatorial types overlap, and where these are carried by the equatorial currents into the western boundary current and encounter a few of the local forms and some of the subarctic forms.

Phytoplankton species appear to have patterns somewhat different from those of the zooplankton. The subarctic and subantarctic gyres contain some bipolar species of zooplankton but few if any of phytoplankton, and the subequatorial zone appears to contain no endemic species of phytoplankton. Phytoplankton species tend to be more widespread and less environmentally specialized than zooplankton species. There is a higher percentage of cosmopolitan species and a higher percentage of circumglobal species in the subarctic environments; however, there are few if any bipolar species. There are no species restricted to environments which are not defined by one of the major circulation gyres. Apparently many of the mechanisms which effectively isolate populations of zooplankton, allowing genetic divergence, are not effective for phytoplankton. The dominance of asexual reproduction among phytoplankton may also be a factor.

INTRODUCTION

In this study we examine the major surface and near-surface circulation of the ocean, both horizontal and vertical, and the effect upon the distribution of heat, nutrients, and life. The gross circulation consists of gyral systems in high and middle latitudes and a system of east-west flows near the equator. Each of these has its own range of light, nutrients, and temperature. And each has its own biomass, with species different from the other systems. Our concern is principally with the upper-level open-ocean zooplankton, as there appears to be more nearly complete information on the distributions of these animals than on the deeper animals or the phytoplankton. We cannot deal here with the coastal, island, or shelf forms, or those inhabiting smaller seas, such as the Mediterranean, North Sea, Caribbean, Gulf of Mexico, or Sea of Japan. Such areas may have important local processes in addition to those imposed by the gross circulation, and the conjectures we make for the open ocean may not be applicable, nor the concepts derived there suitable to the open ocean.

CIRCULATION

There are some aspects of the circulation of the open ocean that seem obviously important in defining or regulating the biological provinces. We would like to call attention to these quite briefly (Fig. 1). These aspects of the circulation of the great ocean include not only the larger systems of horizontal surface flow but also the subsurface and vertical flow that appear to be continuous in time and large in spatial scale.

The upper layers of the ocean are affected by several processes. There is the wind-driven system of trade-wind drift at low latitudes and west-wind drift at high latitudes. The continental boundaries and the wind patterns result in north-south flow at the eastern and western boundaries, and these currents form large gyral systems within the great ocean. In each of the oceans there is an anticyclonic gyre with its axis roughly along the tropic circle, a cyclonic gyre in higher latitudes, and a system of east-west flows near the equator.

Below the surface layer there are a few notable differences (Fig. 2). The anticyclonic gyres lie farther poleward at increasing depth. What we call the North Equatorial Current between about 10°N and 20°N at the surface is found between 20°N and 35°N at 800 m depth. An even more extreme shift is seen in the South Pacific. Comparable features are seen in the Atlantic (MONTGOMERY and POLLAK, 1942). Another difference is the presence of countercurrents flowing poleward along the eastern boundaries; the subarctic and subantarctic gyres extend farther equatorward in the east.

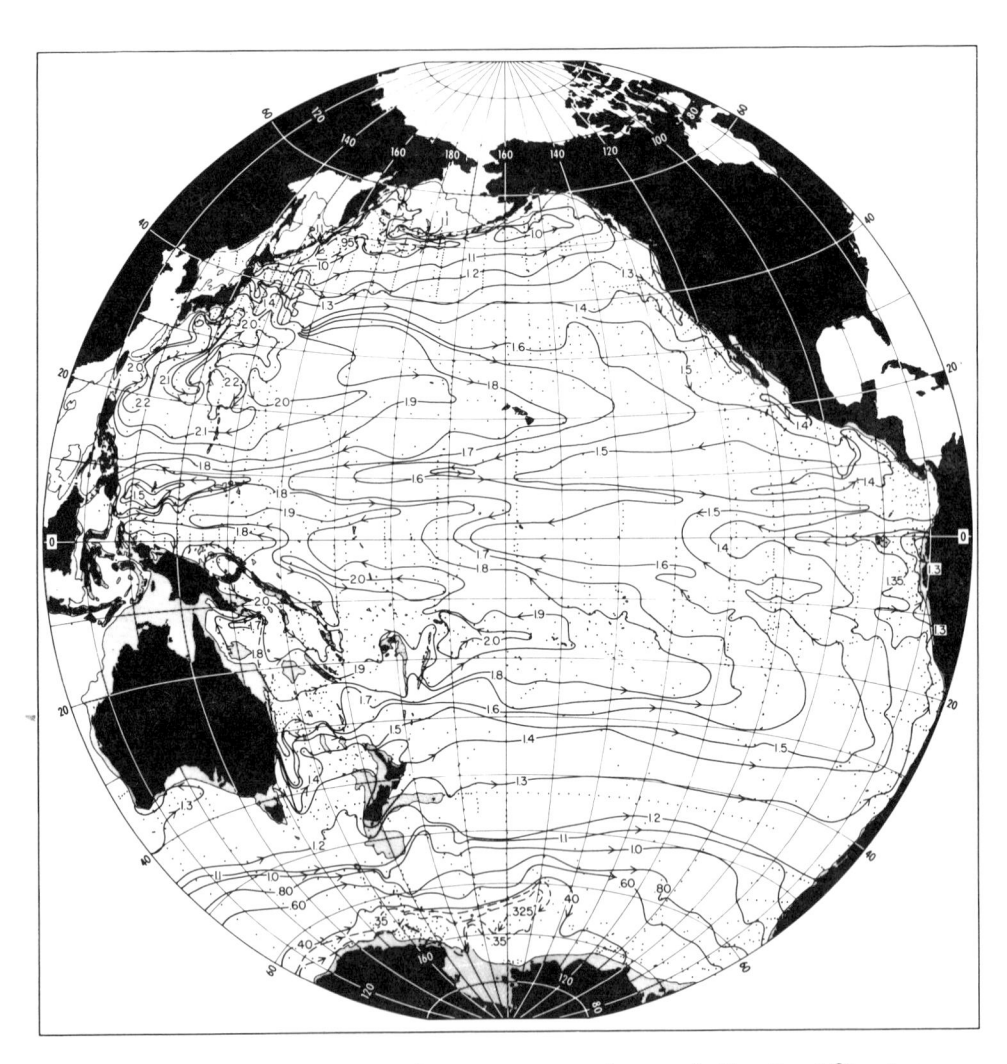

Fig. 1. The circulation of the upper waters of the Pacific Ocean, represented by the geopotential anomaly at the sea-surface relative to the 1000-decibar surface, in dynamic meters (10 J/kg). In the shaded areas the ocean depth is less than 1000 m (REID and ARTHUR), Fig. 1).

OCEAN CIRCULATION AND MARINE LIFE 69

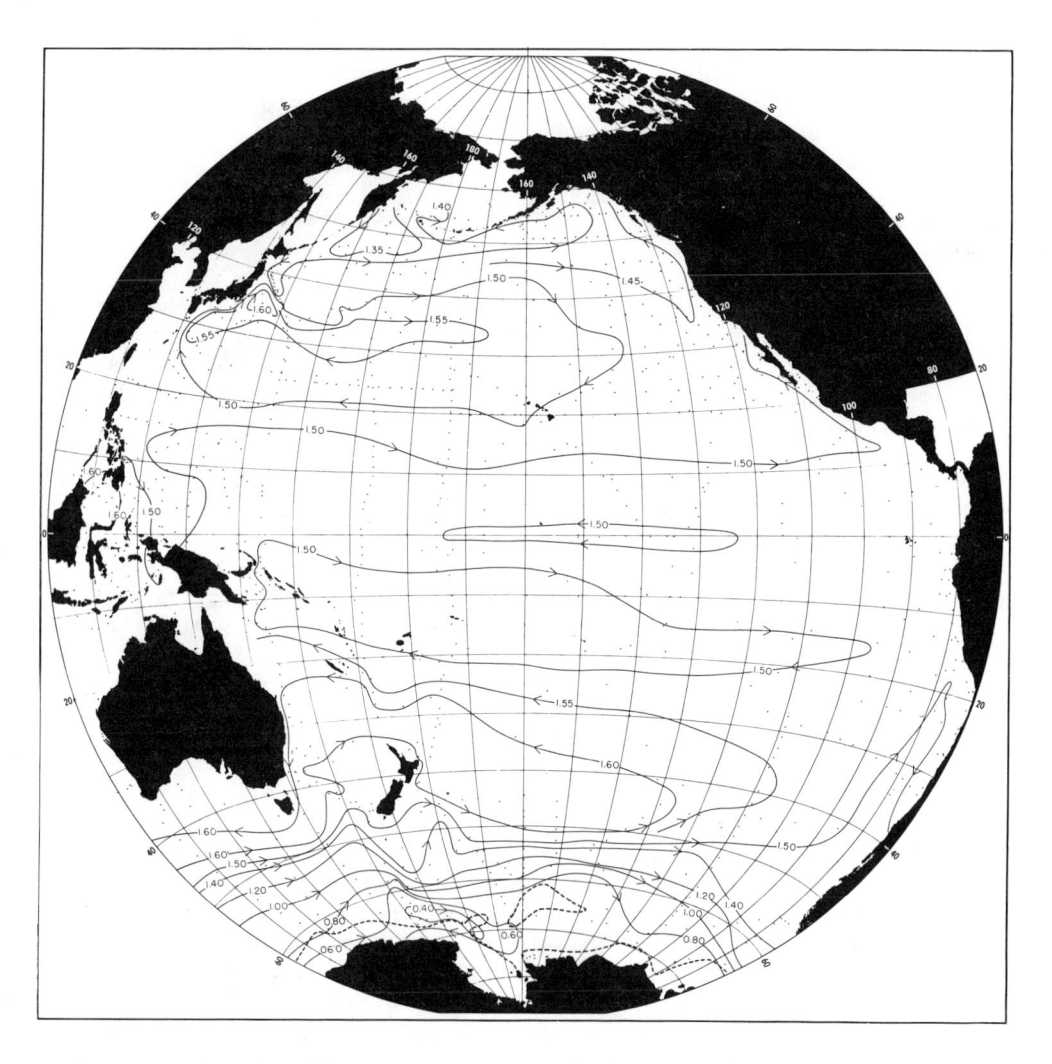

Fig. 2. The subsurface circulation in the Pacific Ocean, represented by the acceleration potential (from whose gradient the geostrophic flow relative to 2000 decibars may be calculated) on the surface where $\alpha_T = 80$ cl/ton, mean depth about 800 m. The dash line near Antarctica represents the intersection of the 80-cl/ton surface with the sea surface in southern summer. South of the dash line the quantity mapped is the geopotential anomaly at the sea surface with respect to the 2000-decibar surface (REID, 1965, Fig. 23).

The winds also transport surface waters from the zones of the westerlies and trades into the subtropical regions between. Thus, the anticyclonic gyres, which are within the subtropical latitudes, are areas of surface convergence, and the cyclonic gyres and the subequatorial zones are areas of surface divergence. This is a continuous effect of the trade winds and westerlies. As a result, surface waters are carried from the cyclonic gyres and the equatorial system into the subtropical gyres, where they sink and return to the equatorial system and cyclonic gyres and rise again toward the surface.

Within this system of convergences and divergences the upper layer is made thin in the high latitudes and near the equator, and thick within the subtropical regions (Fig. 3). At depths below the upper layer (at 200 m, for example) the temperature is higher near the tropic circles than it is at the equator.

Because the circulation is approximately geostrophic, the density also varies horizontally. The dashed lines on Fig. 3 represent two isopycnals, and they are seen to be deepest near the tropic circles and shallowest near the equator and in the high-latitude cyclonic gyres, reflecting the geostrophic nature of the circulation. Subsurface waters move and mix principally along isopycnal surfaces, and may thus change in depth as they move with the currents. One of the isopycnals illustrated here (Fig. 3) lies at about a thousand meters in the subtropical gyres but lies near or at the surface in the subantarctic gyre.

THE CHARACTERISTICS OF THE WATER

The conservative quantities such as salinity (Fig. 4) are concentrated within the convergent gyres, where evaporation exceeds precipitation, and diluted near the equator and in high latitudes, where rainfall dominates. It can be seen on a vertical section (Fig. 5) that both the lowest and the highest values of salinity are found at the sea surface.

If the nutrient phosphate were a conservative concentration, its distribution would be affected by only the same circulation and exchange processes that affect salinity, and its distribution might be much the same as the salinity. But instead, its distribution (Fig. 6) is vastly different--almost opposite, in fact. Values near the surface are low in the convergent areas and high in the divergent areas. The very highest values are found beneath the sea surface.

It is, of course, the activity of the plants and animals that causes this marked difference from the conservative characteristics.

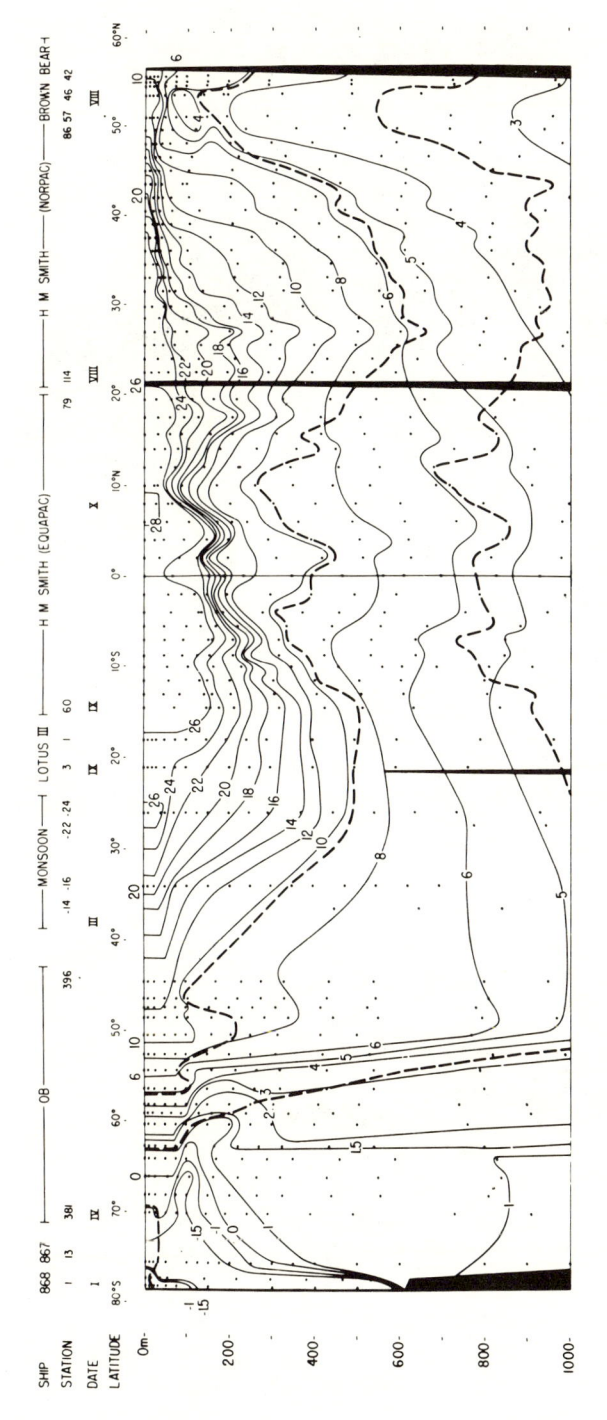

Fig. 3. Vertical section of temperature (°C) in the Pacific Ocean along approximately 160°W, from Antarctica to Alaska. The heavy dash lines on this and the other vertical sections (Figs. 5 and 6) indicate the depths of the 125-cl/ton and 80-cl/ton isopleths (REID, 1965, from Fig. 2).

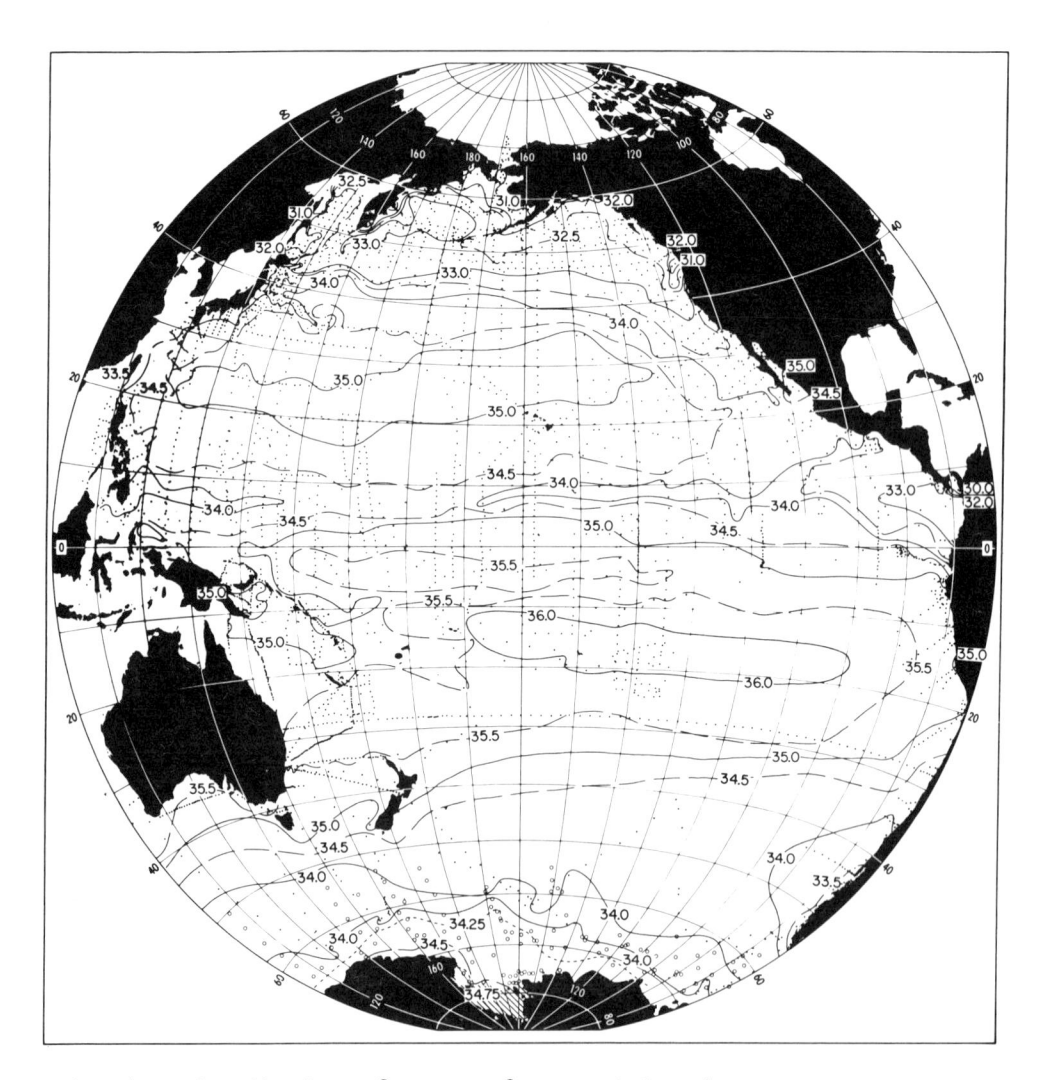

Fig. 4. Distribution of sea-surface salinity (parts per mil) in the Pacific Ocean in northern summer (REID, 1969, Fig. 3).

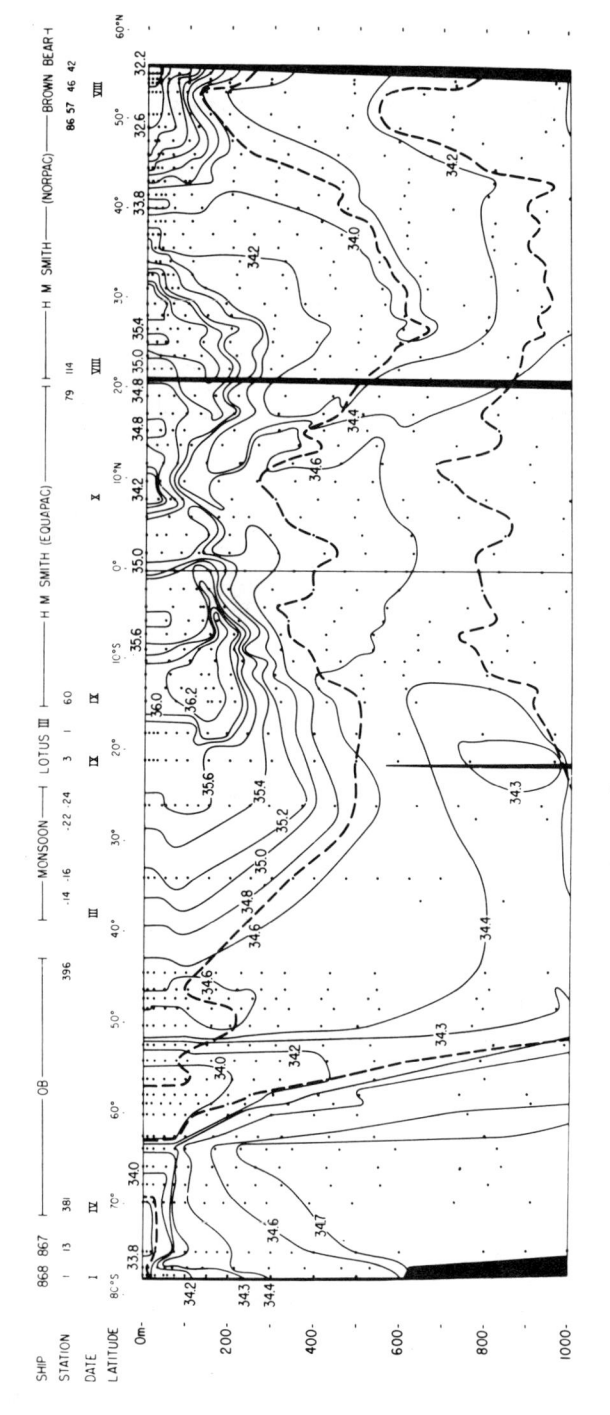

Fig. 5. Vertical section of salinity (parts per mil) in the Pacific Ocean along approximately 160°W, from Antarctica to Alaska (REID, 1965, from Fig. 5).

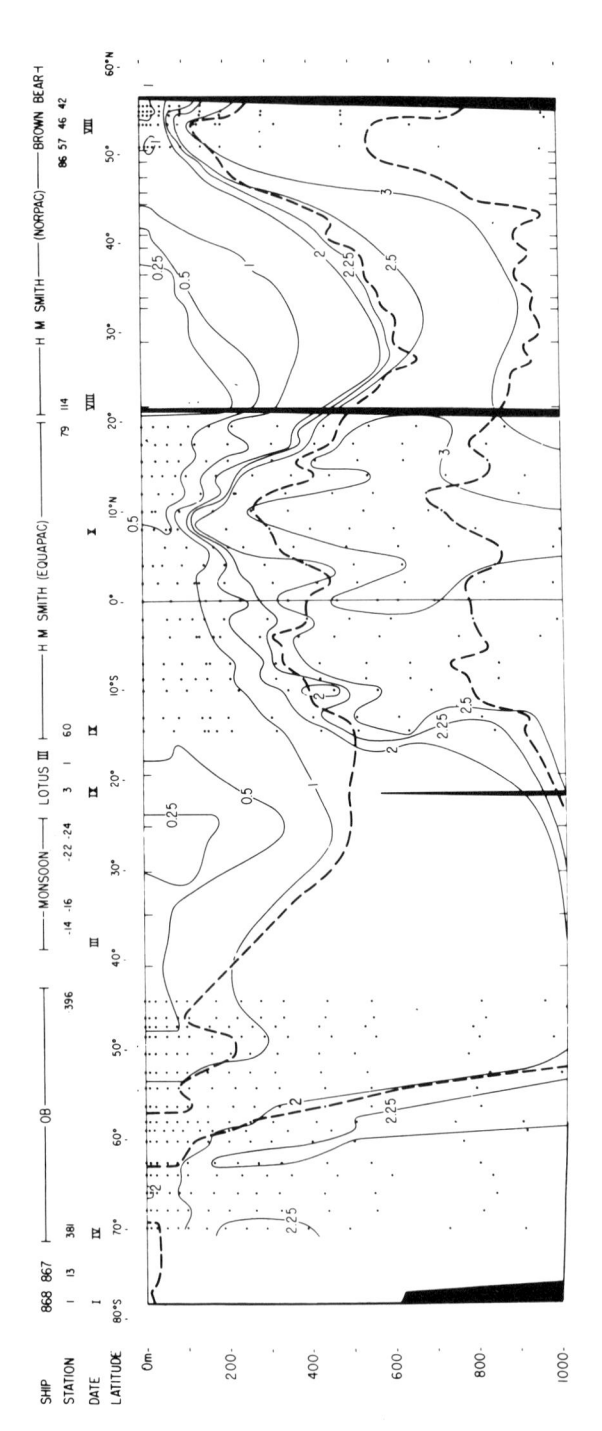

Fig. 6. Vertical section of inorganic phosphate-phosphorus (µg-at/ℓ) in the Pacific Ocean along approximately 160°W, from Antarctica to Alaska (REID, 1965, from Fig. 5).

Nutrients are used at the surface and regenerated in the upper few hundred meters. As a result, the upper, lighted layer is everywhere lower in nutrient concentration than the underlying waters, which show maximum values at different depths in different places. As the deeper waters circulate, the enriched waters from beneath the subtropical gyres are passed back to the equatorial and subarctic zones, where they lie at shallower depths because of their lower density. The effect is to move the enriched water below the lighted zone in the subtropics back to the lighted zones of the equatorial currents and the cyclonic gyres. Under this regime the equatorial and the high-latitude systems are continuously supplied with nutrients, and their concentrations remain high in spite of the use of nutrients within the lighted zone. Within the subtropical gyres the only replenishment is from below by vertical diffusivity and by horizontal convergence from the adjacent richer systems. Such processes are not rapid enough to maintain high concentrations of nutrients within the well-lighted upper waters of the subtropical gyres.

Because of this system of flow, the cyclonic gyres are always rich in nutrients (Fig. 7), and the major seasonal variations in productivity are more a result of changes in illumination than of seasonal variation in nutrients. The subtropical regions are always much lower in nutrient concentration. The equatorial current system is much richer than the subtropical gyres, and is almost as rich in the eastern tropical areas as in the high-latitude cyclonic gyres. There is some seasonal variation, but not nearly so much as in the higher latitudes, and the nutrient concentrations are never low.

The effect of the circulation upon the phosphate is seen even more clearly at 100 m (Fig. 8), where the distribution defines not only the areas of upper-layer divergence, but also reflects the principal features of the horizontal circulation as well. These features can also be seen in the Atlantic Ocean and the Indian Ocean; note that in the northern Indian Ocean the absence of a large-scale anticyclonic flow and the limited overturn lead to quite high values in the Bay of Bengal and the Arabian Sea. Note also that in the subarctic gyre of the North Pacific the concentrations are generally higher than in the corresponding part of the Atlantic. Other differences between these areas will be considered farther on.

Dissolved oxygen (Fig. 9) is maintained near saturation in the surface layer by exchange with the atmosphere and by photosynthesis, but is reduced beneath the upper layer by respiration and the regeneration of the nutrients. Where surface waters move to greater depths, they carry their high

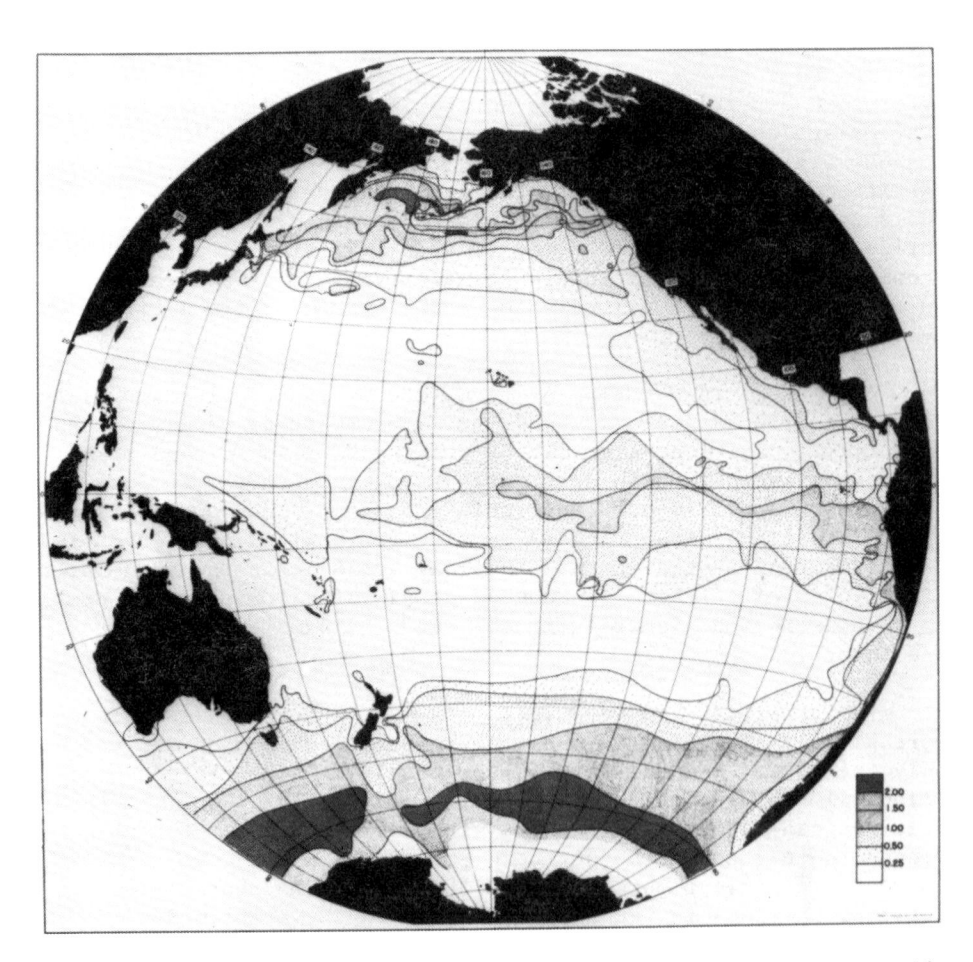

Fig. 7. Distribution of inorganic phosphate-phosphorus (µg-at/ℓ) at the surface of the Pacific Ocean (REID, 1962, Fig. 2b).

Fig. 8. Distribution of inorganic phosphate-phosphorus (μg-at/ℓ) at 100 m in the world ocean; Indian Ocean part adapted from WYRTKI (1971, Fig. 73), Pacific from REID (1962, Fig. 3a): Goode's Projection copyright by the University of Chicago Department of Geography.

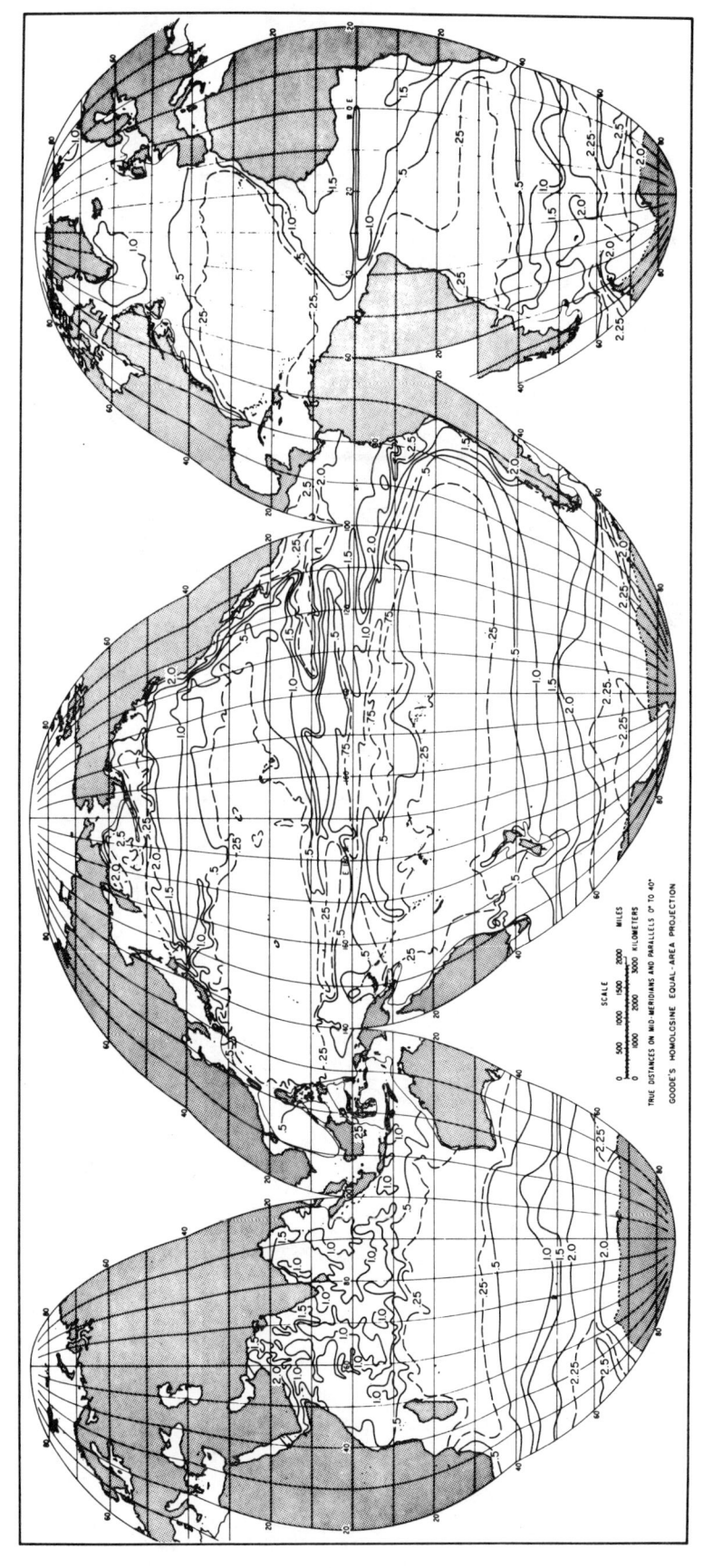

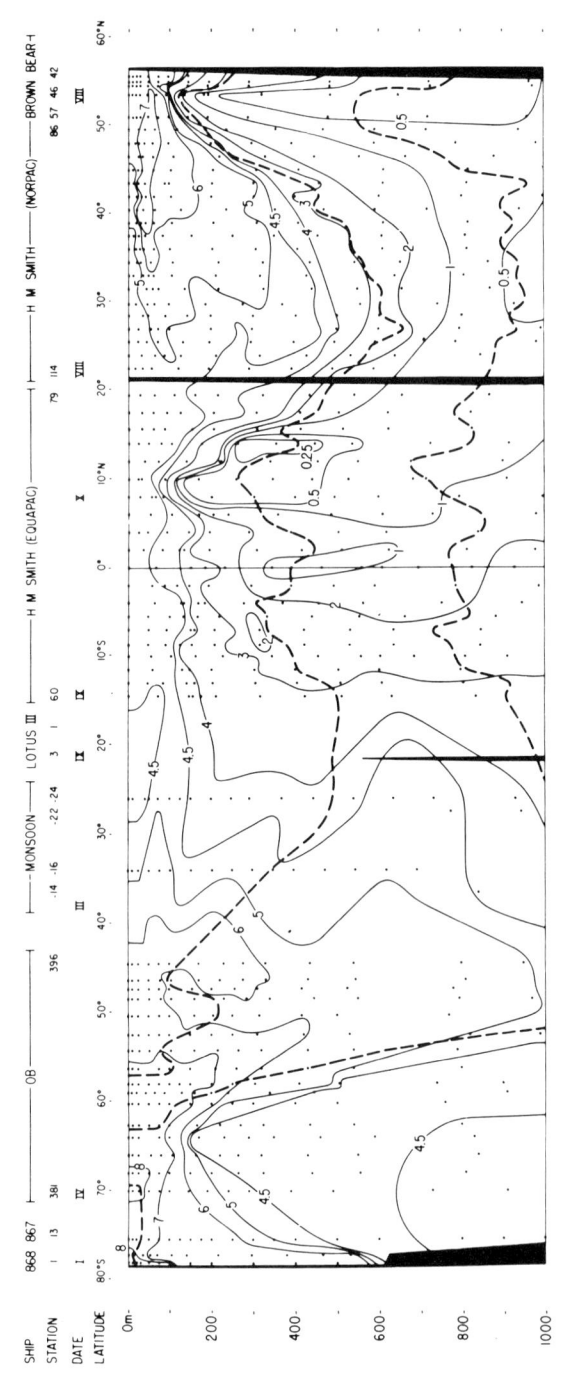

Fig. 9. Vertical section of dissolved oxygen (mℓ/ℓ) in the Pacific Ocean along approximately 160°W, from Antarctica to Alaska (REID, 1965, from Fig. 4).

oxygen concentrations with them, and these waters remain fairly well aerated for some distance from the sources. However, the general circulation, both vertical and horizontal, is such that the equatorial zone receives its subsurface waters only after they have travelled long distances around the gyres. As a result, the concentrations are very low in the waters beneath the upper layer in the equatorial zone.

The distribution of dissolved oxygen at 100 m depth (Fig. 10) reflects several principal features. One is the thickness of the upper layer. Where it extends to or below 100 m the oxygen is high, at or near saturation. Where it is thinner, near the equator and around the periphery of the anticyclonic gyres, the oxygen concentration is lower. The effect of the amount of primary productivity near the surface and the corresponding respiration and decay slightly deeper is seen in the eastern tropical area and along the eastern boundaries. A third effect, that of horizontal advection, is seen in the eastward extension of higher concentration along about 5°N into the eastern tropical Pacific.

Both this figure and the previous figure of 100-m phosphate (Fig. 8) indicate another important difference in the North Indian Ocean. There is no significant supply of oxygen-rich surface water to depths of 100 m there, and the oxygen is low and the phosphate high at 100 m throughout the entire North Indian Ocean.

THE SYSTEMS OF CIRCULATION AND THE BIOMASS

SVERDRUP (1955) presented a schematic map of probable relative productivity of ocean areas, based upon his concept of the rate at which the nutrients at the surface are renewed by physical processes. He was careful to state that it was not known whether his conclusions were valid, but his map seems to have held up very well to the subsequent findings, both as to regions of high nutrient concentrations, which is what in effect he based his map upon, and to the relative levels of primary productivity, which he assumed would follow. KOBLENTZ-MISHKE, VOLKOVINSKY, and KABANOVA (1970) have compiled the available measurements of primary productivity and prepared a world map, which resembles the map of 100-m phosphate (Fig. 8) to a marked degree and amply supports SVERDRUP's (1955) general concepts.

The preceding maps and sections have shown that the major circulation systems of the near-surface ocean include subarctic cyclonic gyres, subtropical anticyclonic gyres, and a system of zonal flows along the equator. The pattern of density structure imposed by the wind stress leads to quite different sets of characteristics within these systems. Because of their several climates and the differences in the vertical circulation, the

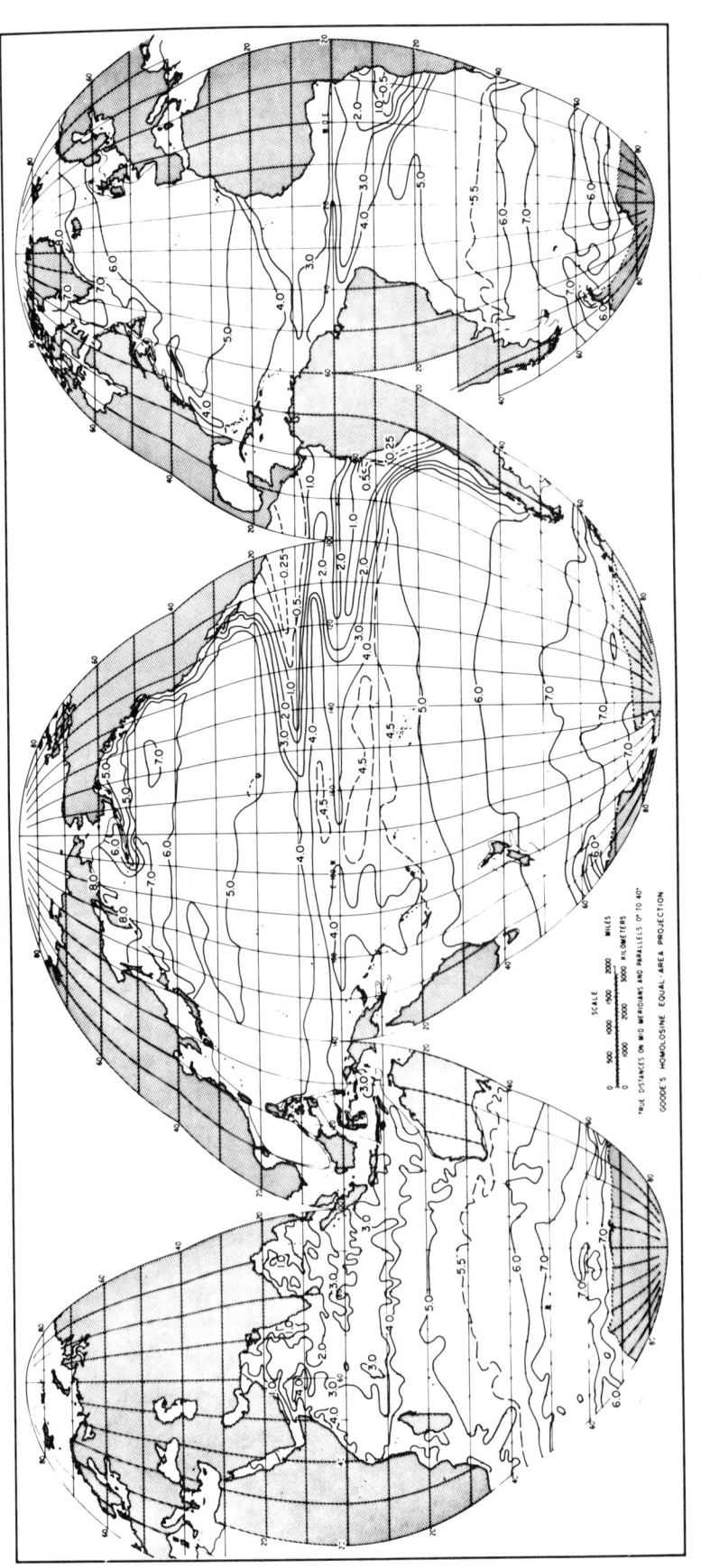

Fig. 10. Distribution of dissolved oxygen (ml/l) at 100 m in the world ocean; Indian Ocean part adapted from WYRTKI (1971, Fig. 72); Goode's Projection copyright by the University of Chicago Department of Geography.

various gyres contain different sets of temperature and nutrient characteristics and show different sorts of seasonal cycles, and as such the gyres may provide separate oceanic habitats (McGOWAN and WILLIAMS, 1973).

The principal cyclonic gyres are in the subarctic and subantarctic zones, and have equatorward extensions along some eastern boundaries. Their upper waters are cold and high in nutrients. Seasonal variation in growth may be expected to be high. It will be shown that a relatively small number of species is found within these gyres, but that the biomass is relatively large.

The principal anticyclonic gyres are in the subtropical zones, and their upper waters are warm and low in nutrients. Seasonal variation is not expected to be as high as within the high-latitude gyres. A larger number of species is found here than in the high-latitude gyres, but the biomass appears to be very much smaller (REID, 1962).

The subequatorial zone contains a series of alternating eastward and westward flow, with associated ridging in thermocline depth, and nutrients are abundant. It is the warmest of the zones at the surface, but is colder beneath the upper layer than are the anticyclonic gyres. It also contains a large number of species and a large biomass, but there are substantial east-west differences; some species are confined to the cooler, nutrient-rich eastern part.

The simplest evidence that these gyres are different oceanic habitats are maps of zooplankton volume (Fig. 11). The upper 150 m of the Pacific shows an enormous variation in zooplankton volume by a factor certainly over 100. Though there is a substantial scatter in the values, particularly in the equatorial and northern Pacific where the data are more numerous, the gross pattern stands out quite clearly. The concentrations within cyclonic systems and the equatorial zone are much higher than those within the anticyclonic gyres.

We can compare this with the transparency map made from Secchi disc readings (Fig. 12). The water transparency should be highest in the regions of low concentrations of plants and should be reduced where plants are abundant. The deepest readings were found within the anticyclonic gyre. The shallowest readings are within the subarctic gyre, from Kamchatka and the Kuril Islands across to North America and in the eastern boundary current and eastern tropical Pacific.

Maps of the upper-level biomass distribution in the Atlantic Ocean (Fig. 13) show the same general features as in the Pacific Ocean. This might be expected from the gross similarity of the

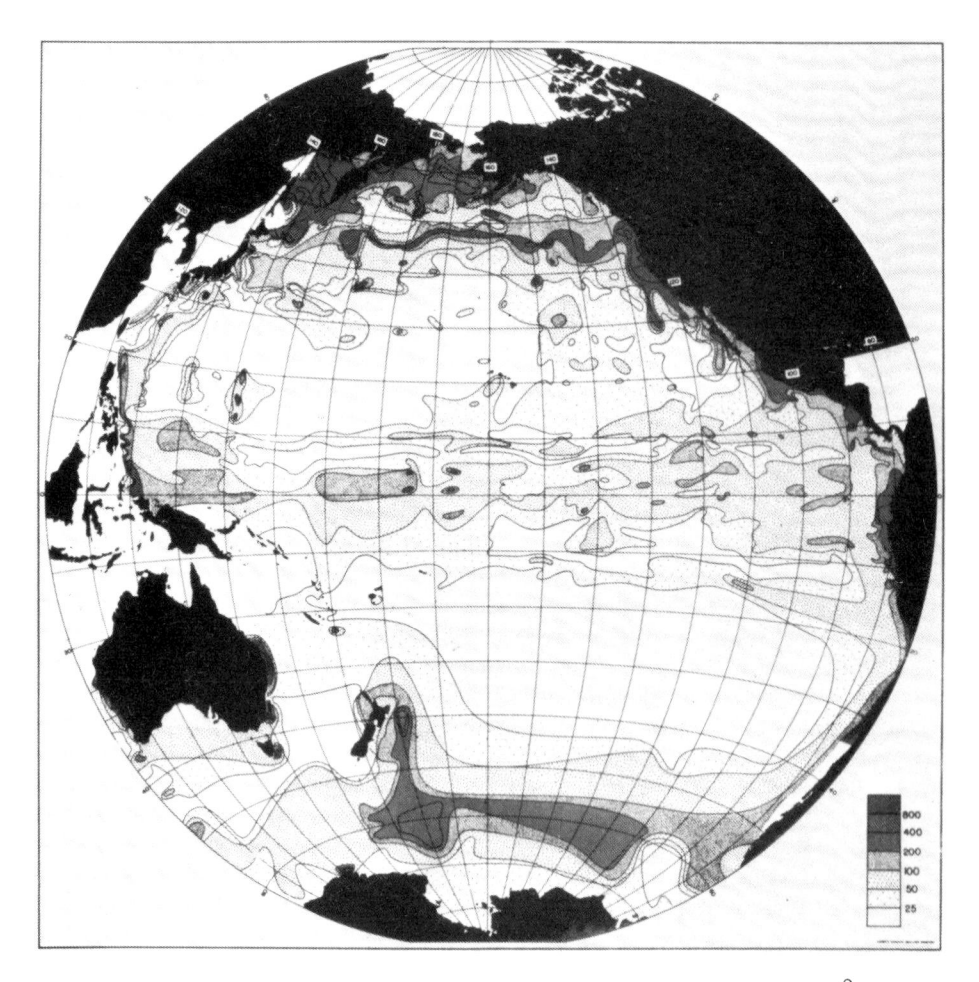

Fig. 11. Distribution of zooplankton volume (parts per 10^9 by volume) in approximately the upper 150 m of the Pacific Ocean (REID, 1962, Fig. 4b).

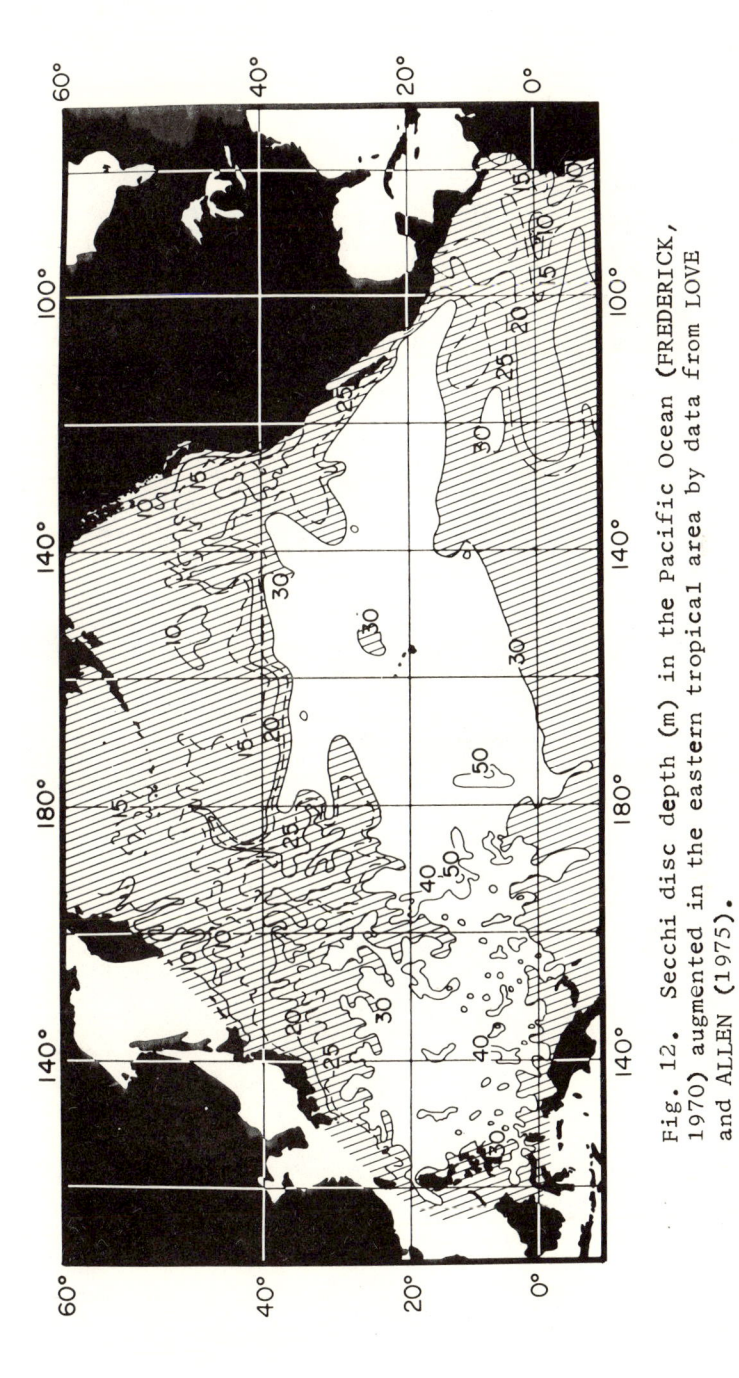

Fig. 12. Secchi disc depth (m) in the Pacific Ocean (FREDERICK, 1970) augmented in the eastern tropical area by data from LOVE and ALLEN (1975).

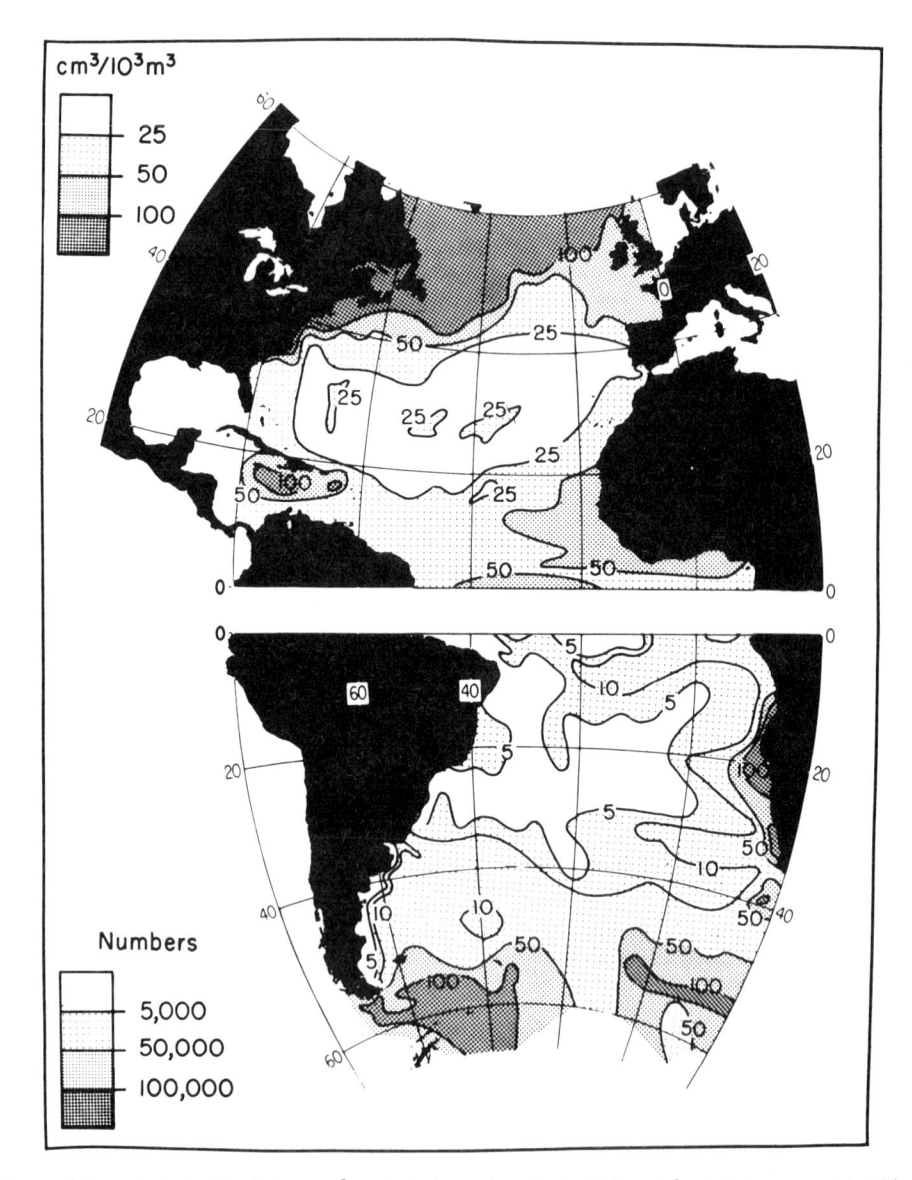

Fig. 13. Distribution of plankton in the Atlantic Ocean. North Atlantic part is zooplankton displacement volume in $cm^3/10^3 m^3$ from plankton tows in the upper 300 m, adapted from BE, FORNS and ROELS (1971). South Atlantic part is total number of plankton (micro- and nannoplankton, in thousands of individuals per liter in the upper 50 m, adapted from HENTSCHEL and WATTENBERG (1930).

upper-level Atlantic circulation (Fig. 14) to that in the Pacific Ocean. Biomass is lowest in the anticyclonic gyres of the North and South Atlantic Ocean, highest in the subarctic and subantarctic gyres, with intermediate values near the equator. Secchi disc measurements for the area north of 20°S mapped by DICKSON (1972) show patterns corresponding to these circulation systems, as does SCHOTT's (1942) map of water color for the entire Atlantic Ocean.

It is worthwhile to point out here the different shape of the North Atlantic and the shift in the position of the anticyclonic gyre, which extends farther north on the eastern side of the ocean than does that of the Pacific Ocean. The North Atlantic current flows northeastward into higher latitudes, and the waters just south of it are part of the subtropical anticyclonic gyre. We have already noted some effect of this shift: the phosphate concentration is lower. But it is in the temperature (Fig. 15) that the greater difference is seen. The northeastward extension of the subtropical gyre in the Atlantic makes the temperatures (at 200 m, for example) 6 to 8 degrees higher than those in the corresponding part of the Pacific.

From consideration of the standing stock of zooplankton, it appears that the equatorial zones can always support a large standing stock, in all seasons. A very large stock is supported by the subarctic cyclonic gyres in summer, and the anticyclonic gyres never show a standing stock as large as the other systems. Thus, the relation of standing stock of zooplankton to nutrients is a simple one: Areas permanently high in nutrients will be productive enough to sustain large standing stocks whenever light is sufficient; areas permanently low in nutrients will never produce or support a high stock of zooplankton. The equatorial current system and the eastern boundary currents are always high in zooplankton, and the cyclonic gyres have abundant zooplankton in summer, with enough stock surviving the winter to start next year's crop as soon as the light is adequate for the phytoplankton to become productive.

The effect of the movement of the water masses upon the biomass itself must be taken into account directly. The movement of the biomass is mostly a consequence of the movement of the water rather than of any horizontal swimming effort of the plants or animal plankton. Without some system of flow that allows a planktonic population to remain within a special environment, the only surviving species would be those forms which can endure the entire range of ocean characteristics--temperature, oxygen, food, for example. Yet we do recognize that most pelagic zooplankton species are not found everywhere, but that different species may occupy different parts of the ocean.

Each of these principal gyral systems spans a limited

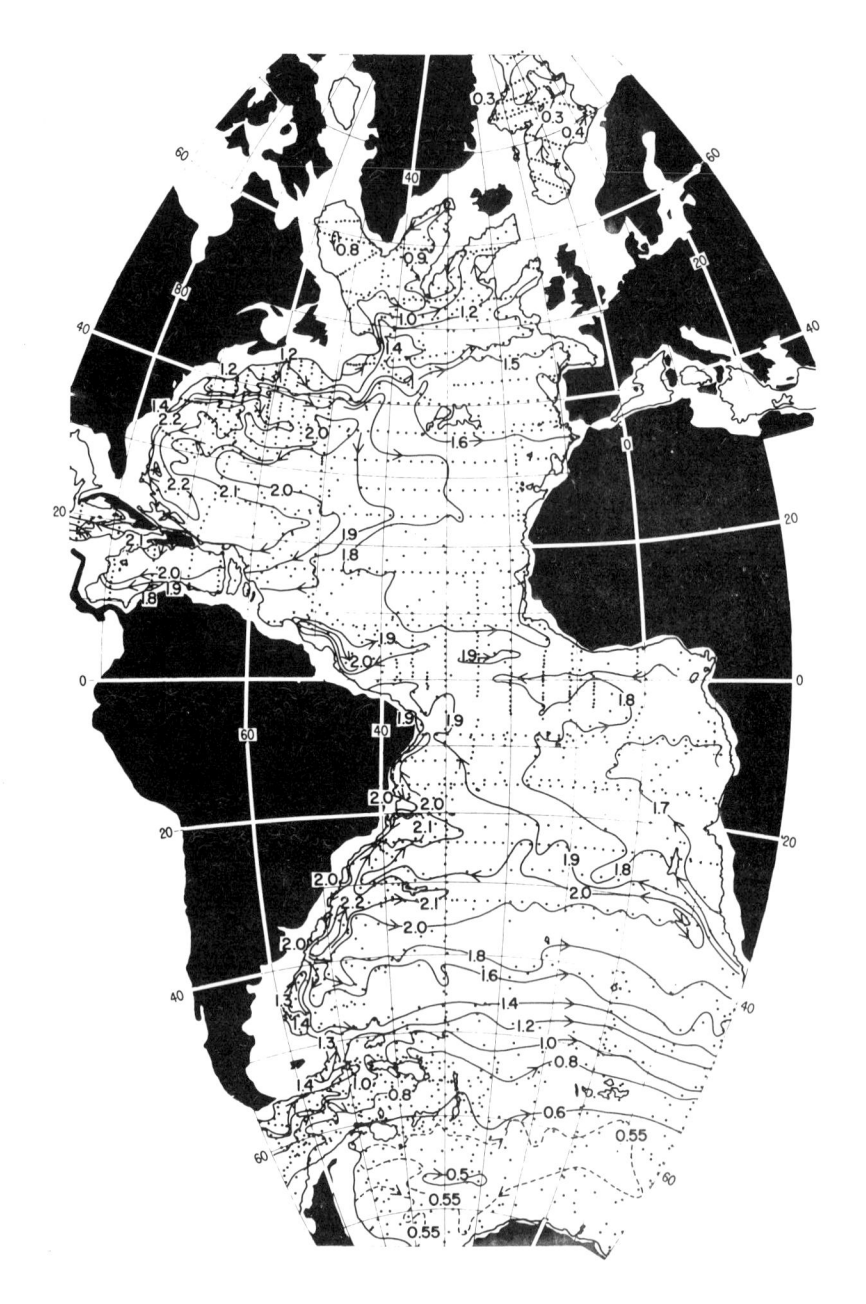

Fig. 14. The circulation of the upper waters of the Atlantic Ocean, represented by the geopotential anomaly at the sea-surface relative to the 2000-decibar surface, in dynamic meters (10 J/kg (REID, NOWLIN and PATZERT, in press).

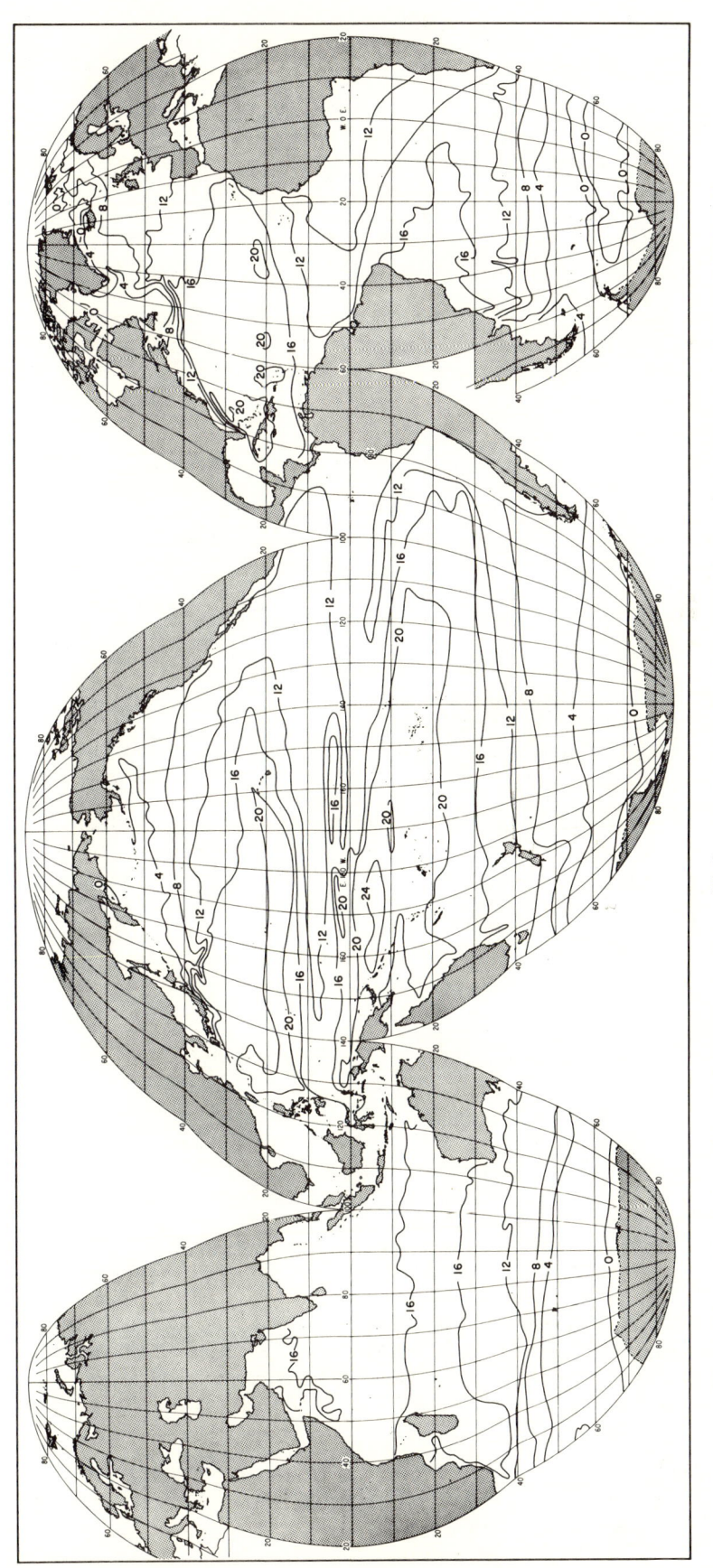

Fig. 15. Distribution of temperature (°C) at 200 m in the world ocean; adapted from SVERDRUP, JOHNSON and FLEMING (1942) with some augmentation from newer data in the Atlantic and Pacific oceans and from WYRTKI (1971, Fig. 76) for the Indian Ocean; Goode's Projection copyright by the University of Chicago Department of Geography.

latitudinal range and has a limited and separate range of characteristics. Each is at least partly protected against loss of organisms. Surface convergence helps to maintain organisms within the anticyclonic gyres. Coastal boundaries help to maintain them within the subarctic and subantarctic cyclonic gyres. We suggest, therefore, that these gyres might be the principal oceanic biological provinces (McGOWAN, 1974). We recognize, of course, that the gyres are interlocked with adjacent gyres along some of the northern and southern edges, and that mixing of the plankton as well as water takes place there. Some overlap is expected and observed (McGOWAN, 1974; HEINRICH, 1975a, b).

A special case is that of the boundary currents. The California Current and Countercurrent, for example, can be considered as an extension of the subarctic gyre, and the colder, richer waters from the subarctic can be carried equatorward through the latitudes of the subtropical gyre. In addition to the endemic, coastal populations, we find not only subarctic forms but also forms from the subtropical gyre mixed into this current, and even some forms from lower latitudes are carried into the system by the Countercurrent (for example, ALVARINO, 1965; BRINTON, 1967; FLEMINGER, 1967; McGOWAN, 1974). Similar distributions are seen in the area of the Peru-Chile Current (for example, BIERI, 1959; BRINTON, 1962), and the fauna of the eastern boundaries is a mixture of forms. Some are not only permanently present but can also maintain themselves by local reproduction. Others are present but in a less suitable environment and exist there only because they are carried in by the flow of water. A similar effect is observed in some of the western boundary currents, which are made up of waters from the subequatorial and subtropical zones. As will be seen, the species are very numerous in the subtropical gyres and the equatorial zones, and they combine in the western boundary currents to give perhaps the highest number of species in any part of the ocean.

DISTRIBUTION OF ZOOPLANKTON SPECIES

Many of the plankton species of the open ocean are found to be restricted to one gyral system: either the cyclonic gyres of high latitudes, the anticyclonic gyres of the subtropics, or to the subequatorial system of zonal currents. Some notable extensions of these patterns are seen in the eastern and western boundary currents. It has been observed also that many species of zooplankton inhabit both the subtropical and equatorial systems.

This is very much the same set of subdivisions that STEUER (1933) chose, based principally upon the copepods. HENTSCHEL (1942) and BEKLEMISHEV (1969) propose many more areas that are essentially subdivisions of STEUER's (1933). We find, however, that some of the texts and papers still make a simple division of warm-water and cold-water species. They note that the biomass of the cold water appears to be larger than that of the warm-water areas, and some of them still appear to assume that the warm-water biomass between about 40°N and 40°S is small everywhere, not recognizing the highly productive equatorial area with its large biomass. We note also that the lower-latitude species are not necessarily in warmer water. Those living just below the upper layer--below 75-100 m--in the eastern equatorial zone are in colder water near the equator than along the tropic circles. The depth range from 200 to 400 m is not warmest at the equator but near the tropic circles.

The subarctic and subantarctic gyres are cold and nutrient-rich. In the northern hemisphere they are partly land-locked in both the Atlantic and Pacific. The euphausiid *Euphausia pacifica* (Fig. 16) extends through the entire subarctic gyre of the North Pacific Ocean, including its extension southward along North America. And a pteropod, *Limacina helicina*, has about the same pattern (Fig. 17), also extending southward in the eastern boundary extension of the subarctic gyre (McGOWAN, 1963). Similar examples are to be found in other taxa of planktonic animals. For example, the chaetognath *Sagitta elegans* (BIERI, 1959), in addition to ranging southward in the California Current, is found in the subsurface extension of the Oyashio (Fig. 2) off Sagami Bay, Japan (MARUMO, 1966).

The much-studied copepod *Calanus finmarchicus* (Fig. 18) is such an Atlantic species, occurring throughout the subarctic gyre of the North Atlantic and extending somewhat southward along the eastern boundary. Another *Calanus*, *C. glacialis* (Fig. 19), inhabits the Arctic and extends into the colder parts of the subarctic gyres and the Okhotsk Sea. Only a few subarctic species have distributions which indicate connections between Atlantic and Pacific oceans through the Arctic: for example, the copepod *Calanus glacialis* and perhaps the euphausiid *Thysanoessa inermis* (Fig. 20). In the Pacific, another euphausiid, *T. longipes*, shows almost the same pattern, inhabiting only the northernmost part of the gyre (BRINTON, 1962). The ranges of the Atlantic-Pacific subarctic species are analogous with the circumglobal antarctic species.

The plankton fauna of the Antarctic (Southern) Ocean is also made up of species having distributional ranges which are linked with gyral circulations. Two copepods, *Calanus simillimus* and *C. propinquus* (Fig. 19), are found within the subantarctic cyclonic gyre. The pelagic tunicate *Salpa thompsoni* lives in the zone of 40°S to 70°S, which encompasses most of the Antarctic Ocean (FOXTON, 1961). And three species of a lineage within the genus *Euphausia*, that is, *E. crystallorophias*, *E. frigida*, and *E.*

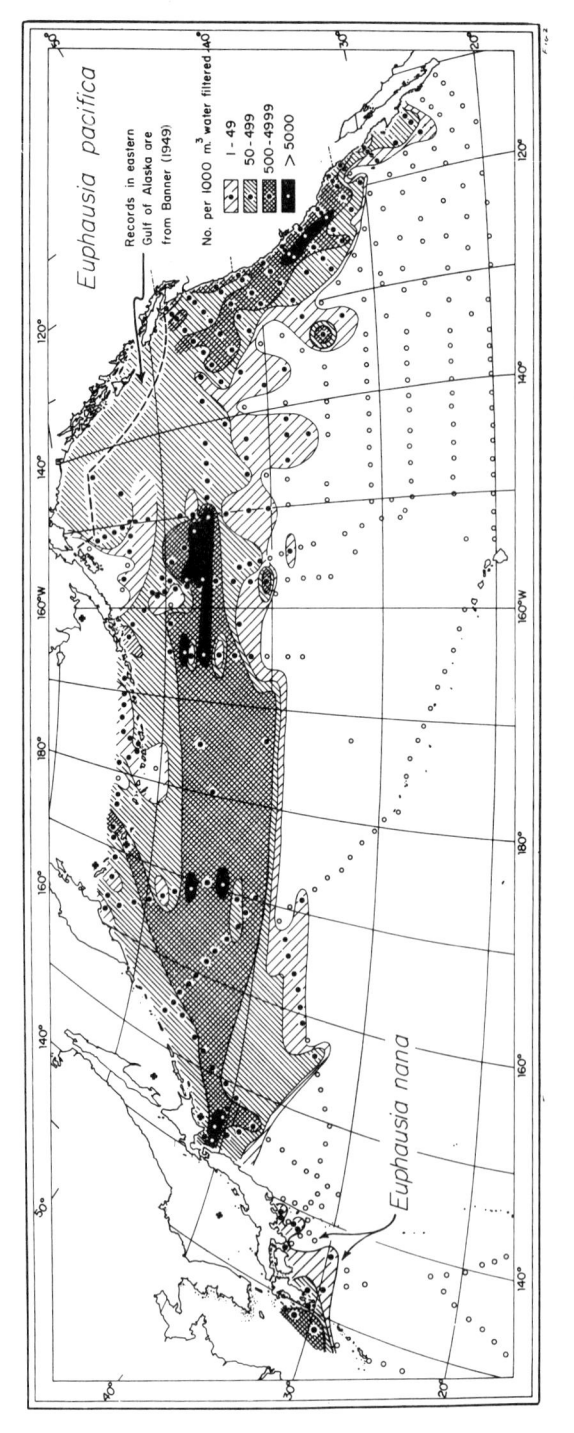

Fig. 16. Geographical distributions of Euphausia pacifica and E. nana in the Pacific Ocean (BRINTON, 1962, Fig. 28). Reprinted by permission of the University of California Press.

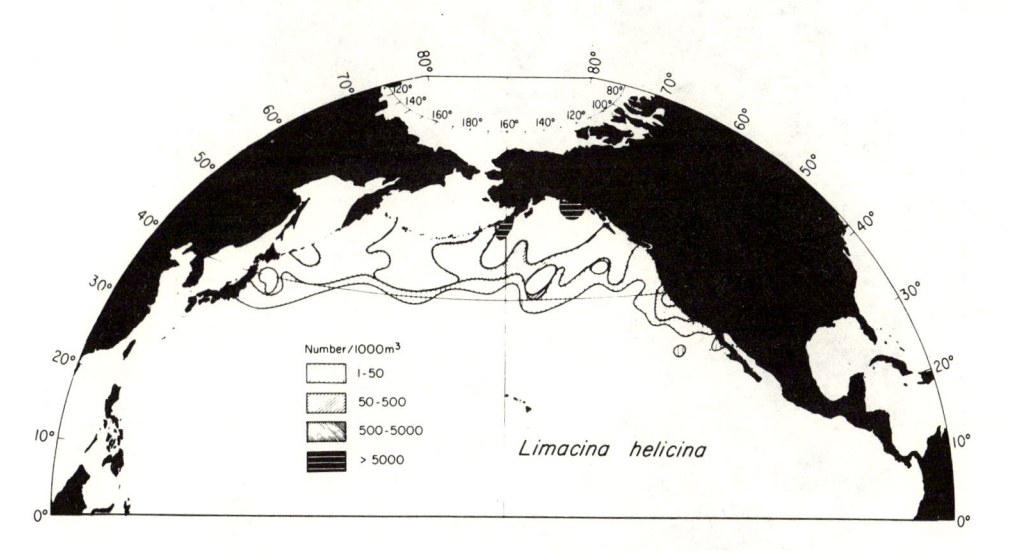

Fig. 17. Geographical distribution of *Limacina helicina* in the Pacific Ocean (McGOWAN, 1963, Fig. 5).

Fig. 18. Geographical distribution of the subarctic calanoid copepod Calanus finmarchicus in the Atlantic Ocean (adapted from FLEMINGER and HULSEMANN (in press).

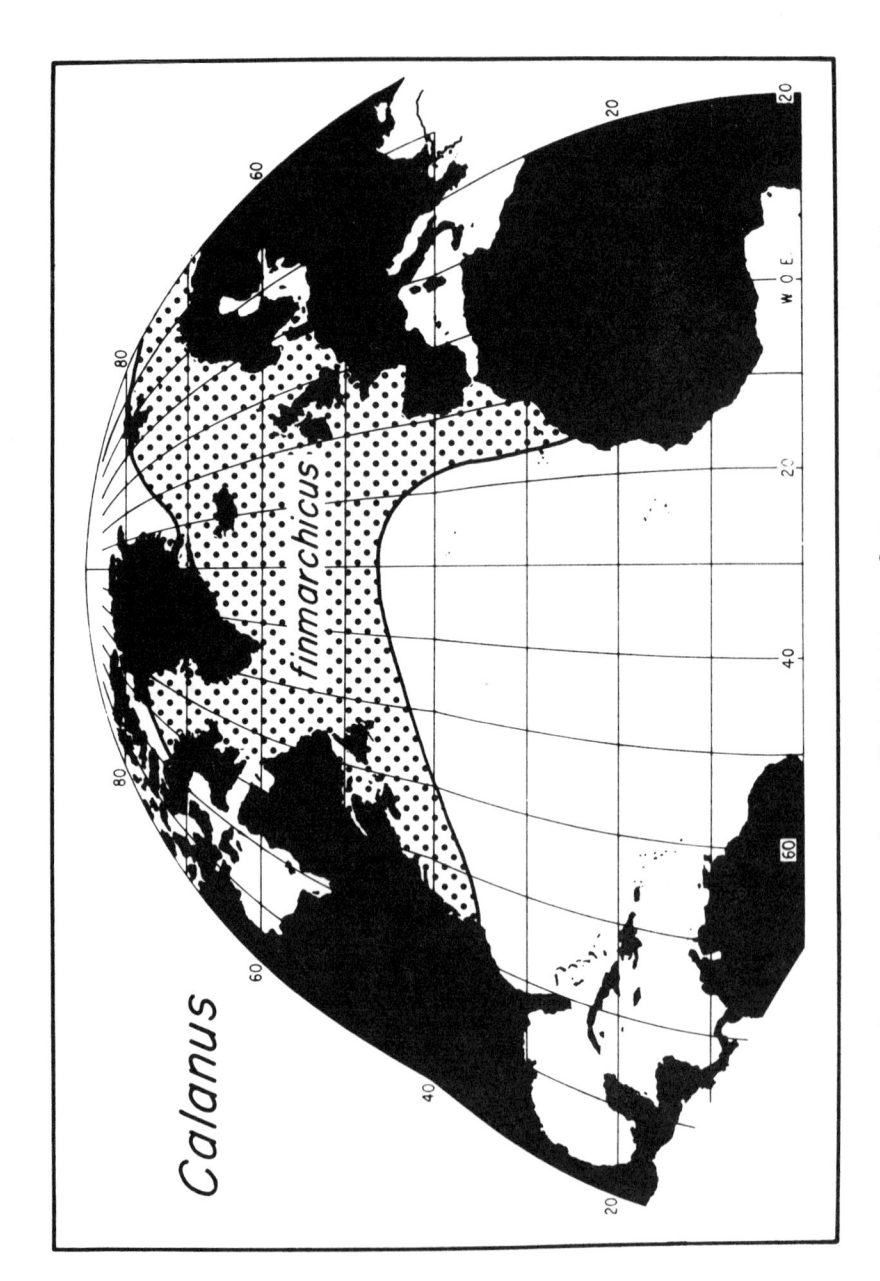

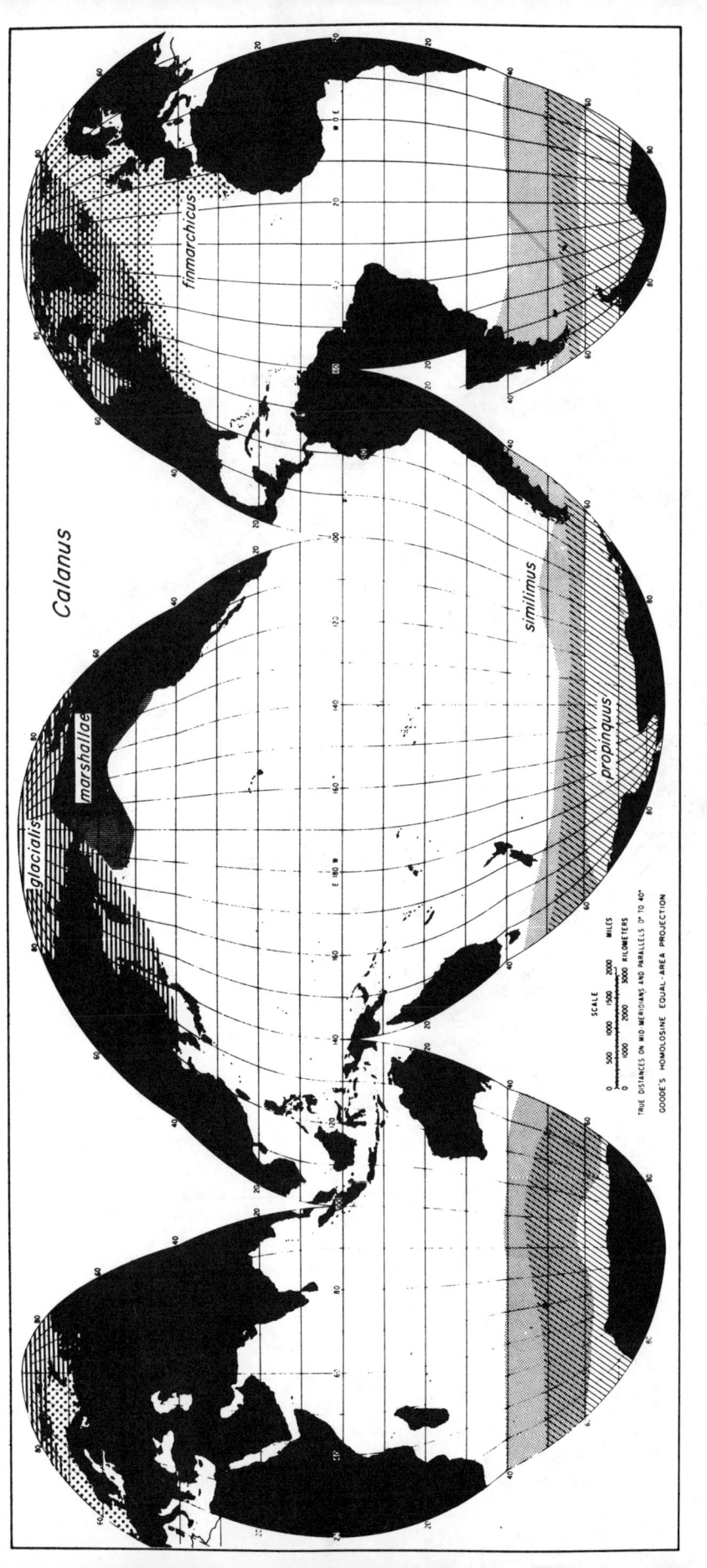

Fig. 19. Geographical distributions of subarctic and subantarctic calanoid copepods *Calanus finmarchicus*, *C. glacialis*, *C. marshallae*, *C. similimus*, and *C. propinquus* in the world ocean; adapted from data of BRODSKY (1964), JASCHNOV (1970), FROST (1974) and FLEMINGER and HULSEMANN (in press); Goode's Projection copyright by the University of Chicago Department of Geography.

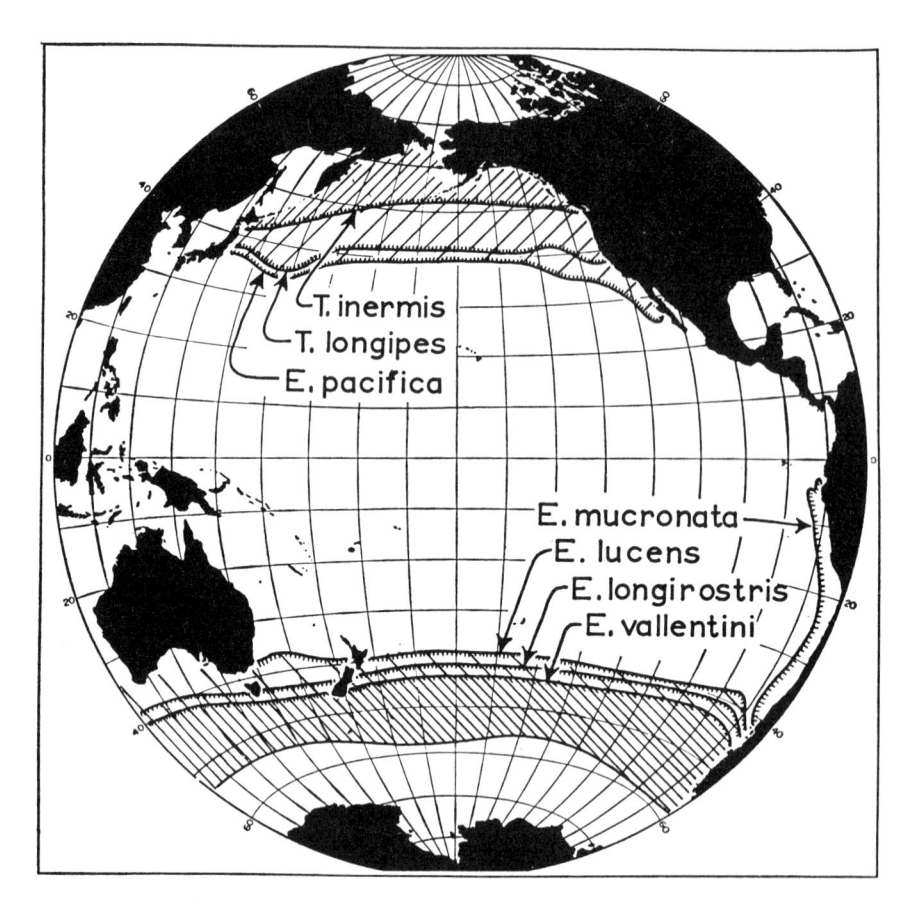

Fig. 20. Geographical distributions of subarctic and subantarctic euphausiid species and Euphausia mucronata, a Peru Current species (BRINTON, 1962, Fig. 103). Reprinted by permission of the University of California Press.

superba (Fig. 20), are found only within the subantarctic gyre (JOHN, 1936).

Zooplankters may have various means of remaining within a circulating system with a wide seasonal range of characteristics. The simplest examples are found in the Southern Ocean, where a continuous circumglobal circulation maintains a unique planktonic fauna (BAKER, 1954). Stocks may remain within the subantarctic gyres by changing their depths at different life stages. Such changes in depth may permit a number of species carried northward by the divergent surface flow to return southward with the deeper flow (MACKINTOSH, 1937; VERVOORT, 1965; VORONINA, 1968).

In the North Atlantic high-latitude species of the genus Calanus are widely known to undertake seasonal vertical migrations (DAMAS, 1905; DAMAS and KOFOED, 1907; SØMME, 1933, 1934; OSTVEDT, 1955) whereby subadults (stage V copepodids) containing large quantities of stored lipids leave the mixed layer with the onset of winter conditions: deep mixing, short day length, and a marked fall in phytoplankton concentrations. The overwintering stocks submerge to depths of 500 to 1000 m, existing largely on their stored reserves and possibly at reduced metabolic rates. Although the mechanism of surfacing the following spring is not known, reemergence of stage V copepodids occurs with the onset of the spring phytoplankton production. Depending on the latitude, spring and early summer broods may not leave the mixed layer, but with the onset of fall, shortened day length, and seasonal overturn, late juveniles once again submerge.

The subtropical gyres are the most extensive of the major oceanic habitats and have a larger proportion of their fauna in common. One euphausiid, Euphausia brevis (Fig. 21), exists globally in five subpopulations that at present appear to be separated by the equatorial zone. In another case, Euphausia hemigibba (Fig. 22) occupies four of the subtropical gyres, but a sibling species, E. gibba, occupies the fifth, in the South Pacific Ocean (BRINTON, 1975). A copepod, Clausocalanus lividus (Fig. 23), has roughly the same pattern (FROST and FLEMINGER, 1968; FROST, 1969). Other subtropical species are less rigidly bound to separate northern and southern gyres in both the Atlantic and the Pacific oceans, like the euphausiid Stylocheiron suhmii, which evidently transgresses the equatorial zones of both oceans, principally the Atlantic, while nevertheless maintaining separate populations among the three oceans (BRINTON, 1975). The pteropods Limacina lesueri, Styliola subula, and Cavolinia inflexa are limited to the subtropical gyres in the Pacific Ocean (McGOWAN, 1971), but apparently not in the Atlantic Ocean (VAN DER SPOEL, 1967).

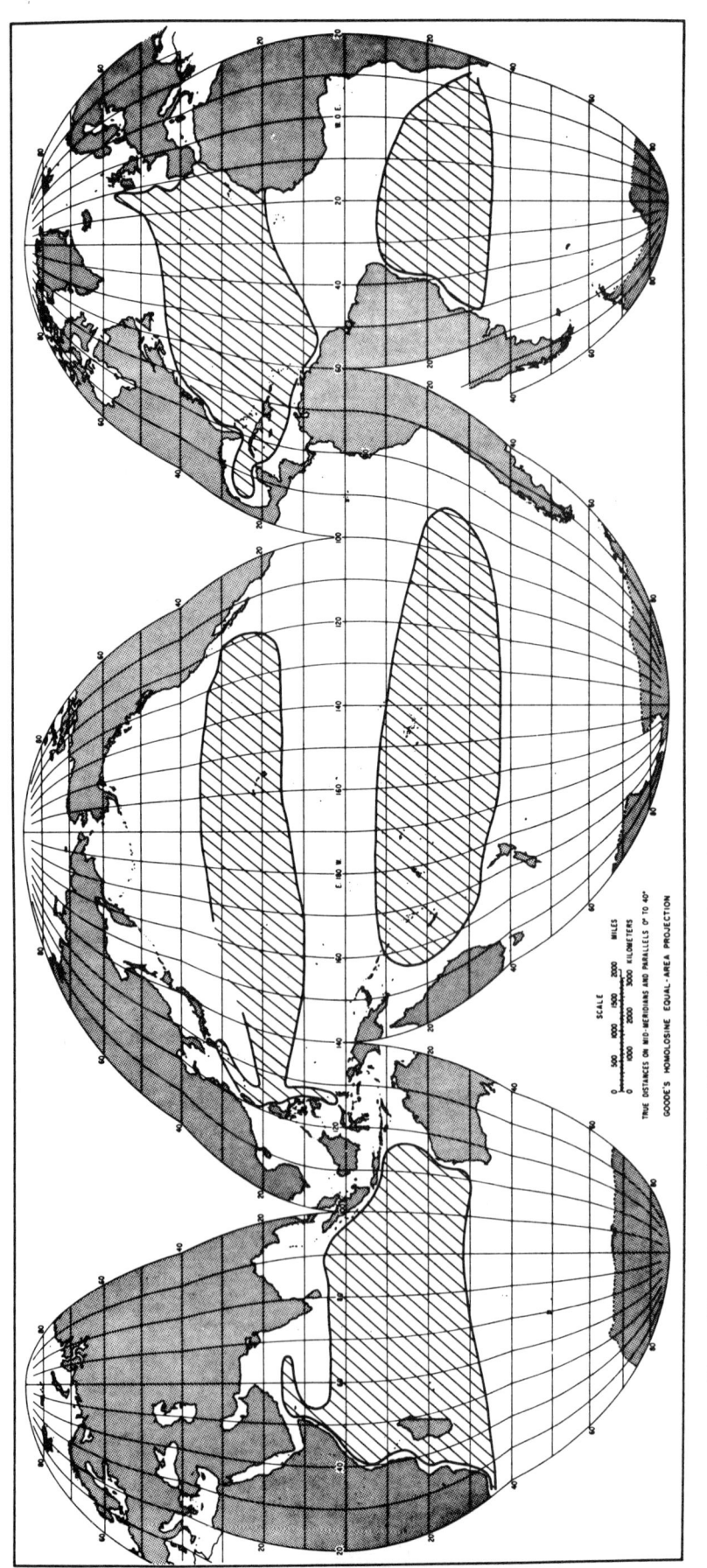

Fig. 21. Geographical distribution of Euphausia brevis in the world ocean (BRINTON, 1975, from Fig. 53a); Goode's Projection copyright by the University of Chicago Department of Geography.

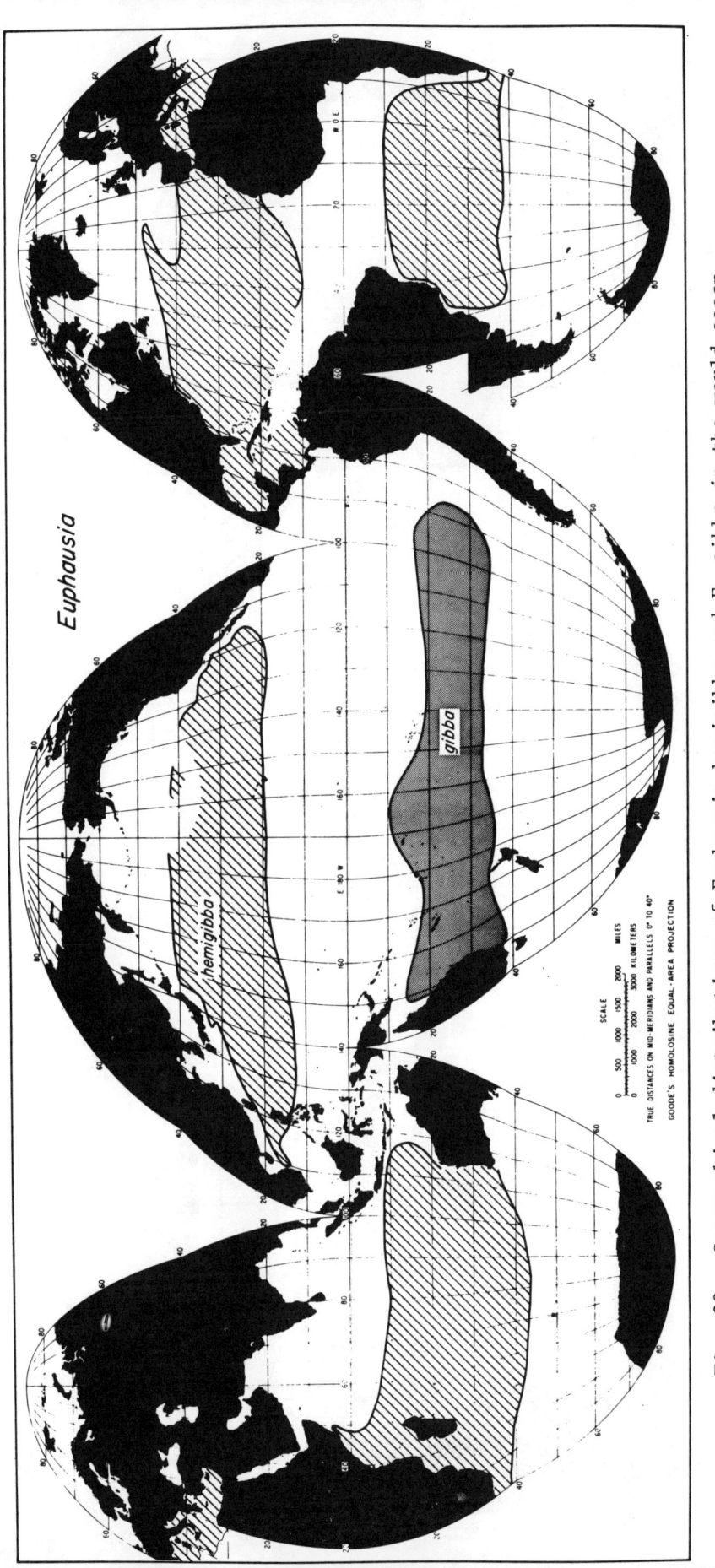

Fig. 22. Geographical distributions of Euphausia hemigibba and E. gibba in the world ocean (BRINTON, 1975, from Fig. 57a); Goode's Projection copyright by the University of Chicago Department of Geography.

Fig. 23. Geographical distributions of Clausocalanus lividus and C. laticeps in the world ocean; adapted from FROST and FLEMINGER (1968) and FROST (1969); Goode's Projection copyright by the University of Chicago Department of Geography.

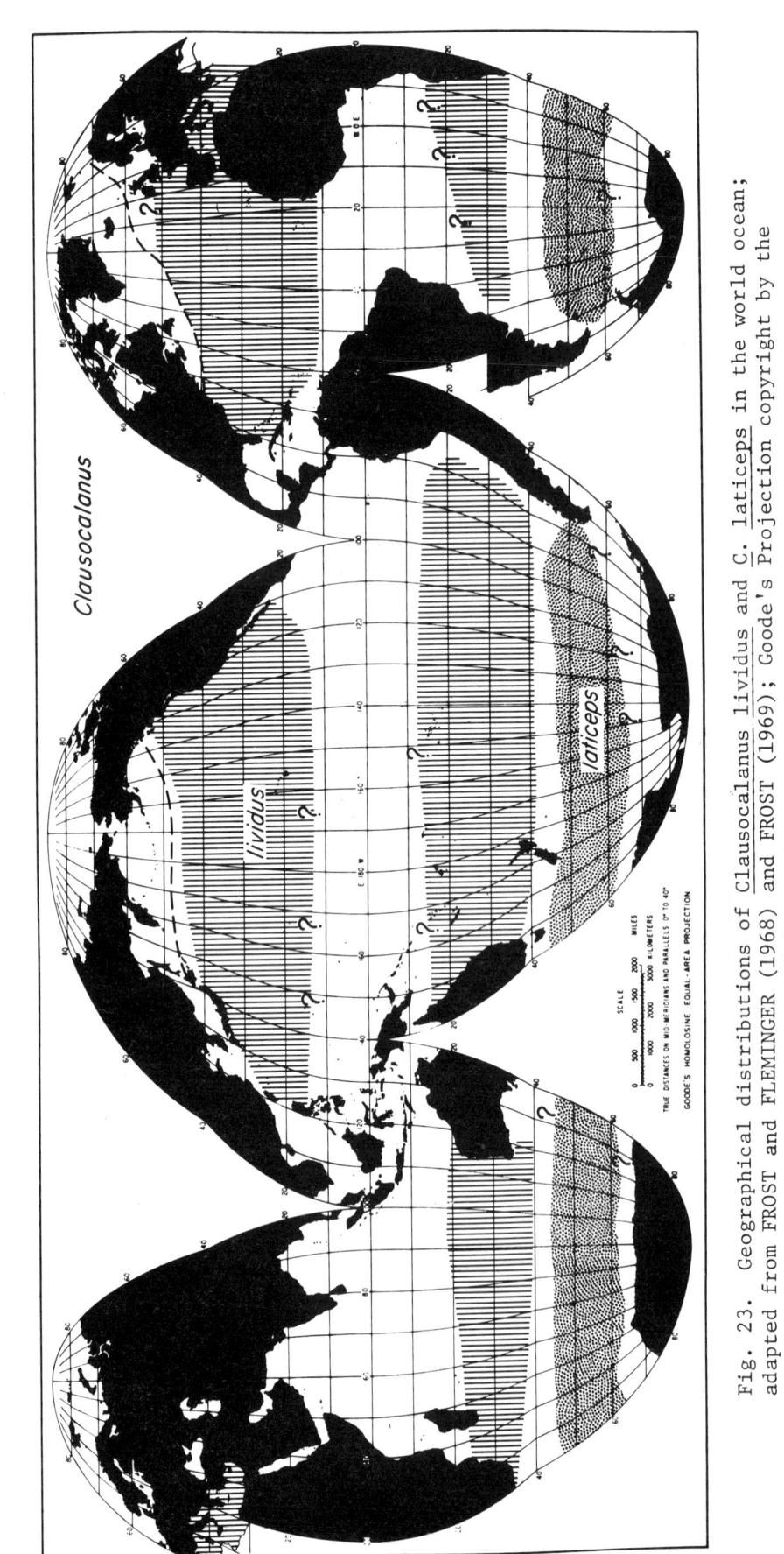

The equatorial zone tends to subdivide into Atlantic and Indo-Pacific provinces with a separate eastern tropical Pacific sector. The copepod genus Pontellina (Fig. 24) illustrates this. One, Pontellina platychela, occupies the equatorial Atlantic; another, P. morii, occupies the Indo-Pacific zone in which interocean confluence is evident north of Australia; and a third P. sobrina, replaces P. morii in the low-oxygen area of the eastern tropical Pacific (FLEMINGER and HULSEMANN, 1974). The pteropod Cavolinia uncinata (McGOWAN, 1971) and the chaetognath Sagitta ferox (BIERI, 1959) are additional examples of species which in the Pacific are limited to the equatorial zone.

Species of subequatorial distribution may in some cases reach higher latitudes by means of western boundary currents. Some even co-occur with species of the temperate zone in the Atlantic, evidently a consequence of ramifications of the Gulf Stream, for example, Euphausia pseudogibba (Fig. 25). In the Indian Ocean, the Agulhas Current system provides a South African habitat for tropical and subtropical species until it converges with the Benguela Current west of the Cape, for example, Euphausia paragibba (Fig. 25) and Pontellina morii (Fig. 24). And some species, such as Euphausia diomedeae (Fig. 26) and Nematoscelis gracilis (not shown), are seen to extend slightly poleward with the eastern boundary countercurrents and quite strongly poleward with the western boundary current at least in the north; data are not sufficient to test their extensions into the Tasman Sea.

Another group of species is found in both the subtropical and equatorial systems. This group is illustrated by two euphausiids, Stylocheiron carinatum (Fig. 27) and S. maximum (Fig. 28). Many species of copepods, chaetognaths, pteropods, and heteropods also fall into this group (BIERI, 1959; McGOWAN, 1971). The geographical range of S. maximum, a mid-depth species, is quite large. It is missing from the extremely low-oxygen (< 1 mℓ/ℓ at some depth) areas of the eastern tropical Pacific and the Indian oceans and perhaps also from that part of the Atlantic. It extends much farther north in the Atlantic than in the Pacific: the North Atlantic subtropical gyre extends farther north in the northeastern Atlantic (Fig. 14), which is notably warmer than the northeastern Pacific, as we have illustrated earlier (Fig. 15).

A number of other species, such as the euphausiid Nematoscelis microps (Fig. 29) and the pteropod Limacina inflata (McGOWAN, 1960), are not found within the areas of low oxygen (< ca. 1.0 mℓ/ℓ).

However, there are some species which seem to be limited to those areas, such as Euphausia distinguenda (Fig. 30), which

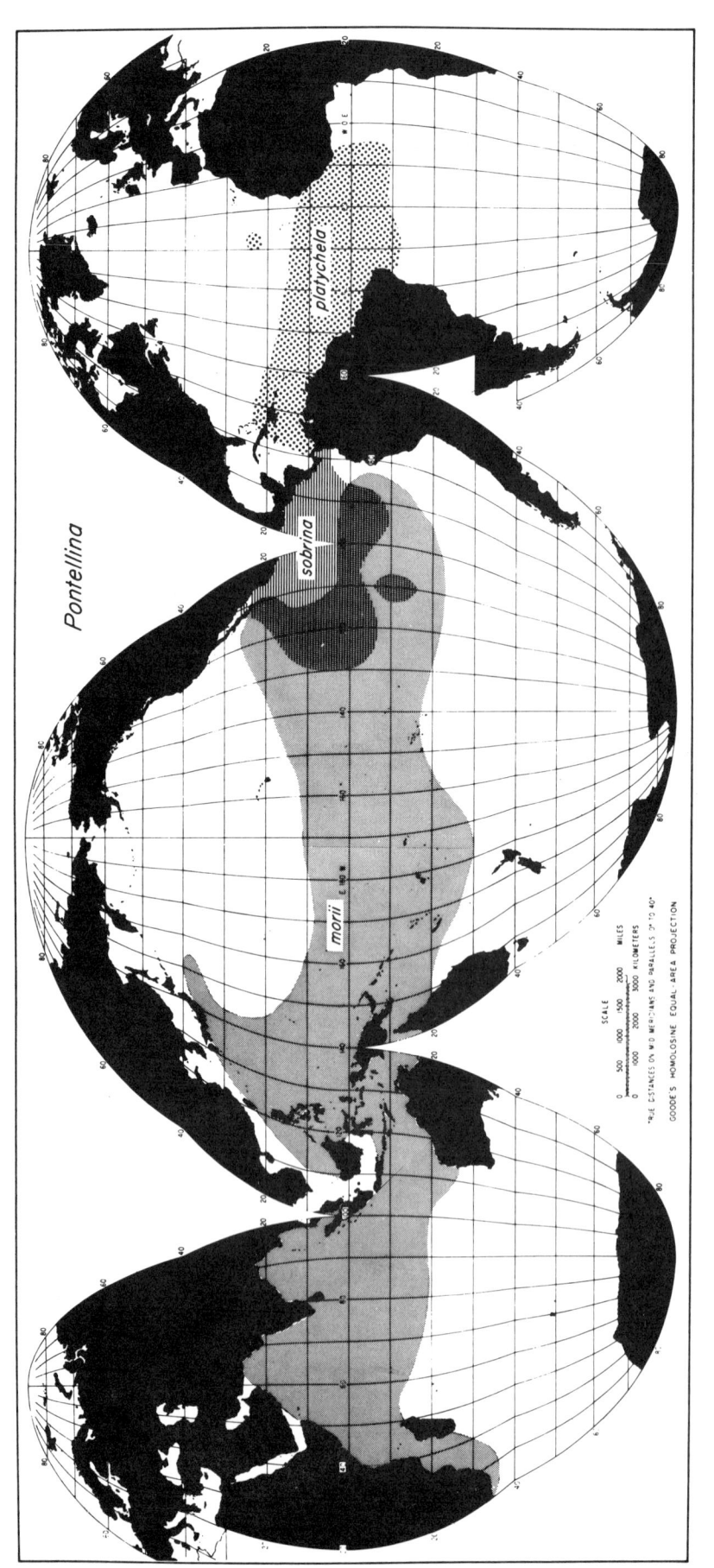

Fig. 24. Geographical distributions of Pontellina platychela, P. morii, and P. sobrina in the world ocean; adapted from FLEMINGER and HULSEMANN (1974); Goode's Projection copyright by the University of Chicago Department of Geography.

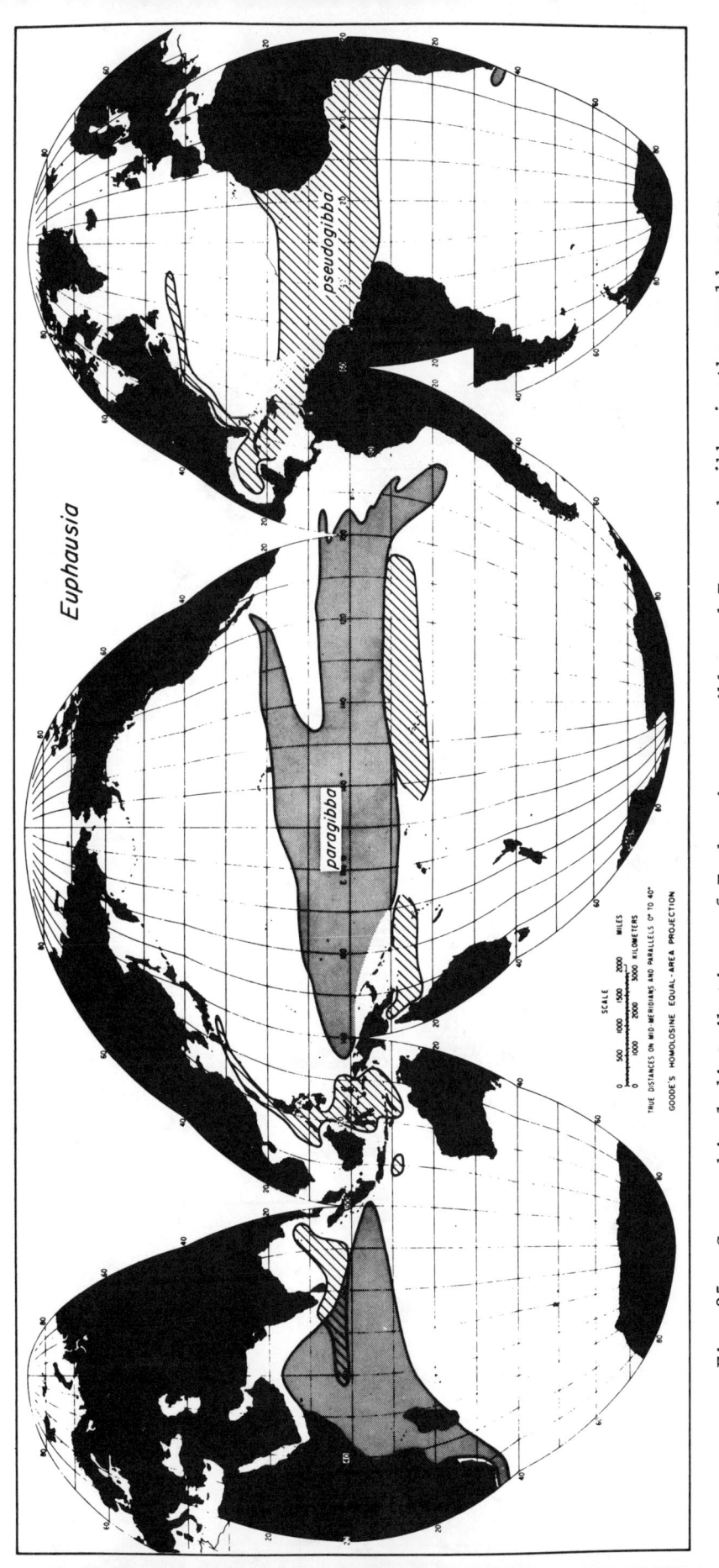

Fig. 25. Geographical distributions of Euphausia paragibba and E. pseudogibba in the world ocean (BRINTON, 1975, from Fig. 57b); Goode's Projection copyright by the University of Chicago Department of Geography.

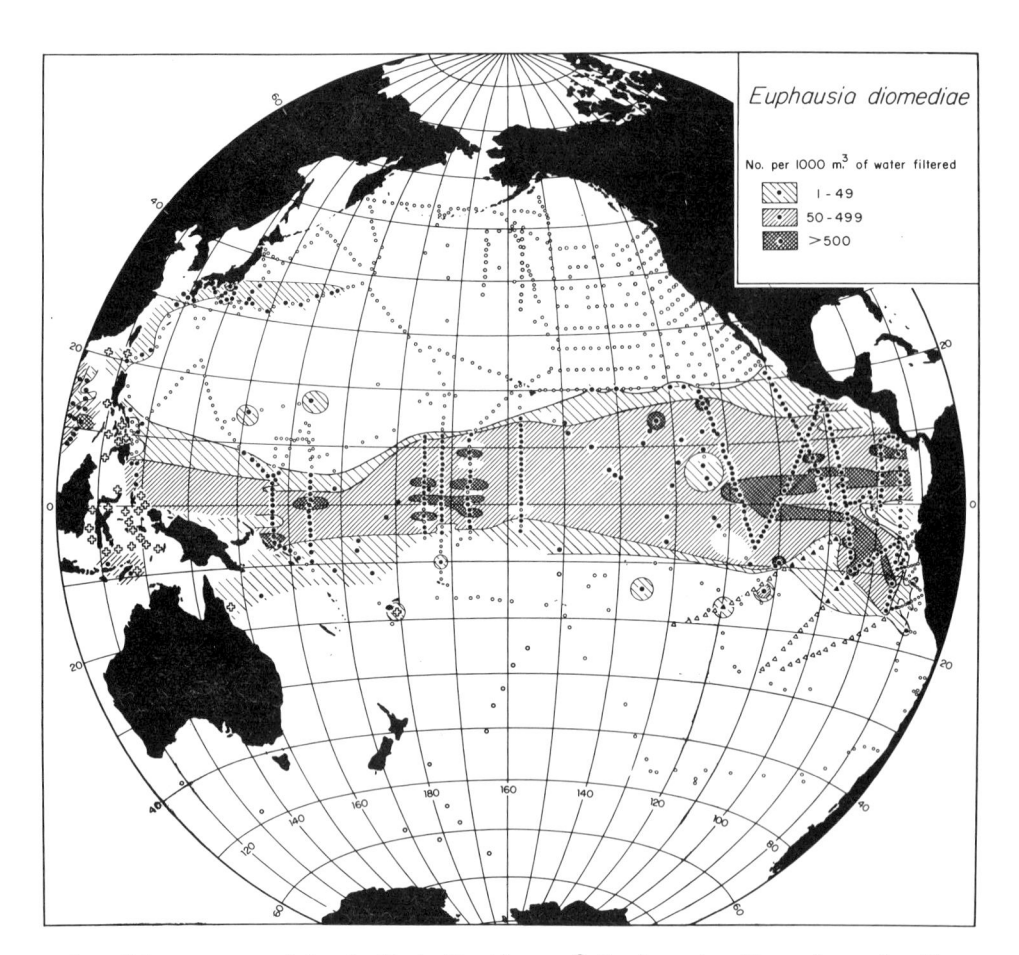

Fig. 26. Geographical distribution of Euphausia diomedeae in the Pacific Ocean (BRINTON, 1962, Fig. 42). Reprinted by permission of the University of California Press.

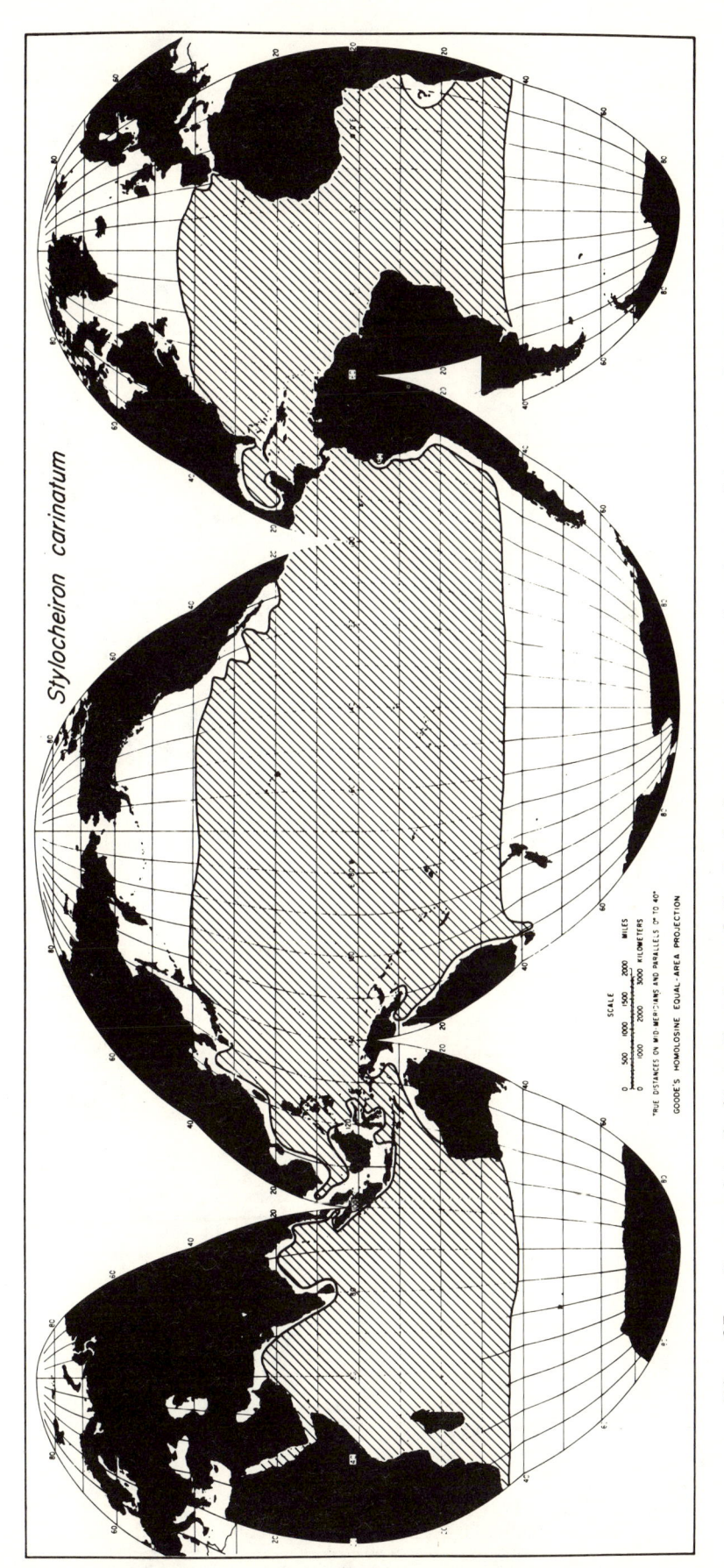

Fig. 27. Geographical distribution of Stylocheiron carinatum in the world ocean (BRINTON, 1975, from Fig. 65a); Goode's Projection copyright by the University of Chicago Department of Geography.

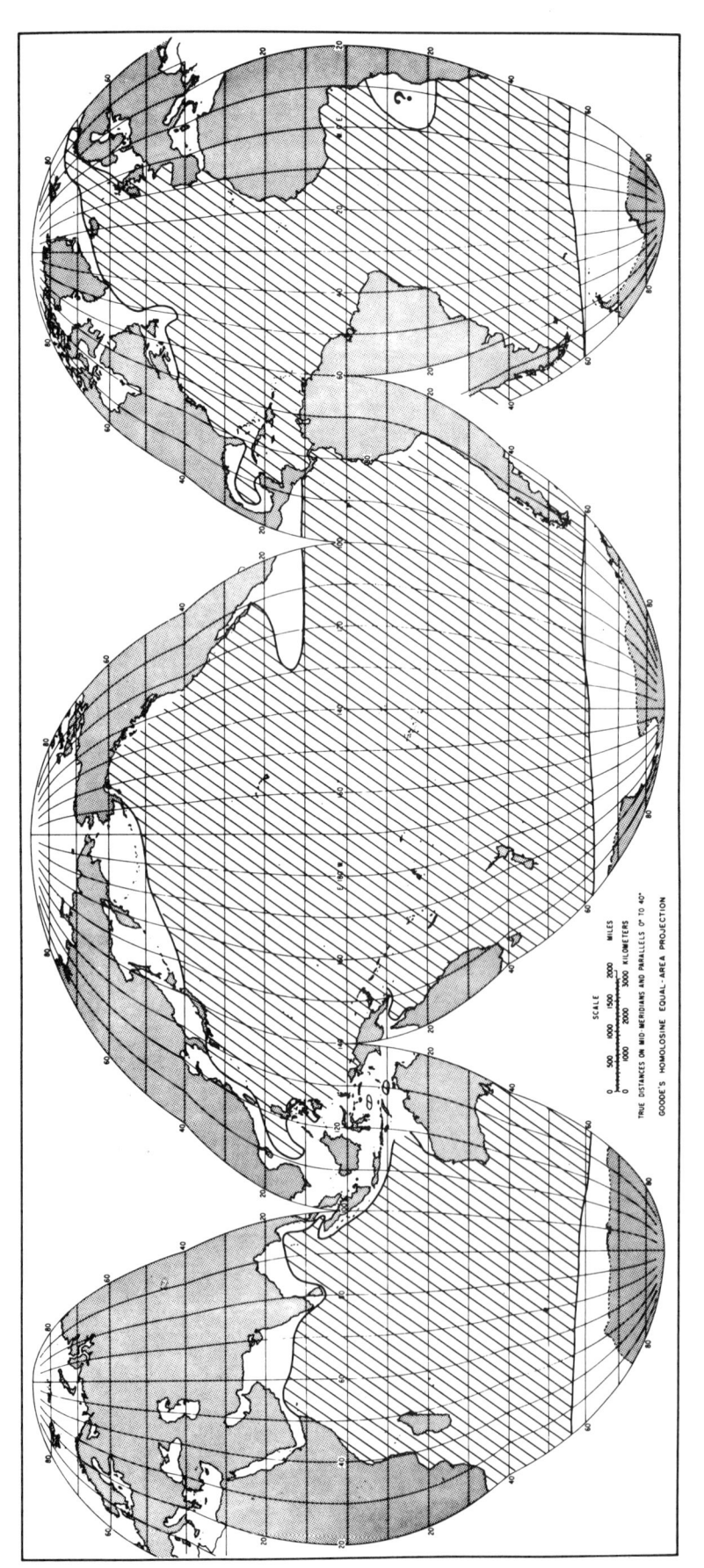

Fig. 28. Geographical distribution of Stylocheiron maximum in the world ocean (BRINTON, 1975, from Fig. 74a); Goode's Projection copyright by the University of Chicago Department of Geography.

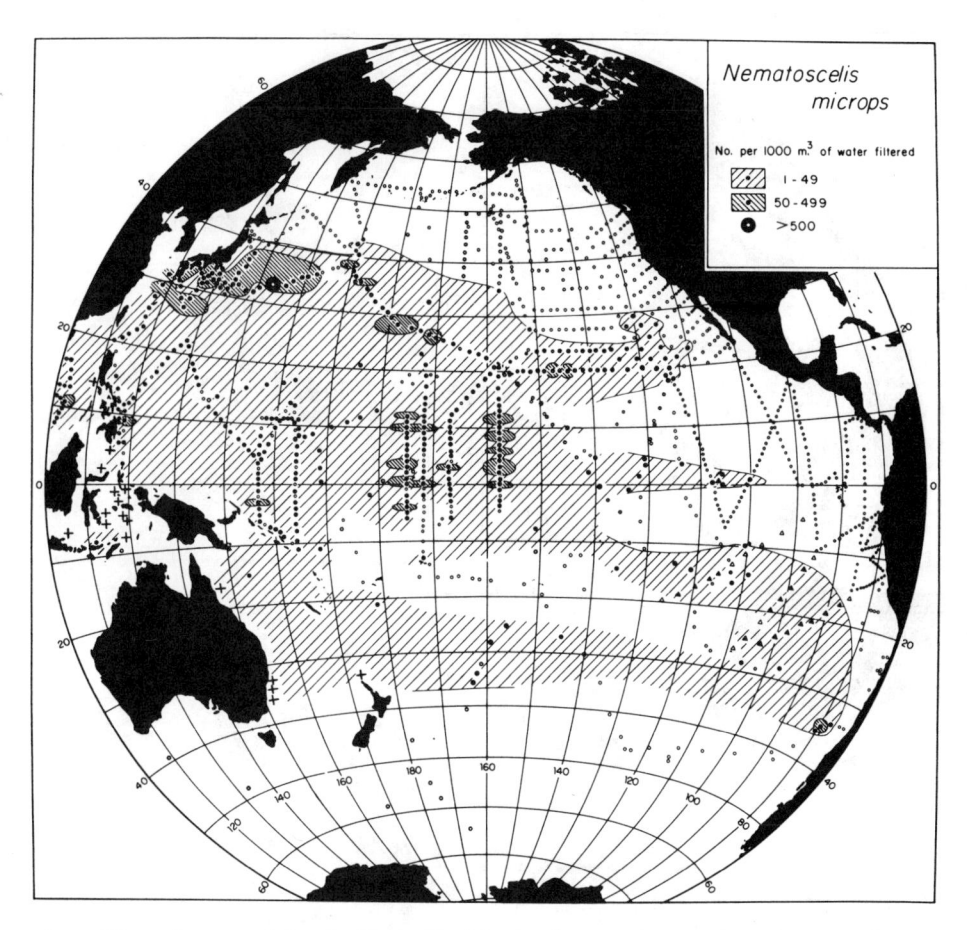

Fig. 29. Geographical distribution of Nematoscelis microps in the Pacific Ocean (BRINTON, 1962, Fig. 66). Reprinted by permission of the University of California Press.

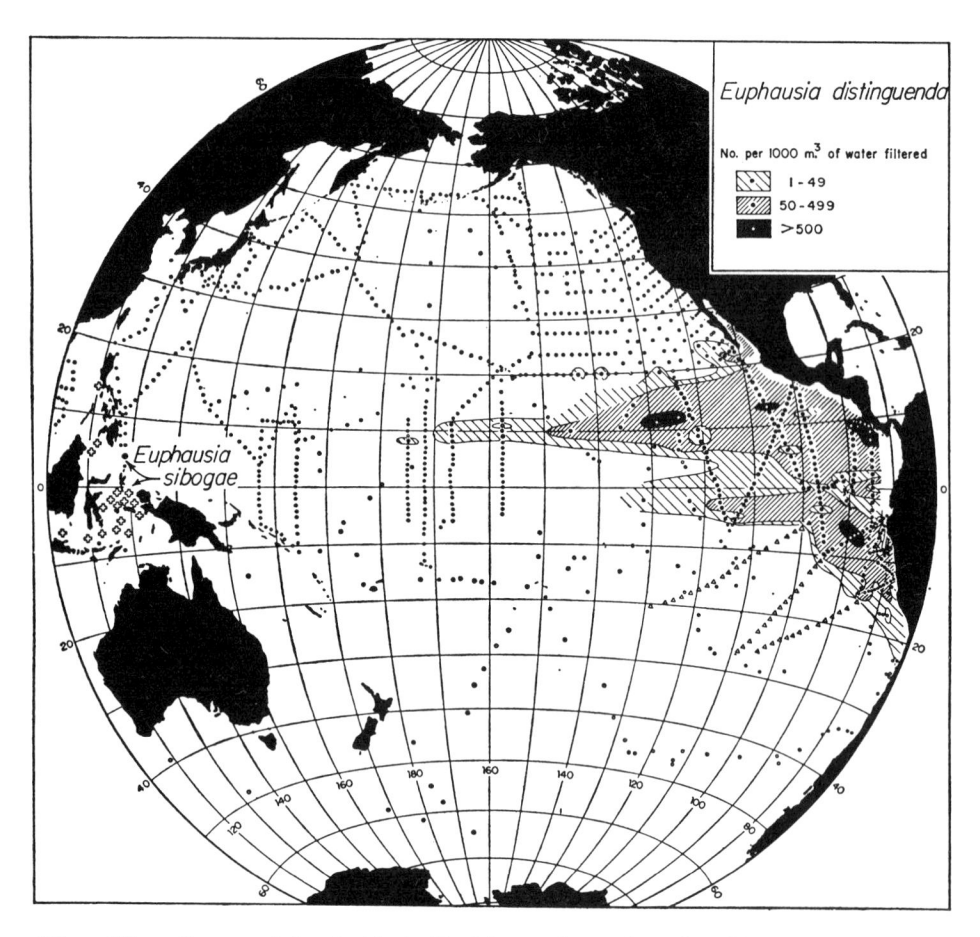

Fig. 30. Geographical distribution of Euphausia distinguenda in the Pacific Ocean (BRINTON, 1962, Fig. 47). Reprinted by permission of the University of California Press.

migrates daily into and out of the low-oxygen layers of the eastern tropical Pacific Ocean. There are other species, such as Limacina trochiformis, which while existing at very low levels of abundance in most of the lower latitudes of the Pacific Ocean are found at very high levels of abundance only in the eastern tropical area (McGOWAN, 1960).

In contrast to the rather simple distributions seen so far, there is one group whose pattern does not always fit this scheme, and whose maintenance we do not understand. Some species occur only along the west-wind drift, in water moving eastward, and they are not known yet to exist in a lower-latitude westward flow that would complete a circuit and account for their pattern. Some of these are found centered at about 40°N to 45°N and some at about 40°S to 45°S. They live partly within the subtropical system and partly within the subarctic or subantarctic system. They have been called species of transition zones, or transition species (JOHNSON and BRINTON, 1963; McGOWAN, 1971), following the use of "transitional" for the hydrographic characteristics of the California Current by SVERDRUP, JOHNSON and FLEMING (1942) and of the North Pacific Ocean near 40°N to 45°N by DODIMEAD, FAVORITE and HIRANO (1963).

Nematoscelis difficilis (Fig. 31) is characteristic of this group. Its sibling, N. megalops (Fig. 31), has analogous ranges in the North Atlantic and southern hemisphere. This species is a vertical migrant, but it rises only to thermocline depth in the night (except in places of strong upwelling, where it surfaces); hence, it avoids relatively rapid easterly or southerly surface transport in its transition zone and boundary current range segments. The large heteropod Carinaria japonica has a very similar distribution in the North Pacific (McGOWAN, 1971) as does the chaetognath Sagitta scrippsae (ALVARINO, 1962).

Another euphausiid species, Thysanoessa gregaria (Fig. 32), occurs in the North Atlantic and is nearly circumglobal in the subantarctic transitional zone (BRINTON, 1975). In the mid-North Pacific it occurs only within the west-wind drift, about 35°N to 45°N, with no obvious path for returning westward and no obvious permanent breeding ground at the western end. It is not a vertical migrant, but its depth range is centered at the seasonal thermocline, tending to submerge offshore in marginally subtropical waters of the California Current.

Eucalanus californicus has almost the same horizontal distribution as T. gregaria in the North Pacific but is known to have a population in the Sea of Japan (FLEMINGER and HULSEMANN, 1973); its presence in the transition zone could stem from that source, but no such source is known for T. gregaria or Carinaria japonica.

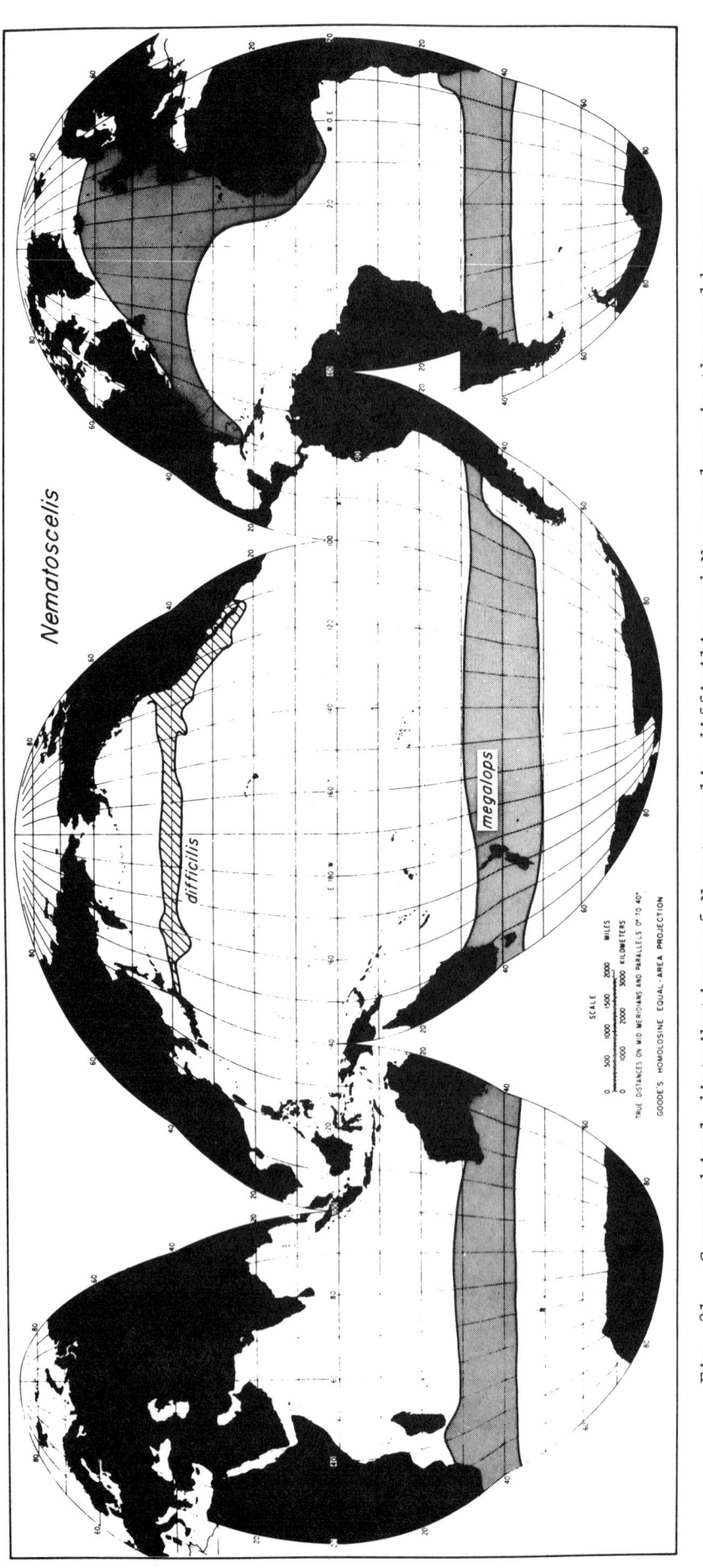

Fig. 31. Geographical distributions of Nematoscelis difficilis and N. megalops in the world ocean; adapted from GOPALAKRISHNAN (1974, Fig. 67b); Goode's Projection copyright by the University of Chicago Department of Geography.

OCEAN CIRCULATION AND MARINE LIFE 109

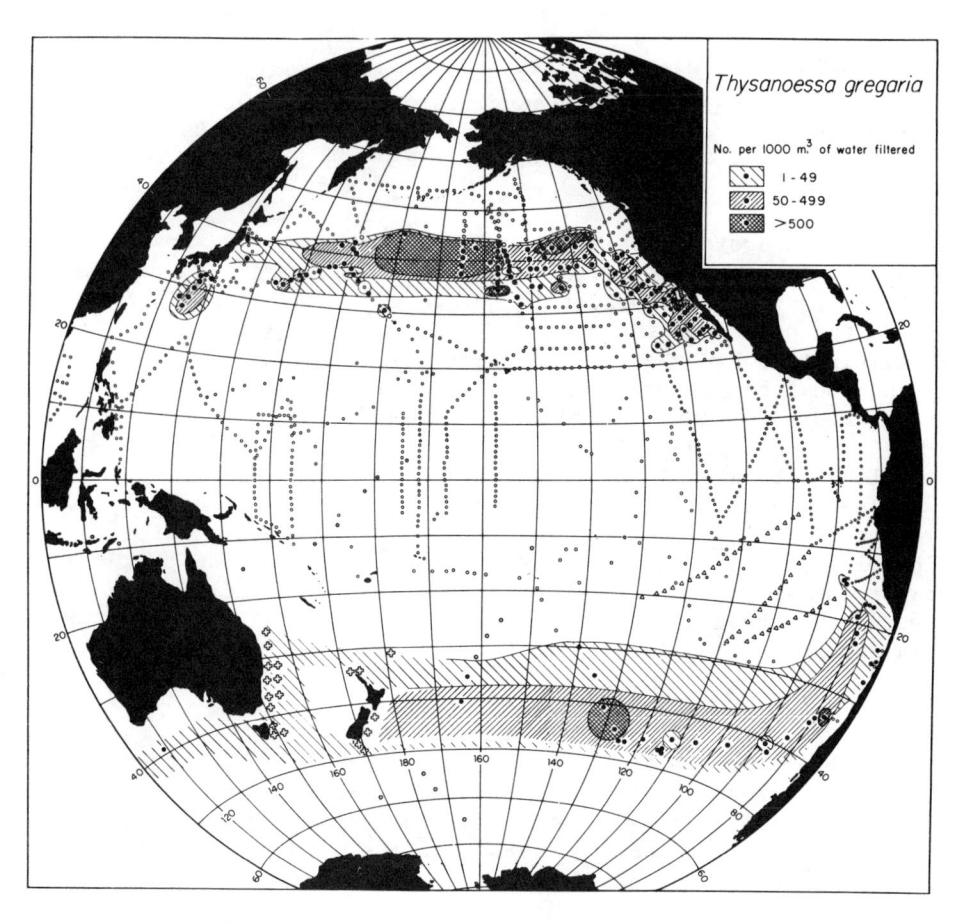

Fig. 32. Geographical distribution of Thysanoessa gregaria in the Pacific Ocean (BRINTON, 1962, Fig. 57). Reprinted by permission of the University of California Press.

Such maps are made mostly from zooplankton hauls reaching depths no greater than 200 m or 400 m. If the species were to extend farther south at a greater depth (as yet undetected), the westward flow there would make it possible to complete the circuit. At depths of 500 m and below, the westward flow lies farther north than it does at the surface. For some species this means of transport may be possible; for others, the data indicate it is not possible.

MACKINTOSH (1937) proposed that some of the most common Antarctic species maintain themselves by seasonal vertical migrations, drifting equatorward in summer in the upper layer, and sinking into the deeper warm layer in winter, which will return them poleward.

If this system applied within the North Pacific subarctic gyre as well, then these transition species might well be carried westward during their winter submergence as well as northward toward the center of the cyclonic gyre. McGOWAN (1971) proposed an alternative hypothesis, that seasonal and meridional differences in the zonal component of wind-driven transport might maintain the transition zone populations.

To review global patterns, Eucalanus (Fig. 33) provides examples of subarctic (bungii), transitional (californicus), subtropical (hyalinus), equatorial (elongatus), and an equatorial form (inermis) limited somewhat like Euphausia distinguenda to the area lowest in oxygen. There is some doubt as to whether hyalinus is present as more than a transient in the equatorial Atlantic: the data are not so complete as we would like.

NUMBER OF ZOOPLANKTON SPECIES BY AREA

We have shown that biomass is smallest within the subtropical gyres and largest within the subarctic gyres and the equatorial system. We have shown also that these systems are biological provinces for many species. How does the number of species vary from province to province? We cannot, of course, make complete lists of all zooplankton by area, but there are a few groups that are sufficiently known to make at least a start.

The euphausiids (Fig. 34) show fewest species in the subarctic and subantarctic gyres. Large numbers of species are found within the subtropical gyres, but there is not a uniform increase to the equator. These high numbers between 45°N and 45°S are made up of three sorts. Some species are found throughout this range. Some are restricted to the subtropical system, and most of these are vertical migrants. Others are restricted to the subequatorial system.

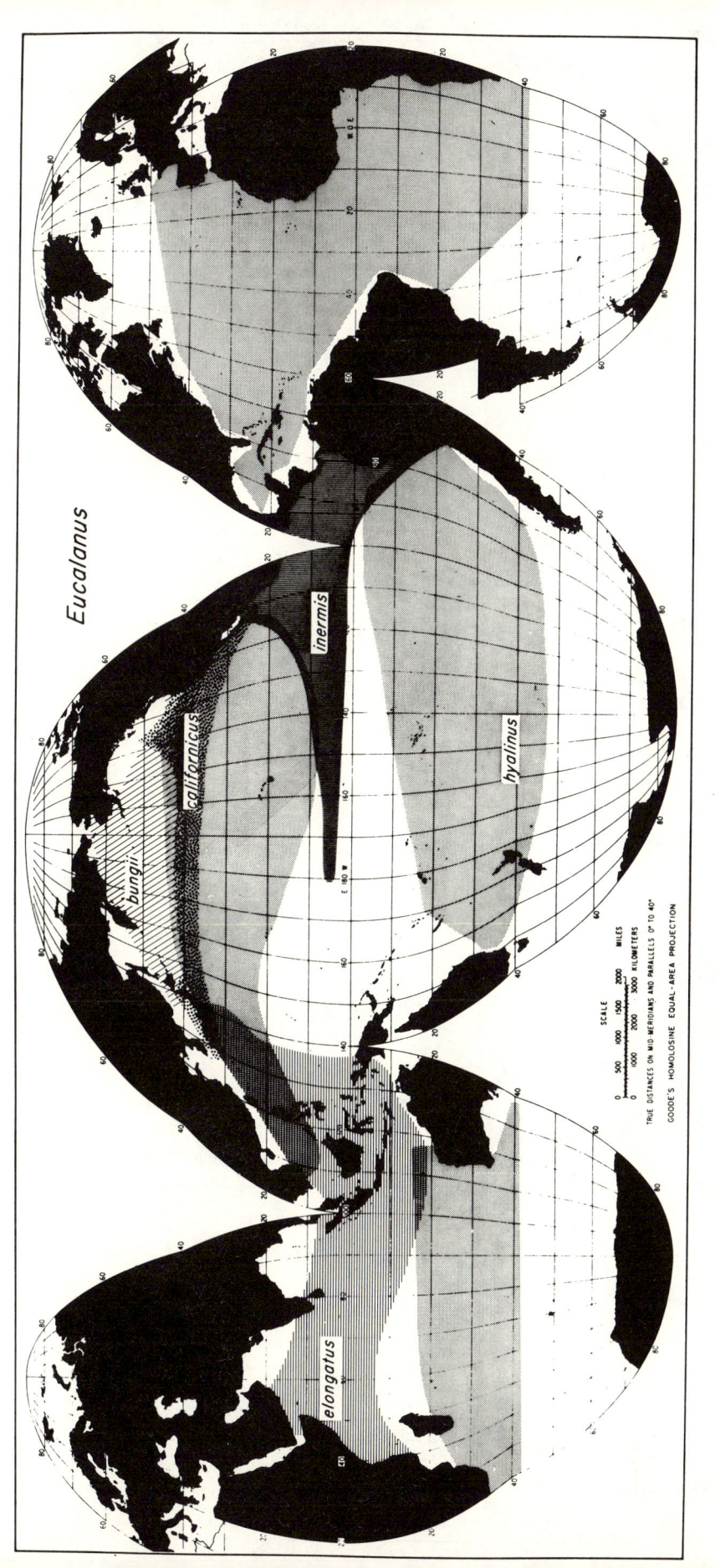

Fig. 33. Geographical distributions of Eucalanus bungii, E. californicus, E. elongatus, E. inermis, and E. hyalinus in the world ocean; adapted from FLEMINGER and HULSEMANN (1973); Goode's Projection copyright by the University of Chicago Department of Geography.

Fig. 34. Distribution of euphausiids by numbers of species in areas 5° in latitude and longitude in the world ocean; Goode's Projection copyright by the University of Chicago Department of Geography.

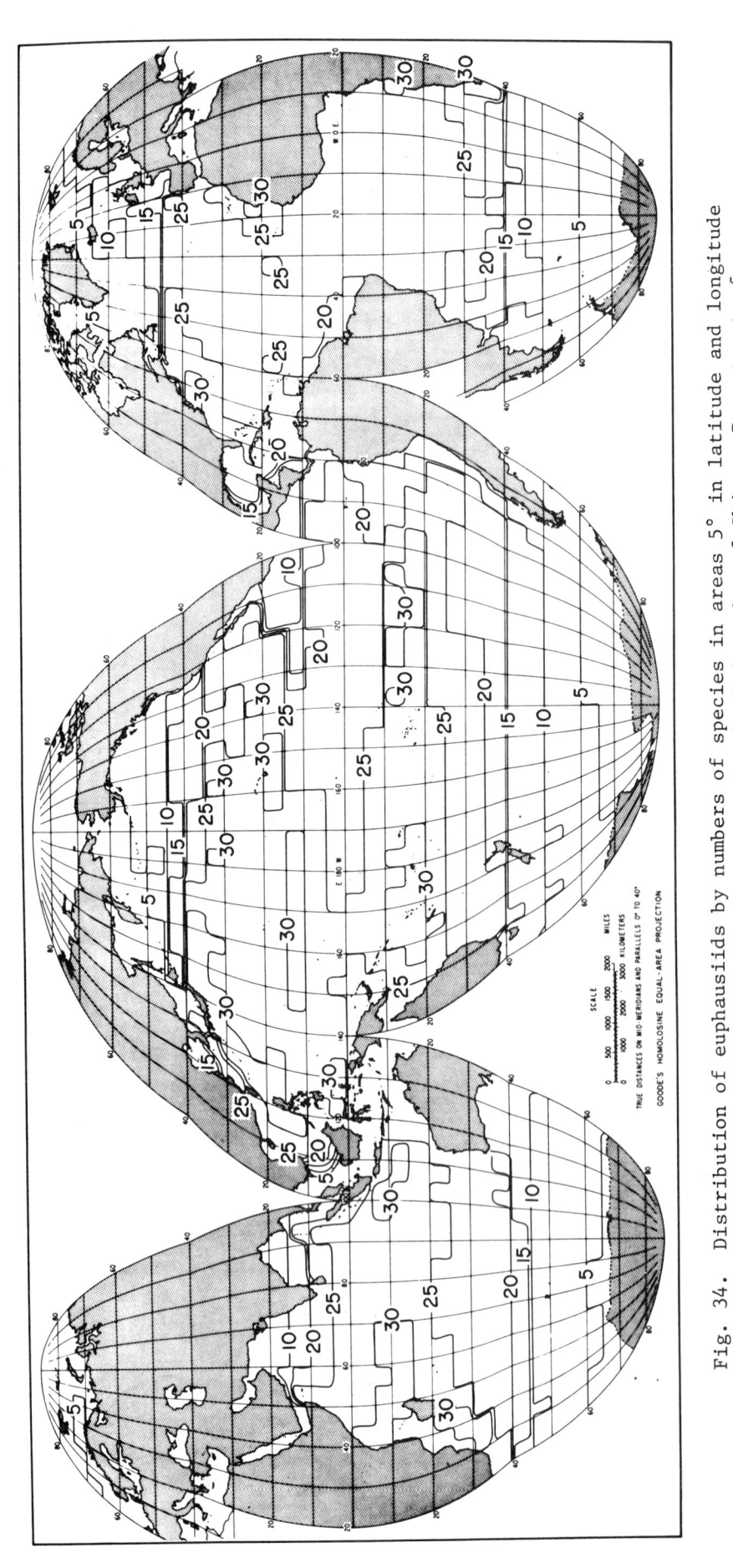

The subequatorial zone shows slightly fewer species along the equator in the Indian and Atlantic oceans, and substantially fewer in the Pacific. The low numbers in the eastern tropical Pacific and northern Indian Ocean appear to correspond with the extrema in subsurface oxygen values found there (Fig. 10).

Large numbers of species are found within the western boundary currents. Although the biomass is small there, the subtropical forms, the equatorial forms carried poleward by the strong flow, a small number of subarctic forms carried equatorward, and the local forms provide perhaps the largest total number of euphausiid species.

The northeastern Atlantic Ocean shows considerably more species than the corresponding part of the Pacific. We recall that the subtropical gyre of the North Atlantic extends farther north in the east than does that of the Pacific and that the northern Atlantic is notably warmer. The extreme northern distribution of Stylocheiron maximum (Fig. 28) has already been noted.

Pteropod species (Fig. 35) are distributed much like the euphausiids, except the effect of the oxygen minimum in the eastern tropical Pacific is not so obvious.

Chaetognaths (Fig. 36) are much like euphausiids and pteropods in the central and western Pacific, but the number seems to increase rather than decrease in the eastern tropical Pacific regions of low oxygen concentration. They show large numbers in the western boundary currents.

Thus, there is not a simple decrease in numbers of species from the equator to the high latitudes. Some habitats support many more species than do others, for example, the subtropical anticyclonic gyres as compared to the subarctic gyres. Large-scale mixing at the boundaries of faunal provinces, as in the western boundary currents, can cause large numbers of species to be present. The subarctic and subantarctic gyres are provinces of high biomass but few species. The subtropical gyres have the lowest biomass but the largest number of species of these three systems. Mixing of species from different systems occurs within the western boundary currents, and perhaps the highest number of species occurs there.

DIFFERENCES IN CHARACTERISTICS OF THE ORGANISMS IN THE SEPARATE GYRES

We have tried to show that the circulation of the ocean provides physical systems that can contain plankton and that have relatively narrow ranges of physical characteristics, and that these systems provide large-area habitats for specialized planktonic

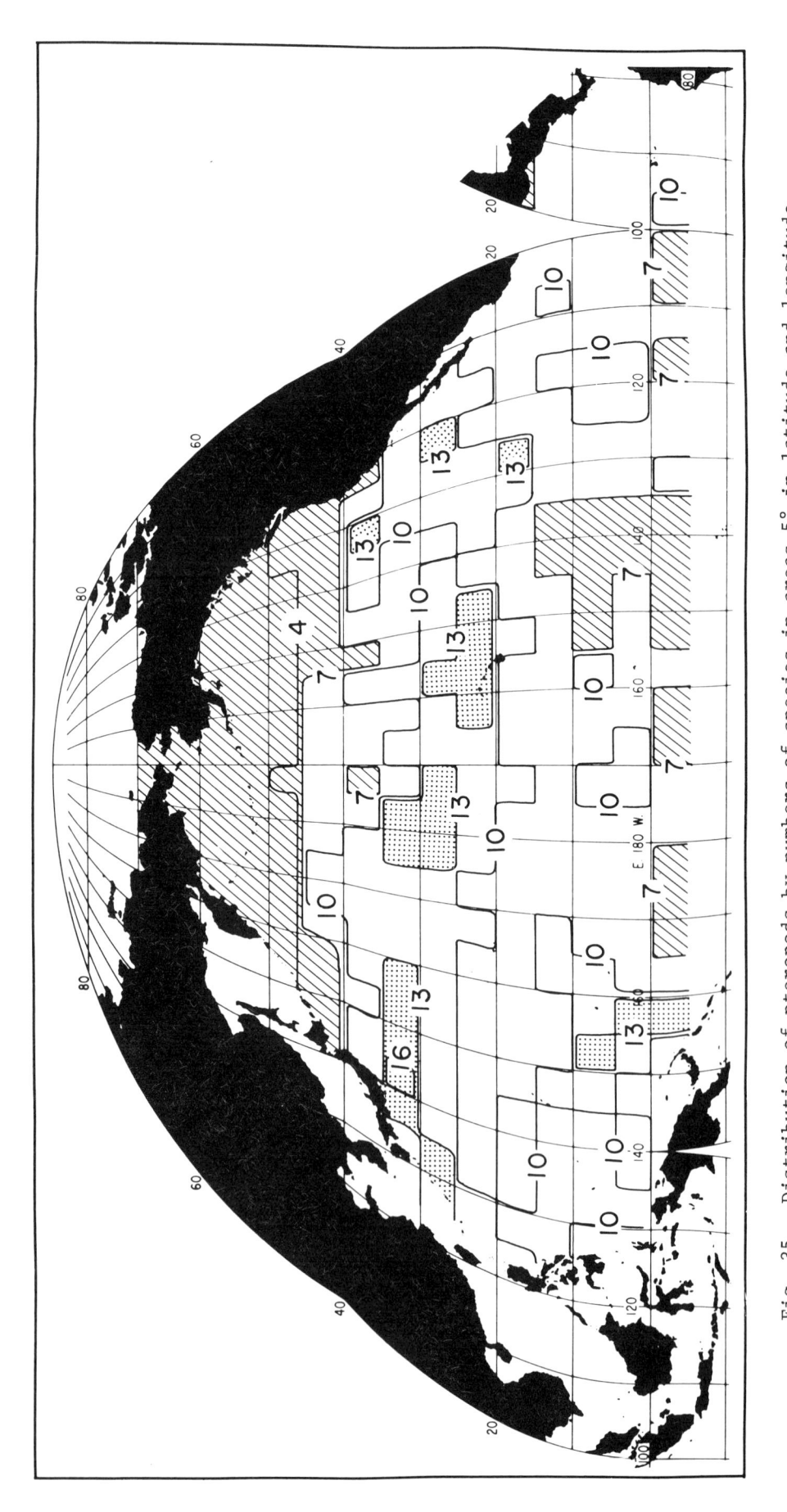

Fig. 35. Distribution of pteropods by numbers of species in areas 5° in latitude and longitude in the Pacific Ocean; Goode's Projection copyright by the University of Chicago Department of Geography.

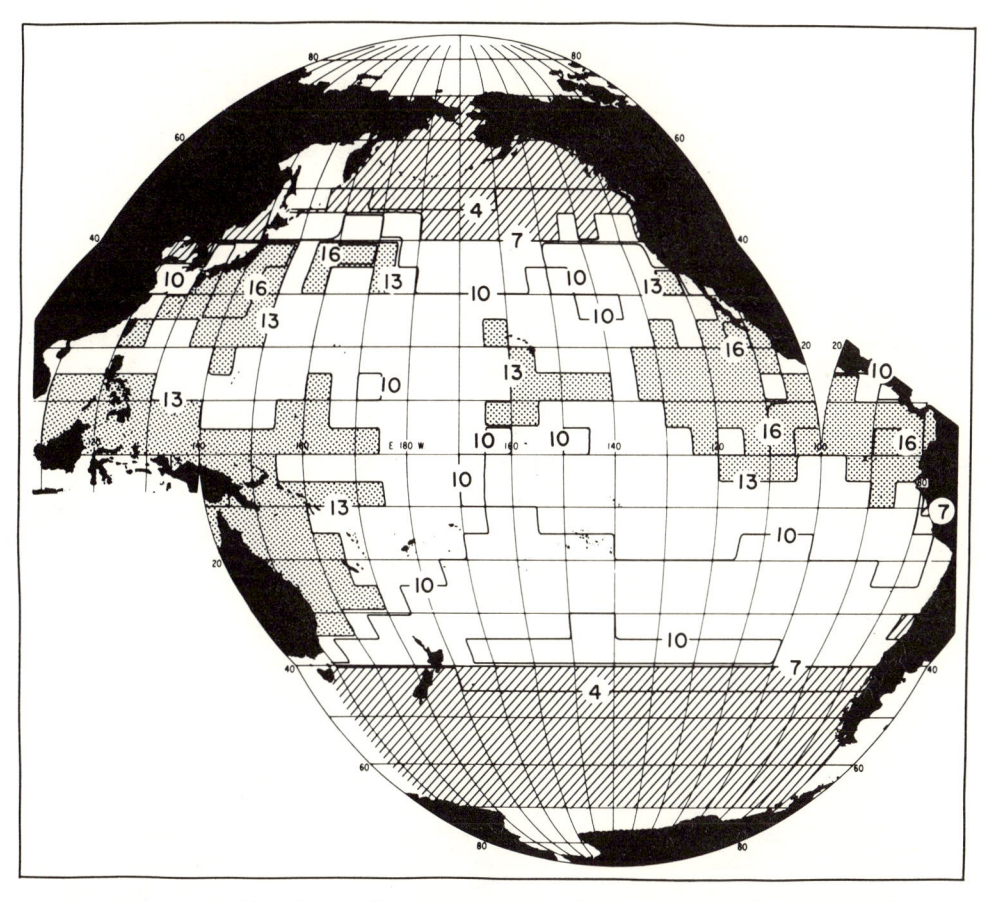

Fig. 36. Distribution of chaetognaths by numbers of species in areas 5° in latitude and longitude in the Pacific Ocean; Goode's Projection copyright by the University of Chicago Department of Geography.

forms. What are the differences in the forms that inhabit the different sorts of systems? What can we say, if anything, about their appearance and behavior?

Secchi disc measurements have shown that the subtropical systems of low nutrient and low biomass are more transparent than the subarctic and equatorial systems. In the more transparent waters of the subtropical systems many species are much more transparent and are often tinted to shades of blue, tending to match the water color shown by SCHOTT's (1942) map. This includes such diverse animals as pontellid copepods, surface-living mysids, siphonophores, nudibranchs, and the mollusc Janthina. In the richer, lower-transparency systems many species tend to be more opaque and pigments, when present, are commonly orange to red.

Within crustacean genera the subarctic and subantarctic zooplankton tend to be largest in size, the subtropical species are smallest, and the equatorial species tend to be intermediate in size (for example, in genus Euphausia, subarctic and transition zone E. pacifica is 20 mm, subantarctic euphausiids are 20-50 mm, subtropical E. brevis is 8 mm, equatorial E. diomedeae is 10-15 mm). Certain crustacean species living beneath the upper layer, especially in equatorial and subtropical areas, tend to be largest, particularly within the groups Copepoda, Ostracoda, Amphipoda, Euphausiacea and Decapoda. Thus, in these forms there appears to be some correlation between large body size and low temperature, but this is probably not independent of food supply.

Crustacean herbivores dominate the zooplankton biomass in the subarctic and subantarctic systems, notably the euphausiid Euphausia superba and the copepod Calanus propinquus (Fig. 19), but omnivores and carnivores are relatively more abundant in the subtropical and equatorial biomass. For example, in the euphausiids, most species of the genera with raptorial limbs, Nematobrachion, Nematoscelis, and Stylocheiron, are restricted to subtropical and equatorial regions (exceptions are two deep-living species of Stylocheiron, maximum and elongatum, Figs. 28 and 37 and Nematobrachion boopis whose range has been seen to extend beyond the transition zones, at least in the North Atlantic). Evidently such predaceous forms are adapted to areas where food stocks are low and filtering is possible only in a thin surface layer.

There seems to be some difference in the depth ranges of the organisms as related to the system they inhabit. Many crustacean species that extend throughout the subtropical and subequatorial systems appear to have vertical distributions bounded by the permanent thermocline. Species such as Stylocheiron elongatum (Fig. 37) and Nematoscelis microps (Fig. 38) live in or beneath the permanent pycnocline. These species are excluded from the eastern tropical Pacific and the northern Indian Ocean, where the

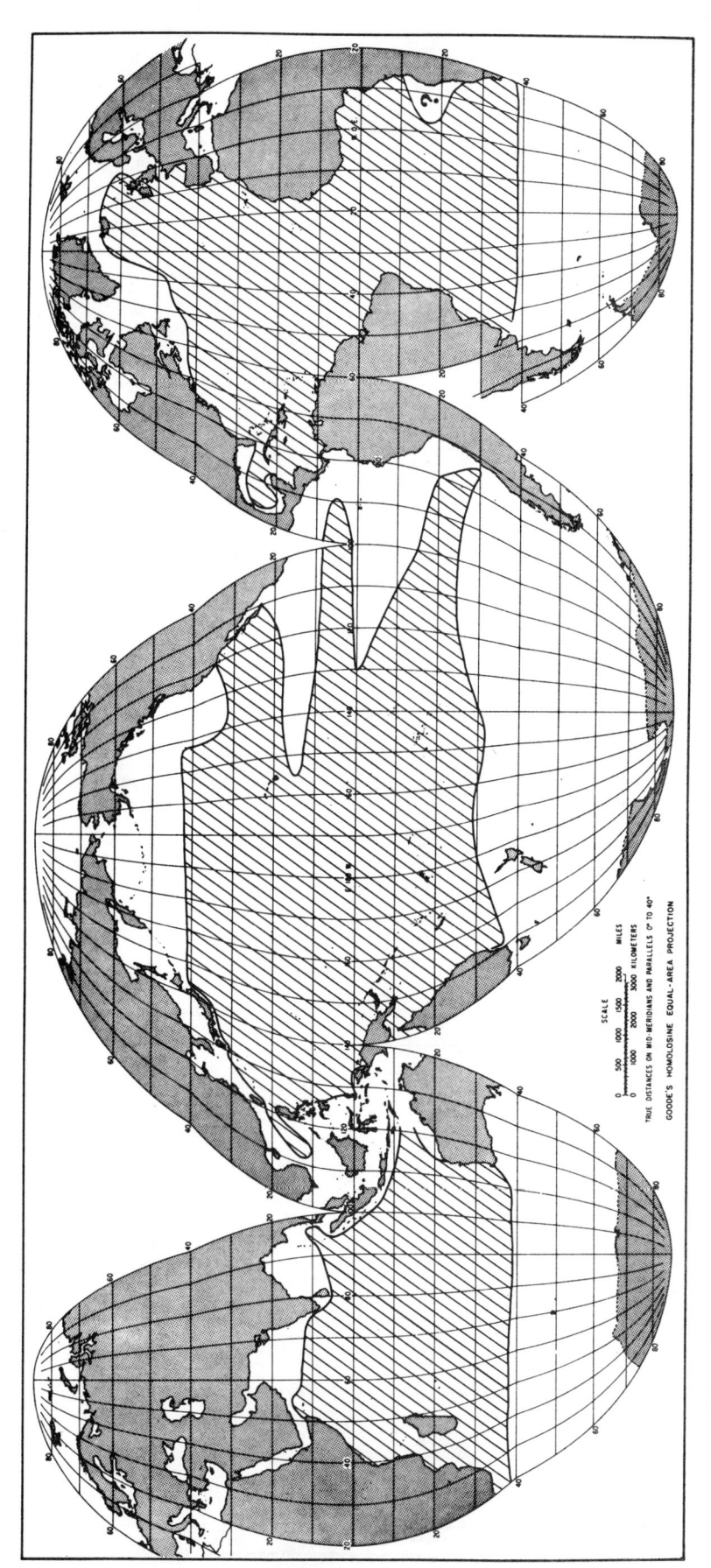

Fig. 37. Geographical distribution of *Stylocheiron elongatum* in the world ocean (BRINTON, 1975, from Fig. 72a); Goode's Projection copyright by the University of Chicago Department of Geography.

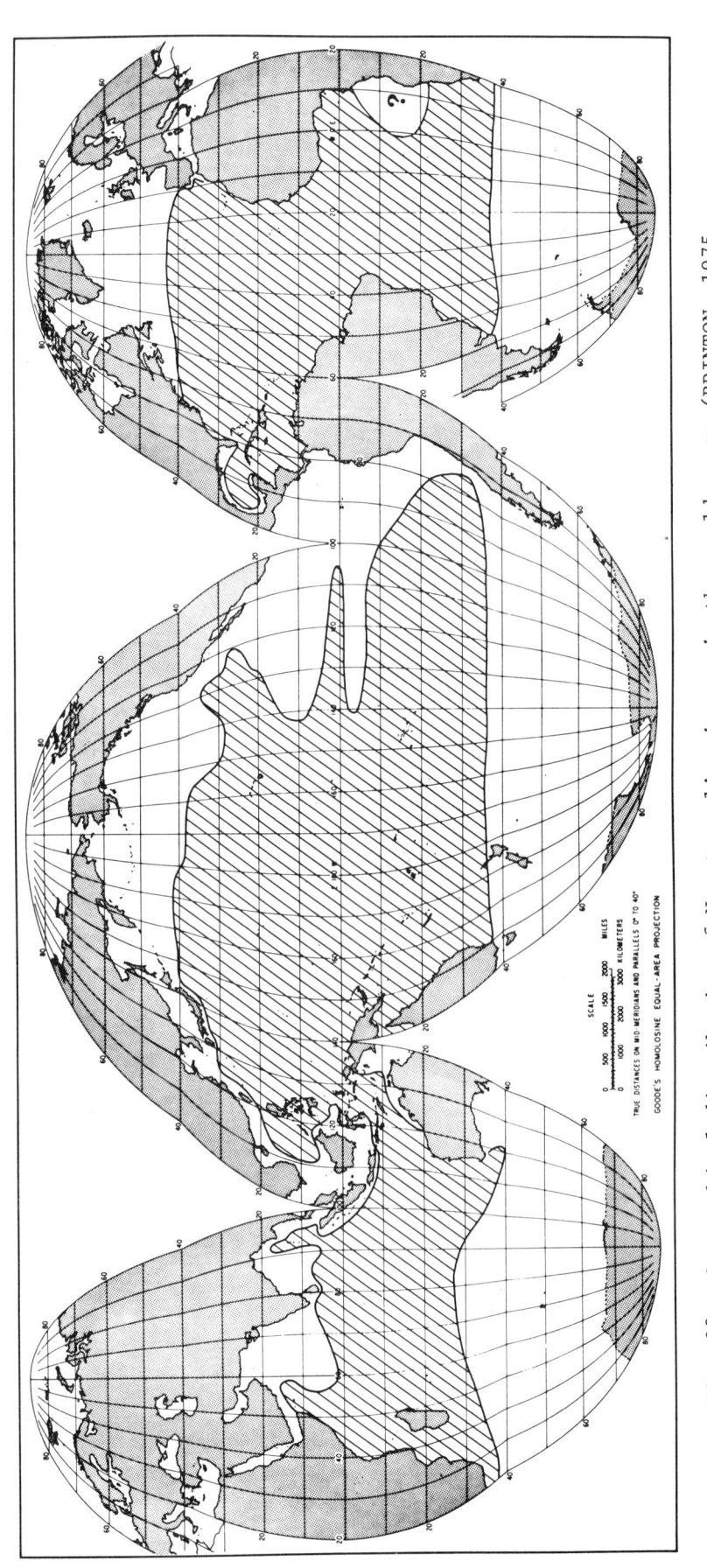

Fig. 38. Geographical distribution of Nematoscelis microps in the world ocean (BRINTON, 1975, from Fig. 61a); Goode's Projection copyright by the University of Chicago Department of Geography.

low-oxygen layers intercept their depth ranges. Other species inhabiting both systems live always above the permanent thermocline, such as Stylocheiron carinatum (Fig. 27) and many copepods.

On the other hand, certain species that live both above and in the thermocline, including those which daily migrate from the surface layer to beneath the thermocline, appear to be bound to only one of the two systems: they are limited either to the subtropical gyre or the equatorial system. The converse--that species living only in or beneath the seasonal thermocline range from 40°N to 40°S--is true for most species of euphausiids, except in the genus Nematoscelis where certain species which migrate between the seasonal thermocline and deeper layers appear to be restricted to one province or the other (N. atlantica is subtropical, N. gracilis is equatorial). Thus, the dual surface-deep habitats occupied by migrators tend to regulate horizontal range more restrictively than either depth range does for the upper-level inhabitants or for those living permanently at depth. We do not know of any zooplankton species that is excluded from the equatorial system but is found within both the subtropical and subarctic gyres.

The basis for cosmopolitanism, once considered a fairly common phenomenon, has virtually disappeared following closer taxonomic scrutiny of widespread oceanic zooplankton species. For example, the salp Salpa fusiformis, once thought to range from the subarctic to the antarctic continent, is now seen to be made up of four species, either sexually or asexually reproducing, which has a subarctic-to-subantarctic range in the upper layers of the sea.

The most familiar of planktonic copepods, Calanus finmarchicus, was also formerly regarded as ranging from the Arctic to the Antarctic. Combining the efforts of a number of workers shows that C. finmarchicus sensu lato is comprised of at least 11 different species (BRODSKY, 1959, 1965, 1972; FROST, 1974; FLEMINGER and HULSEMANN, in press). Warm-water circumglobal distributions are also diminishing in number and biogeographic significance, as critical interocean comparisons of planktonic crustaceans reveal that Atlantic populations are specifically distinct from their Indian and Pacific Ocean counterparts and that eastern subequatorial species may be specifically distinct from their subequatorial-subtropical counterparts (for example, Figs. 24, 30, and 33; JONES, 1965; BOWMAN, 1967; FLEMINGER, 1973; FLEMINGER and HULSEMANN, 1973, 1974; BRADFORD, 1974).

PHYTOPLANKTON

Most of our information about the global distributions of phytoplankton species has come from the early expeditions during the last half of the 19th and the first half of the 20th centuries. Since then, the use of electron microscopes has led to modification of species and genera, for example, the diatom genera Nitzschia and Fragilariopsis (HASLE, 1964, 1965a, b); and Thalassiosira (HASLE and HEIMDAL, 1970) and the coccolithophorid genus Syracosphaera (GAARDER and HEIMAL, in press). This in turn has modified our phytogeographical concepts, and this revision may be expected to continue.

The distributions of phytoplankton species have usually been described in terms of climate zone (WIMPENNY, 1966) or temperature and salinity (BRAARUD, 1962; SMAYDA, 1958), but the patterns appear to relate to the circulation systems as well as to the physical or chemical characteristics alone. When phytoplankton and zooplankton distributions are compared, some similarities are apparent, and some striking differences, and these differences may give insight into the mechanisms by which both phytoplankton and zooplankton species maintain their patterns.

For instance, a significant number of phytoplankton species occur throughout all of the major oceanic environments--subarctic and subantarctic, subtropical and equatorial. The pennate diatom, Thalassionema nitzschioides, has been reported from 60°S to 75°N and at temperatures from below 0°C to above 30°C (SMAYDA, 1958). Other "cosmopolitan species" include the coccolithophorid Emiliania huxleyi (SMAYDA, 1958) and the dinoflagellate Ceratium furca (GRAHAM and BRONIKOVSKY, 1944). The percentage of these cosmopolitan species is difficult to document, but appears to be of the order of 15%. Of the diatom species described from the California Current in 1943 (CUPP, 1943), approximately 15% were described as cosmopolitan (or widespread or ubiquitous, etc.). The same percentage of cosmopolites were reported within the dinoflagellate genus Ceratium collected by the Carnegie Expedition (GRAHAM and BRONIKOVSKY, 1944). Although these are rough approximations, they illustrate the differences from the zooplankton, for which such broadly distributed species are restricted to depths greater than a thousand meters.

Of the phytoplankton species found only in the subarctic or subantarctic systems, only two or three are believed to be bipolar (HASLE, 1975). The development of bipolar distributions by submergence or by transport in subsurface nearshore countercurrent, two mechanisms proposed for zooplankton, seems unlikely for phytoplankton species, because they are unable to perform the necessary vertical migrations. Once in the deeper waters below the depth of penetration of light sufficient for photosynthesis, they may be unable to rise again. Indeed, the paucity of bipolar distributions among phytoplankton may give indirect support to the theories of submergence and subsurface transport.

We have shown some examples of species distributions that indicate these circulation systems provide major and separate open-ocean habitats for large numbers of organisms.

We have noted some of the gyre-to-gyre differences in the zooplankton organisms: that is size, color, depth, and migratory range.

And we have made a few remarks upon the similarities and differences of the patterns of phytoplankton and zooplankton.

We realize, of course, that much of the information we have provided here is not original, but, in some cases, of quite long standing. However, we found it interesting to bring this information and these conjectures together.

Acknowledgments--The work reported here was supported by the National Science Foundation, the Office of Naval Research, and by the Marine Life Research Program, the Scripps Institution's component of the California Cooperative Fisheries Investigations, a project sponsored by the Marine Research Committee of the State of California.

REFERENCES

ALVARINO A. (1962) Two new Pacific chaetognaths. Bulletin of the Scripps Institution of Oceanography of the University of California, 8(1), 1-50.

ALVARINO A. (1965) Distributional atlas of chaetognatha in the California Current region. CalCOFI Atlas No. 3, State of California Marine Research Committee, i-xiii, plates 1-291.

BAKER A. de C. (1954) The circumpolar continuity of Antarctic plankton species. 'Discovery' Reports, 27, 201-218.

BANNER A.H. (1949) A taxonomic study of the Mysidacea and Euphausiacea (Crustacea) of the northeastern Pacific. Part III, Order Euphausiacea. Transactions of the Royal Canadian Institute, 28(58), 1-63.

BÉ A. W. H., J.M. FORNS and O. A. ROELS (1971) Plankton abundance in the North Atlantic Ocean. In: Fertility of the Sea, Vol. 1, J. D. COSTLOW, JR., editor, Gordon and Breach, pp. 17-50.

BEKLEMISHEV C. W. (1969) Ecology and biogeography of the open ocean. Akademia Nauk SSSR, Publishing House "NAUKA", 291 pp. (In Russian)

BIERI R. (1959) The distribution of the planktonic Chaetognatha in the Pacific and their relationship to the water masses. Limnology and Oceanography, 4(1), 1-28.

BOWMAN T. E. (1967) The planktonic shrimp, Lucifer chacei sp. nov. (Sergestidae: Luciferinae), the Pacific twin of the Atlantic Lucifer faxoni. Pacific Science, 2 (2), 266-271.

BRAARUD T. (1962) Species distribution in marine phytoplankton. Journal of the Oceanographical Society of Japan, 20th Anniversary Volume, 628-649.

BRADFORD J. M. (1974) Euchaeta marina (Prestandrea) (Copepoda, Calanoida) and two closely related new species from the Pacific. Pacific Science, 28(2), 159-169.

BRINTON E. (1962) The distribution of Pacific euphausiids. Bulletin of the Scripps Institution of Oceanography of the University of California, 8 (2), 51-270.

BRINTON E. (1967) Distributional atlas of Euphausiacea (Crustacea) in the California Current region, Part I. CalCOFI Atlas No. 5, State of California Marine Research Committee, i-xi, plates 1-275.

BRINTON E. (1975) Euphausiids of southeast Asian waters. Naga Report, Vol. 4, Part 5, Scientific Results of Marine Investigations of the South China Sea and the Gulf of Thailand 1959-1961, Scripps Institution of Oceanography of the University of California, 287 pp.

BRODSKY K. A. (1959) On phylogenetic relations of some Calanus (Copepoda) species of northern and southern hemispheres. Zoologicheskii Zhurnal, Moscow, 38(10), 1537-1553. (In Russian)

BRODSKY K. A. (1964) Variability and systematics of the species of the genus Calanus (Copepoda). Explorations of the fauna of the seas II(X): Results of biological investigations of the Soviet Antarctic Expedition (1955-1958, 2, Akademia Nauk SSSR Zoologicheskii Institut, 189-251. (In Russian)

BRODSKY K.A. (1965) Variability and systematics of the species of the genus Calanus(Copepoda). 1. Calanus pacificus Brodsky, 1948, and C. sinicus Brodsky, n. sp. Explorations of the fauna of the seas III(XI): Marine fauna of the northwest Pacific Ocean, Publishing house "Nauka", pp. 22-71. (In Russian)

BRODSKY K. A. (1972) Phylogeny of the family Calanidae (Copepoda) on the basis of comparative-morphological analysis of its characters. Explorations of the fauna of the seas XII(XX): Geographical and seasonal variability of marine planktonic organisms, Publishing house "NAUKA", pp. 5-110. (In Russian)

CUPP E. E. (1943) Marine plankton diatoms of the west coast of North America. Bulletin of the Scripps Institution of the University of California, 5(1), 1-237.

DAMAS D. (1905) Notes biologiques sur les Copépodes de la mer norvégienne. Publications de Circonstance, Conseil Permanent International pour l'Exploration de la Mer, 22, 3-23.

DAMAS D. and E. KOEFOED (1907) Le plankton de la mer du Grönland. In: Croisière Oceanographique dans la Mer du Grönland, 1905, Duc d'Orleans, 357-453.

DICKSON R.R. (1972) On the relationship between ocean transparency and the depth of sonic scattering layers in the North Atlantic. Journal du Conseil Permanent International pour l'Exploration de la Mer, 34(3), 416-422.

DODIMEAD A. J., F. FAVORITE AND T. HIRANO (1963) Review of the oceanography of the subarctic Pacific region. Part II of Salmon of the North Pacific Ocean, Bulletin, International North Pacific Fisheries Commission, No. 13, 195 pp.

FLEMINGER A. (1967) Distributional atlas of calanoid copepods in the California Current region, Part II. CalCOFI Atlas No. 7, State of California Marine Research Committee, i-xv, plates 1-213.

FLEMINGER A. (1973) Pattern, number variability and taxonomic significance of integumental organs (sensilla and glandular pores) in the genus Eucalanus (Copepoda, Calanoida). Fishery Bulletin, U.S. National Marine Fisheries Service, 71(4), 965-1010.

FLEMINGER A. and K. HULSEMANN (1973) Relationship of Indian Ocean epiplanktonic calanoids to1the world oceans. In: Ecological Studies, Analysis and Synthesis, Vol. 3, B. ZEITZSCHEL, editor, Springer-Verlag, pp. 339-347.

FLEMINGER A. and K. HULSEMANN (1974) Systematics and distribution of the four sibling species comprising the genus Pontellina Dana (Copepoda, Calanoida). Fishery Bulletin, U.S. National Marine Fisheries Service, 72(1), 63-120.

FLEMINGER A. and K. HULSEMANN (in press) Geographical range and taxonomic divergence in North Atlantic Calanus (C. helgolandicus, C. finmarchicus, C. glacialis). Marine Biology.

FOXTON P. (1961) Salpa fusiformis Cuvier and related species. 'Discovery' Reports, 32, 1-32.

FREDERICK M. A. (1970) An atlas of Secchi disc transparency measurements and Forel-Ule color codes for the oceans of the world. M.S. thesis, U.S. Naval Postgraduate School, Monterey, California, 179 pp.

FROST B. (1969) Distribution of the oceanic epipelagic copepod genus Clausocalanus with an analysis of sympatry of North Pacific species. Ph.D. dissertation, University of California, San Diego, La Jolla, California, 319 pp.

FROST B. W. (1974) Calanus marshallae, a new species of calanoid copepod closely allied to the sibling species C. finmarchicus and C. glacialis. Marine Biology, 26, 77-99.

FROST, B. and A. FLEMINGER (1968) A revision of the genus Clausocalanus (Copepoda, Calanoida) with remarks on distributional patterns in diagnostic characters. Bulletin of the Scripps Institution of Oceanography of the University of California, 12, 1-235.

GAARDER K. R. and B. R. HEIMDAL (in press) A revision of the genus Syracosphaera Lohmann (Coccolithineae).

GOPALAKRISHNAN K. (1974) Zoogeographic of the Nematoscelis crustacea Euphausiacea. Fishery Bulletin, U.S. National Marine Fisheries Service, 72(4), 1039-1074.

GRAHAM H. W. and N. BRONIKOVSKY (1944) The genus Ceratium in the Pacific and North Atlantic Oceans. Scientific Results. Cruise VII of the Carnegie 1928-1929, Biology, 5, 1-209.

HASLE G. R. (1964) Nitzschia and Fragilariopsis species studied in the light and electron microscopes. I. Some marine species of the groups Nitzschiella and Lanceolatae. Skrifter utgitt av det Norske videnskaps-akademi i Oslo, I. Mat.-Naturv. Klasse, Ny Serie, 16, 1-48.

HASLE G. R. (1965a) Nitzschia and Fragilariopsis species studied in the light and electron microscopes. II. The group Pseudonitzschia. Skrifter utgitt av det Norske videnskaps-akademi i Oslo, I. Mat.-Naturv. Klasse, Ny Serie, 18, 1-45.

HASLE G. R. (1965b) Nitzschia and Fragilariopsis species studied in the light and electron microscopes. III. The genus Fragilariopsis. Skrifter utgitt av det Norske videnskaps-AKADEMI i Oslo, I. Mat.-Naturv. Klasse, Ny Serie, 21, 1-49.

HASLE G. R. (1975) The biogeography of some marine planktonic diatoms. Deep-Sea Research, 23(4), 319-338.

HASLE G. R. and B. R. HEIMDAL (1970) Some species of the centric diatom genus Thalassiosira studied in the light and electron microscopes. Nova Hedwigia, Beiheft, 31, 543-581.

HEINRICH A. K. (1975a) The significance of the expatriated species in the structure of the planktonic tropical communities of the Pacific Ocean. Okeanologiia, 15(4), 721-725. (In Russian)

HEINRICH A. K. (1975b) On the boundaries of oceanic planktonic communities. Okeanologiia, 15(6), 1097-1100. (In Russian)

HENTSCHEL E. (1942) Eine biologische Karta des Atlantischen Ozeans. Zoologischer Anzeiger, 137(7/8), 103-123.

HENTSCHEL E. and H. WATTENBERG (1930) Plankton und Phosphat in der Oberflachenschicht des Sudatlantischen Ozeans. Annalen der Hydrographie und Maritimen Meteorologie, 58, 273-277.

HONJO S. and H. OKADA (1974) Community structure of coccolithophores in the photic layer of the mid-Pacific. Micropaleontology, 20, 209-230.

JASCHNOV W. A. (1970) Distribution of Calanus species in the seas of the northern hemisphere. Internationale Revue der gesamten Hydrobiologie und Hydrographie, 55, 197-212.

JOHN D. D. (1936) The southern species of the genus Euphausia. 'Discovery' Reports, 14, 193-324.

JOHNSON M. W. and E. BRINTON (1963) Biological species, water masses and currents. In: The Sea, Vol. 2, M. N. HILL, editor, Interscience, 381-414.

JONES E. C. (1965) Evidence of isolation between populations of Candacia pachydactyla (Dana) (Copepoda: Calanoida) in the Atlantic and the Indo-Pacific Oceans. In: Symposium on Crustacea, Marine Biological Association of India, Part 1, Series 2, 406-410.

KOBLENTZ-MISHKE O. J., V. V. VOLKOVINSKY, J. G. KABANOVA (1970) Plankton primary production of the world ocean. In: Scientific Exploration of the South Pacific, W. S. WOOSTER, editor, Scientific Committee on Oceanic Research (SCOR, pp. 183-193.

LOVE C. M. and R. M. ALLEN, editors (1975) EASTROPAC Atlas, Vol. 10: Biological and nutrient chemistry data from principal particpating ships, third survey cruise, February-March 1968. National Oceanic and Atmospheric Administration, National Marine Fisheries Service, Circular 330, i-viii, 137 figures.

MACKINTOSH N. A. (1937) The seasonal circulation of the Antarctic macroplankton. 'Discovery' Reports, 16, 365-412.

MARUMO R. (1966) Sagitta elegans in the Oyashio Undercurrent. Journal of the Oceanographical Society of Japan, 22(4), 129-137.

McGOWAN, J. A. (1960) The Systematics, distribution, and abundance of the Euthecosomata of the North Pacific. Ph.D. dissertation, University of California, San Diego, La Jolla, California, 212 pp.

McGOWAN J. A. (1963) Geographical variation in Limacina helicina in the North Pacific. Speciation in the sea. Systematics Association Publication No.5, British Museum of Natural History, pp.109-128.

McGOWAN J. A. (1971) Oceanic biogeography of the Pacific. In: The Micropaleontology of Oceans, B. M. FUNNELL and W. R. RIEDEL, editors, Cambridge University Press, pp. 3-74.

McGOWAN J. A. (1974) The nature of oceanic ecosystems. In: The Biology of the Oceanic Pacific, C. MILLER, editor, Oregon State University Press, pp. 9-28.

McGOWAN J. A. and P. M. WILLIAMS (1973) Oceanic habitat differences in the North Pacific. Journal of Experimental Marine Biology and Ecology, 12, 187-217.

MONTGOMERY R. B. and M. J. POLLAK (1942) Sigma-T surfaces in the Atlantic Ocean. Journal of Marine Research, 5(1), 20-27.

ØSTVEDT O.-J. (1955) Zooplankton investigations from Weather Ship M in the Norwegian Sea, 1948-49. Hvalradets Skrifter (Scientific Results of Marine Biological Research), No. 40, 93 pp.

REID J. L. JR. (1962) On the circulation, phosphate-phosphorus content and zooplankton volumes in the upper part of the Pacific Ocean. Limnology and Oceanography, 7(3), 287-306.

REID J. L. JR. (1965) Intermediate waters of the Pacific Ocean. The Johns Hopkins Oceanographic Studies, No.2, 85 pp., 32 figures.

REID J. L. (1969) Sea-surface temperature, salinity, and density of the Pacific Ocean in summer and in winter. Deep-Sea Research, 16 (Supplement),215-224.

REID J. L. and R. S. ARTHUR (1975) Interpretation of maps of geopotential anomaly for the deep Pacific Ocean. Journal of Marine Research, 33 (Supplement), 37-52.

REID J. L., W.D. NOWLIN, JR., W. C. PATZERT (in press) On the characteristics and circulation of the southwestern Atlantic Ocean. Journal of Physical Oceanography, 7.

SCHOTT G. 1942 Geographie des Atlantischen Ozeans, C. Boysen, 438 pp., 27 tafeln.

SIMONSEN R. and T. KANAYA (1961) Notes on the marine species of the diatom genus Denticula Kutz. Internationale Revue der gesamten Hydrobiologie und Hydrographie, 46(4), 498-513.

SMAYDA T. J. (1958) Biogeographical studies of marine phytoplankton. Oikos, 9(2), 158-191.

SØMME J. D. (1933) Animal plankton and sea currents. American Naturalist, 67(708), 33-34, 42-43.

SØMME J. D. (1934) Animal plankton of the Norwegian coast waters and the open sea. I. Production of Calanus finmarchicus (Gunner) and Calanus hyperboreus (Krøyer) in the Lofoten area. Fiskeridirektoratets Skrifter, Serie Havundersøkelser, 4(9), 163 pp.

STEUER A. (1933) Zur planmässigen Erforschung der geographischen Verbreitung des Haliplanktons, besonders der Copepoden. Zoogeographica, 1(3), 269-302.

SVERDRUP H. U. (1955) The place of physical oceanography in oceanographic research. Journal of Marine Research, 14(4), 287-294.

SVERDRUP H. U., M. W. JOHNSON and R. H. FLEMING (1942) The oceans: their physics, chemistry and general biology, Prentice-Hall, 1087 pp.

VAN DER SPOEL S. (1967) Euthecosomata - a group with remarkable developmental stages (Gastropoda, Pteropoda). J. Noorduijn En Zoon N.V., 375 pp.

VENRICK E. L. (1971) Recurrent groups of diatom species in the North Pacific. Ecology, 52(4), 614-625.

VERVOORT W. (1965) Notes on the biogeography and ecology of free-living marine Copepoda. Monographiae Biologicae, 15, 381-400.

VORONINA N. M. (1968) The distribution of zooplankton in the Southern Ocean and its dependence on the circulation of water. Sarsia, 34, 277-284.

WIMPENNY R. S. (1966) The plankton of the sea, Faber and Faber, 425 pp.

WYRTKI K. (1971) Oceanographic Atlas of the International Indian Ocean Expedition, The National Science Foundation, 531 pp.

THE BALTIC - A SYSTEMS ANALYSIS OF A SEMI-ENCLOSED SEA

B.-O. JANSSON

DEPARTMENT OF ZOOLOGY AND THE ASKÖ LABORATORY
STOCKHOLM, SWEDEN

ABSTRACT

The Baltic Sea has an area of 365 000 km² and a water of low, stable salinity (about 6-7 ‰ in the surface water) maintained through large freshwater outflows from rivers and salt water inflows over the sills to the North Sea. The absence of tides also contributes to the long residence time of the water, 25-40 years, the stratification of the water and the conversation of toxic substances. The seasonal pulse is pronounced, with ice cover in winter, a spring circulation, a summer thermocline and an autumn circulation.

The Baltic Sea is presented as one huge ecosystem, where the producers dominate the upper box of the stratified water. The phytal subsystem covers large areas in the rocky archipelagos. The structure and function of this diversified system is described in "Odum-models" and computer simulations.

The pelagic producer system has one typical Baltic feature, the blue-green bloom of the nitrogen fixing Nodularia spumigena which is presented in remote sensing photographs and ground truth measurements.

The soft bottoms constitute the main consumer part of the Baltic ecosystem. Below 60 m depth the bottoms switch between aerobic and anaerobic states with nutrient flux and hydrogen sulphide formation as important processes. The role of man's wastes flows and of the pulse-like inflows of North Sea water as forcing functions are discussed and presented in computer simulations. Examples of eutrophication effects such as decrease of bottom fauna and increase of stocks of pelagic fish such as herring and sprat are given.

The feedback-loops between the phytal, the pelagic and the soft bottom subsystems are exemplified by living flows such as migrating fish and hydrodynamic processes such as upwelling and downwelling. The urgent need for a closer co-operation between physicists and biologists is stressed.

1. The evolution of the Baltic ecosystem

From a macroscopic point of view, the Baltic Sea can be described as a huge ecosystem with the vegetation covered coastal areas and the phytoplankton dominated surface waters as producers and the vast soft bottom areas as consumers. These are connected with each other by the biogeochemical cycles, where the water plays the role of a dynamic transport system. As in any other area, the Baltic Sea started as a self-designed system and many types were tried and rejected during the big physical changes after the last glaciation when it was successively an ice-lake, a marine Yoldia Sea, a freshwater Ancylus Lake and a brackish Littorina Sea. Today the Baltic Sea, with an area of 365 000 km^2 and a salinity at the surface of 6-7 ‰ is the largest true brackish water area in the world, large enough to give examples of mesoscale events but with species of organisms few enough to encourage a systems analysis of the whole sea. The Baltic Sea may be young as a marine system but the organisms, usually requiring a large time scale for adaption have actually invaded from old marine and freshwater areas where they were selected to conform to the stress and dynamics of the physical and chemical environment.

2. Morphology of the basin

The Baltic Sea extends in a north-south direction and is divided into a series of basins by the presence of sills (Fig.1). The most important of these, the Oresund and the Strait of Darss, constitute with their 8 and 18 m respectively, the connection of the Baltic proper with the Kattegatt and the shallow Belt Sea, while its northern boundary is created by the shallow Aland Sill and the Archipelago Sea. Further north, the Bothnian Sea extends to the shallow Kvarken, north of which the Bothnian Bay constitutes the northernmost part of the Baltic Sea. The Gulf of Finland and the Gulf of Riga are two eastern bays connected to the Baltic proper. Within the latter, the Bornholm Deep (105 m) and the Gothland Deep (249 m) are two well known basins, often used as "indicators" or "alarm-clocks" for the actual state of the whole Baltic proper. But on the whole, the Baltic is a shallow sea with a mean depth of only 55 m. Roughly 17% of the total Baltic area is shallower than 10 m which means favourable conditions for growth of macroscopic plants within a proportionally large area.

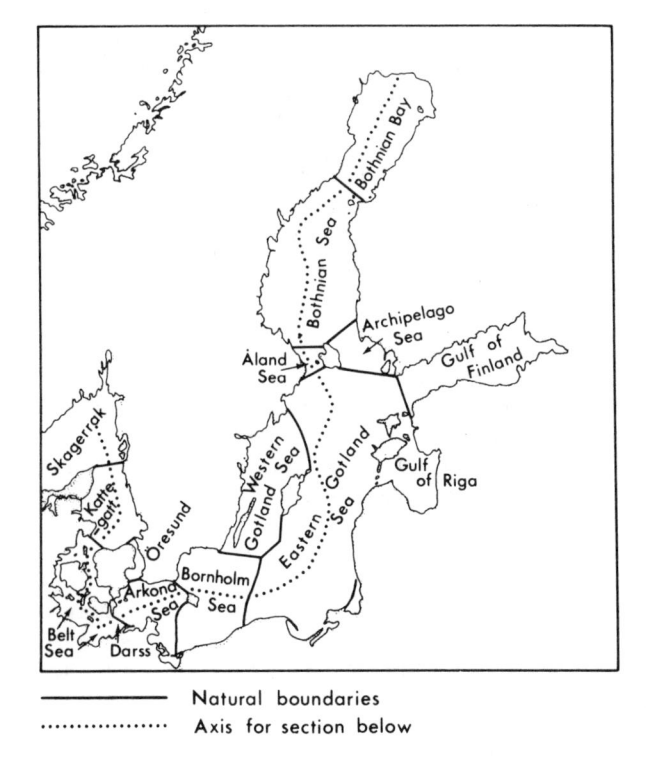

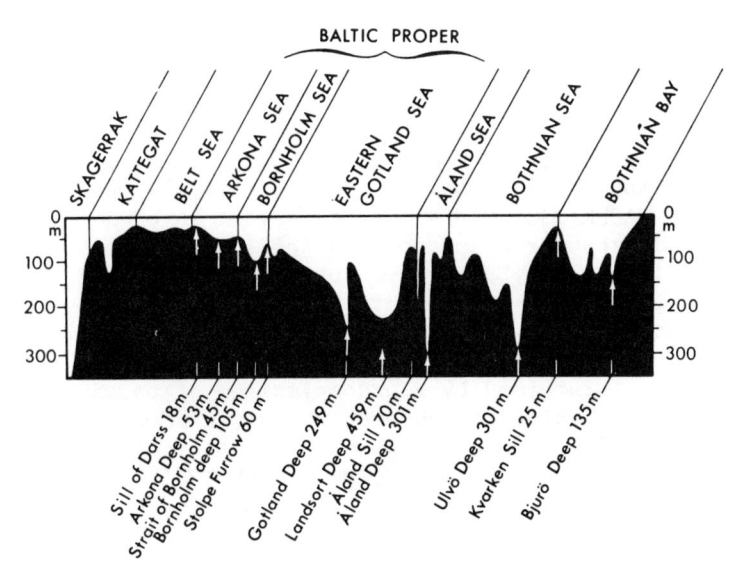

THE BALTIC - A SYSTEMS ANALYSIS OF A SEMI-ENCLOSED SEA

Figure 1. The natural division of the Baltic (after WATTENBERG, 1949) and the main sills and basins (after DIETRICH and KOSLER, 1974).

The bottom sediments clearly show the past history of the Baltic Sea. While the southern bottom areas are dominated by sandy sediments of alluvial origin, the northern parts consist of old bedrock from precambrium and palaeozoicum. On top of this, land runoff from rivers has deposited mineral sediments, particularly in the north where several large rivers reach the sea. The organic fraction comes both from land runoff and from dead Baltic plants and animals. Wave erosion, transportation and sedimentation continuously working on the bottom are responsible for the distribution of the bottom sediments. The southern and southwestern coasts up to the Gulf of Finland are shallow and sandy areas. The Swedish coast of the Baltic proper is mostly rocky, with large archipelagos. They form a rugged band along the Swedish coast, over to Finland through the Archipelago Sea and along the northern coast of the Gulf of Finland. In the Bothnian Sea and Bothnian Bay, stony and sandy sediments again dominate the shallow coastal areas. The deeper parts of the basins are filled with muddy sediments.

Into this series of basins several large rivers, especially in the north, empty their water. This flows southwards over the inflowing salt water from the North Sea and creates a heavily stratified body of water in the Baltic proper. The hydrodynamical forces have evolved a characteristic system which constitutes the basis for the whole Baltic ecosystem.

3. The physical system

The structure of this non-tidal system is determined not only by external forces, important for the water and material budget, but also by internal processes involving advection and mixing. The 1500 km long series of basins has a water volume of ca. 21 000 km^3, of which the Baltic proper constitutes ca. 13 000 km^3 (EHLIN, MATTISSON and ZACHRISSON, 1974). The Baltic is one of the most studied seas in the world and observational series since the beginning of the century have given us a fairly good knowledge of the water balance. This is maintained mainly by the freshwater budget and meteorological forcing. The water balance is worked by processes of vertical and horizontal exchange. The local water budget is maintained by a precipitation of ca. 500 mm·yr^{-1} and an evaporation which, taken over the whole area, balances the precipitation. The horizontal water balance is dominated by the river runoff which constitutes ca. 40% of the total water exchange. The subsurface runoff makes up less than 10% of the total river input (ZEKTZER, 1973). The input from the North Sea is by SOSKIN, (1963) divided into 1) a "continuous" flow of water along the bottom, generated primarily by the horizontal salinity gradient and 2) a more intermittent, intensive inflow, mostly during autumn and winter. The latter is mainly forced by the meteorological conditions such as the distribution of air pressure and wind field over the total area. The morphology of the connection between the Baltic and the Kattegatt is favourable for such inflows since westerly winds tend to elevate the sea level at the Swedish coast while depressing it in the Arkona Basin (DIETRICH and SCHOTT, 1974).

Of the inflowing water from the Kattegatt to the Danish Straits (ca. 1200 $km^3 \cdot yr^{-1}$) only a small part continues into the Baltic proper. The Baltic Sea has a positive water balance with a net outflow of ca. 500 $km^3 \cdot yr^{-1}$ (SOSKIN, 1963) corresponding to the freshwater input. This transport is small compared to the volume of the total Baltic. SVANSSON (1972) calculated that it would take 35 years for a successively accumulating, persistant substance to reach a steady state. The Baltic has therefore more of a stagnant than a throughflow character (FALKENMARK and MIKULSKI, 1975).

The water flowing into the Baltic proper has an average salinity of 17.5‰ (FONSELIUS, 1970), is heavier than the surface water in the Baltic and is therefore injected at a subsurface level determined by its density. It will maintain the salinity of the deep water in the Baltic (ca. 11‰) and sustain the stable pycnocline at ca. 60 m depth (the primary halocline) which is one of the main characteristics of the Baltic (Fig. 2). The slow flow of deep water will proceed into the Baltic proper filling the depressions along its eastern margins. A small amount might penetrate into the Gulf of Bothnia through the narrow deep furrow west of Åland but most of the slow flow will continue westwards, north of Gothland, pass the Landsort Deep, the greatest depth in the Baltic Sea (459 m) and continue southwards through the shallow areas between Gothland and Öland. The diluted Baltic water flows out at the surface through the Belts and the Öresund and continues northwards as a low-salinity wedge against the Swedish westcoast (the Baltic Current).

While the transport of the "continuous" flow is still fairly unknown, the pulse-like injections of North Sea water has drawn more attention. The largest inflow hitherto known occurred in 1951 when ca. 200 km^3 of saline water (22‰) was forced into the Baltic during a couple of weeks. The water of such a heavy inflow creeps along the bottom, filling the deep depressions, the Bornholm Deep, the Gothland Deep, the Landsort Deep flushing the basins of their stagnant water. A secondary halocline is formed, usually at 110-130 m depth (VOIPIO and MÄLKKI, 1972). The big inflow in 1951 reached the Gothland Deep after 5-6 months, whereas more "normal" inflows such as the 30 km^3 inflow in 1969 was detected in the Gothland Basin 8-9 months later (FRANCKE and NEHRING, 1973).

The magnitude of the Baltic coupled with efficient redistribution modes leads to an unusually stable salinity distribution.

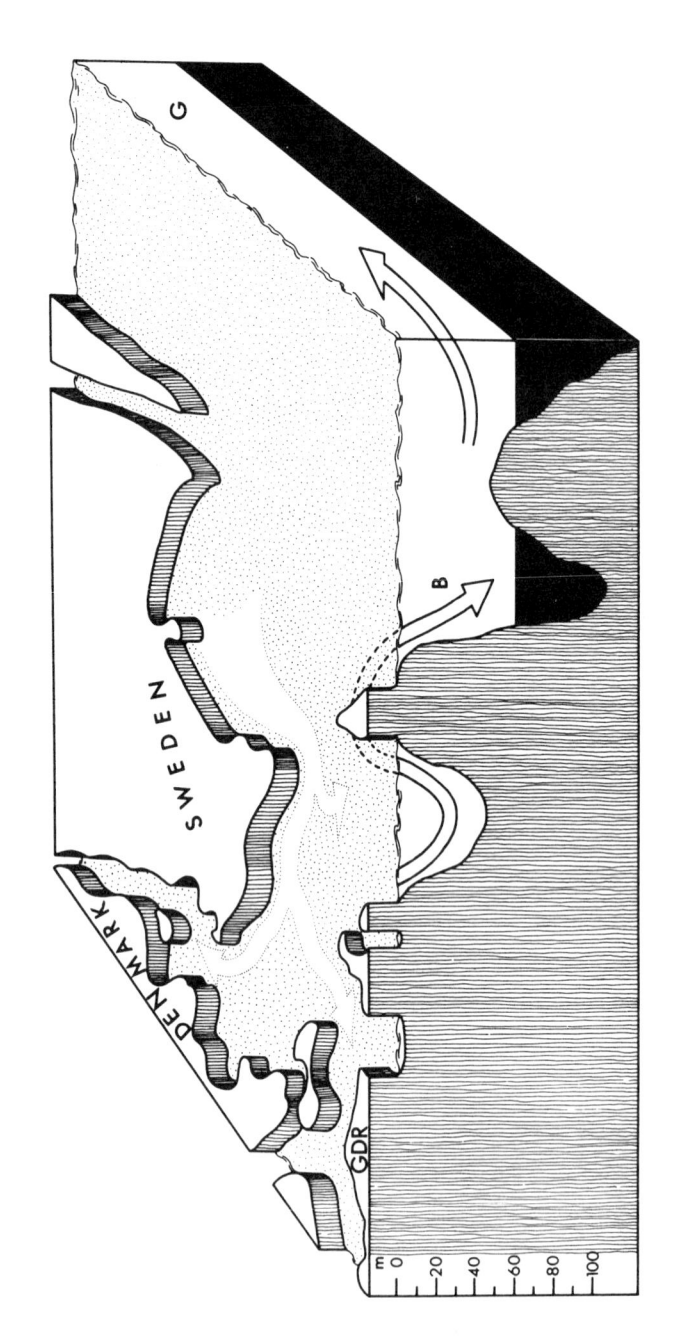

Figure 2. Block diagram showing the outflow of surface water through the Oresund and the Strait of Darss, the inflow of saline water along the bottom and the primary halocline in the Bornholm Deep (B) and the Gothland Basin (G) with underlying bottom water (black).

Except for the "noise" in the Arkona Sea, which acts as a kind of buffer zone, an annual fluctuation of 0.5‰ for the surface water has been calculated (DIETRICH, 1950). The mean salinity of the water is about 6-7‰, of the deep water, 8-12‰ (FONSELIUS, 1972). In the Bothnian Sea and Bothnian Bay, a stable salinity stratification is almost missing except for the deepest part of the Bothnian Sea, which will intermittently receive pulses of saline water from the Baltic proper. The salinity in these northern basins decreases successively with values close to freshwater in the northernmost part.

The residence time of the water in the Baltic has been calculated as 25 years (FALKENMARK and MIKULSKI, 1974). One third of the out-flowing water, however, returns with the inflowing water giving a residence time of 35-40 years. This means that much happens to the water during its stay. Exporting its load of oxygen to the organisms in the deep and bottom layers, the inflowing water receives nutrients such as phosphate, nitrate, ammonia from decomposition processes and, when oxygen is running short, hydrogen sulphide. The primary halocline acts as a barrier between the deep water and the less saline surface water. In the summer a thermocline at 20-30 m depth develops in this layer. Below this, the water is trapped between the thermocline and the primary halocline and is maintained mainly by the winter convection and therefore cold, 0-3°C throughout the year. The water below the primary halocline is warmer, 4-5°C, but except for a thin upper layer, the Baltic must be regarded as a very cold sea. The cooling in winter usually leads to the formation of an ice cover which in the Bothnian Bay persists for nearly 4 months but for the coastal area in the Baltic proper lasts for only 1 month.

Obviously the lid of the primary halocline must open somewhere and a transport of nutrients fertilize the sunlit parts of the watermass, constituting an important forcing function of the Baltic ecosystem. The stability of the stratification is therefore of crucial importance. An experimental approach to this complex problem was adopted by WELANDER, (1974). He showed that if the Baltic is simplified as a two-layer fjord-type estuary driven by freshwater inflow and mixing, a decrease in the freshwater supply would lead to a decrease in stability and show one single steady state, sensitive even to small changes in the statistical characteristics of the forcing functions.

A numerical model, using the time statistics of weather and runoff as basic forcings on the time-dependent salt conservation equation integrated with the vorticity equation in the longitudinal-vertical plane has been developed by WILMOT, (1974). It is especially fitted to explore the requisite conditions for the salt water intrusions and the distribution of salinity in the Baltic.

The nutrients in the deeper layers may penetrate the stable halocline and reach the surface either through turbulent mixing, (e.g. wind entrainment, internal mixing) or vertical advection (upwelling). Few attempts have been made to quantify the exchange through the halocline. KULLENBERG, (1974) calculated the diffusive, vertical exchange, partly from known nutrient gradients, partly from his own field measurements. He concluded that the diffusion through the halocline in the open Baltic, though small, could not be disregarded in comparison with the coastal mixing. Based upon the bi-weekly measurements of surface temperature of the Baltic made by SMHI and the Fishery Board of Sweden, it is possible to recognize coastal regions of divergent temperature, e.g. off the Stockholm Archipelago and the Hano Bight, which correspond with strong SW-W winds (Fig. 3). Intermittent upwelling has recently also been proved for these areas. WALIN, (1972) was the first to point out the importance of internal Kelvin waves for the Baltic coastal areas. He suggested that the nutrient flux from the deeper layers of the Baltic takes place mainly along the coasts through boundary mixing. For the Swedish coast opposite Gothland, upwelling has also been shown to take place intermittently both during summer and winter by SVANSSON, (1975) who has also summerized the coastal-offshore dynamics in the Baltic.

In the Asko-Landsort area, SHAFFER, (1975) was able to show the presence of a wind-driven, downwelling regime in the fall and the importance of coastal jets. This system was capable of a considerable horizontal exchange and particle transport between the coast and the offshore area. Little vertical transport through the halocline was implied however. Mean cross-halocline transports over the rest of the fall, which was characterized by upwelling events, were calculated to be quite intense. A flow of the same magnitude, calculated for a 600 m broad area along the whole Baltic coast would account for the total, vertical uptransport in the Baltic proper (SHAFFER, pers. comm.).

These examples clearly show the importance of the coastal areas for the dynamics of the Baltic proper. The upwelling areas might be regarded as "nutrient windows", which intermittently open up between the surface and the deeper layers, letting through a flow of nutrients, which is of decisive importance for the evolution of the structure and dynamics of the Baltic ecosystem.

4. The biological system

4.1. The biological components

The resources of the Baltic for the establishment and adjustment of the biological components into a large-scale ecosystem may be roughly condensed as:

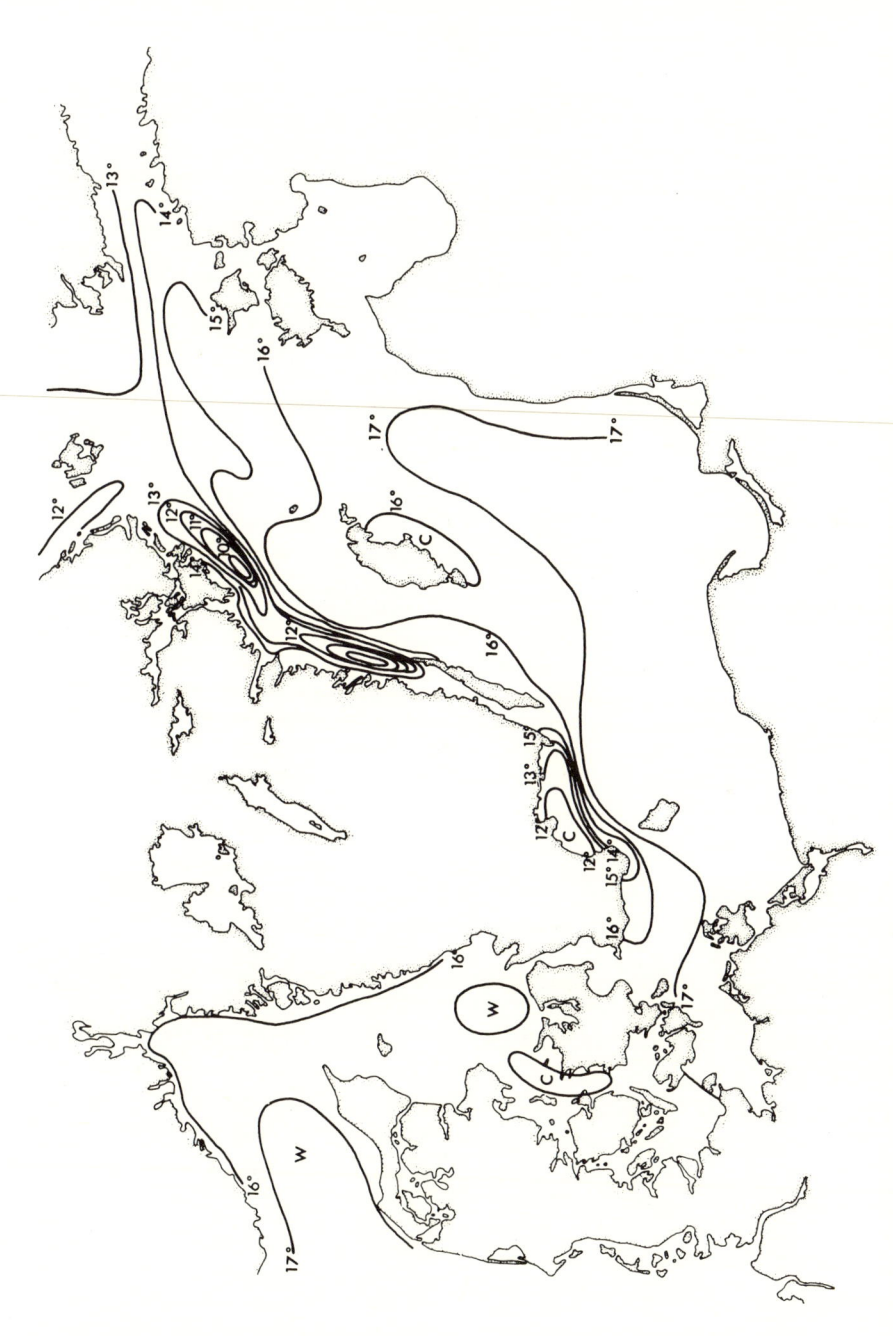

THE BALTIC - A SYSTEMS ANALYSIS OF A SEMI-ENCLOSED SEA

Figure 3. Surface temperature of the Baltic on September 19, from the bi-weekly measurements by the Swedish Meteorological and Hydrological Institute (SMHI).

1. A substratum ranging between hard and soft bottom but where the rocky archipelagos at the coast constitute a typical Baltic feature.

2. A heavy annual pulse of solar energy.

3. A water mass, diluted, but with the same proportions between the main constituents as in marine water and with an unusual stability in the total salinity.

4. A physical system with a heavy stratification of the water, a spring and autumn convection of the surface water and intermittent pulses of inflows from the North Sea and of upwelling along the coasts.

The biological components in the Baltic ecosystem consist of a curious mixture of species, selected through invasions from adjacent seas and lakes. Some arrived during Arctic conditions and were trapped when the climate grew warmer and are not living as glacial relicts in the deep of the Baltic, like the fourhorned sculpin, Oncocottus quadricornis, while others are still on the march following the slow increase in salinity and pushing their innermost limit further to the north, like the common jellyfish, Aurelia aurita.

This heterogeneous flora and fauna offers a whole spectrum of basic problems in marine biology, physiology and autecology which has attracted many scientific workers, and the Baltic Sea is probably the best known sea in the world (cf. REMANE and SCHLIEPER, 1958, SEGERSTRALE, 1957). The main features of the Baltic organisms are:

1. They can withstand low salinities.

2. They are less specialized in the Baltic than members of the same species in fully marine areas. Macoma baltica for example occupies both sandy and soft bottoms in the Baltic but is restricted to shallow sand bottoms on the Swedish west coast. This might be the result of decreased predator pressure or increased number of free niches due to the lower diversity in the Baltic systems (cf. DAHL, 1973).

3. The Baltic specimens are smaller than those living in fully marine waters. A blue mussel, Mytilus edulis on the Swedish west coast is about 4-5 times longer than its Baltic counterpart. The latter has to put more energy into maintaining the basic metabolism in a dilute medium, has lower filtering activity and grows more slowly.

4. The number of species in the Baltic ecosystem is low compared to marine areas. Fig. 4 shows some marine species at their innermost limit. It also shows the low number of macroscopic species in the central Baltic, 88, compared to the 1500 off the Norwegian coast.

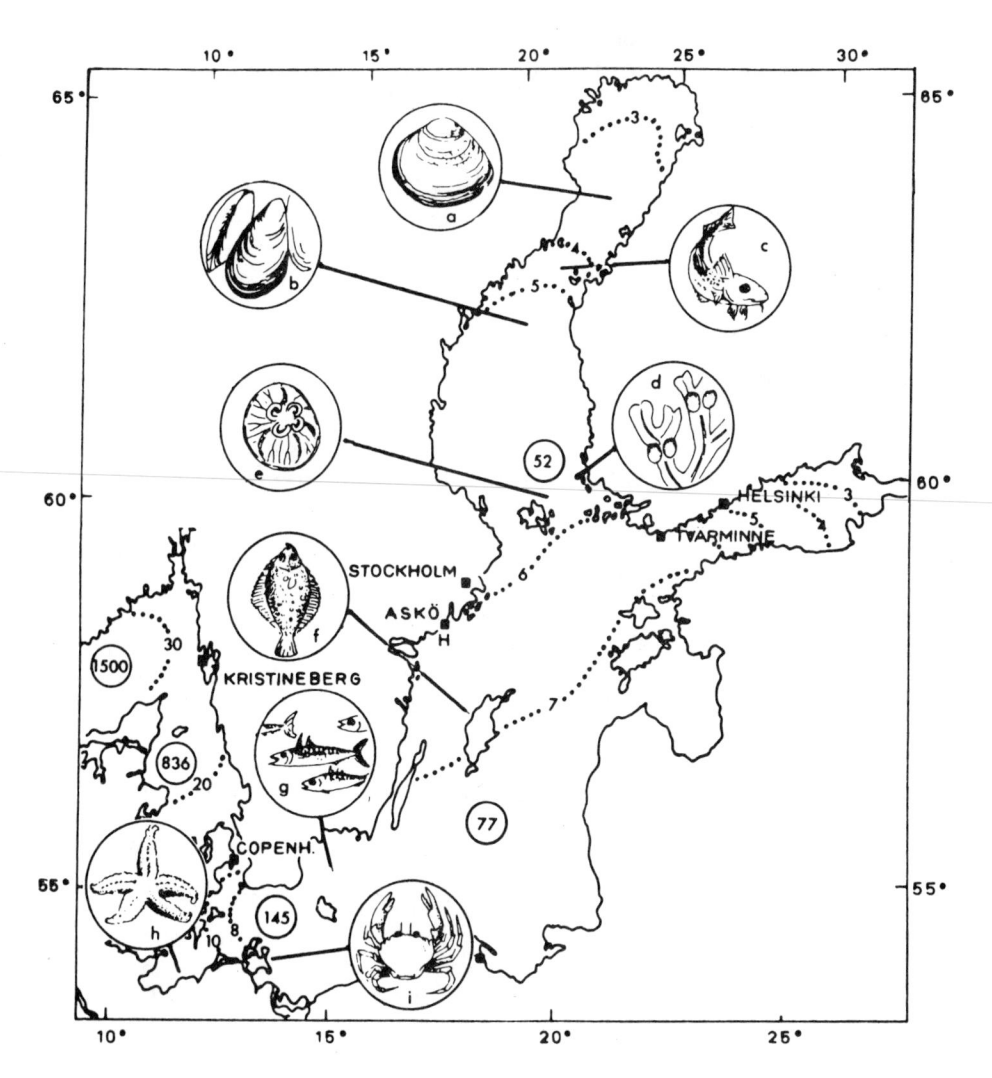

Figure 4. The Baltic is an area of low salinity (dotted lines are isohalines) and low diversity (numbers in circles are numbers of macroscopic animals, according to ZENKEVITCH, (1963). Some marine organisms are shown at their innermost limit (according to SEGERSTRALE, 1957). Dotted lines: surface water isohalines. Numbers within circles: number of macrofauna species. a-i distribution limits for some common marine species: a. Macoma baltica b. Mytilus edulis, c. cod, d. Fucus vesiculosus, e. Aurelia aurita, f. plaice, g. mackerel, h. Asterias rubens, i. Carcinus maenas (B-O. JANSSON, 1972).

Thus the biological components in the Baltic ecosystem represents a selected set of species with a large physiological and ecological potential. These "hardy" species live at the limit of their salinity tolerance however, and are probably sensitive to changes, though the opposite conclusion has been put forward by some authors (JÄRNELÖV and ROSENBERG, 1976). The mixing of freshwater and marine species also shows the Baltic as a system established between the marine and freshwater regimes and utilizing resources from both - a large-scale, edge community.

4.2. The main subsystems of the Baltic

4.2.1. The phytal subsystem

The evolution of a well developed phytal subsystem is partly due to the large shallow area with good light conditions (17% of the bottoms shallower than 10 m), and partly to a coastline greatly increased by the numerous islands and skerries of the rocky archipelagos. Fig. 5 shows part of the area between Finland and Sweden, a submerged rocky landscape. Nutrients from land runoff are mixed with nutrient-rich Baltic bottom water, welling up due to favourable winds (SHAFFER, in print). Reduction in number of species is also valid for the flora, the number of dominating species decreasing from 154 off the Norwegian coast to 24 at the SW Finnish coast (SCHWENKE, 1974).

The main floral components of the rocky coast are sorted out vertically according to their optimum for fixing solar energy at their complementary colour with green algae nearest the surface (mainly the freshwater species (Cladophora glomerata) followed by brown algae (Fucus vesiculosus, Pilayella spp.) and red algae (Furcellaria fastigiata, Phyllophora spp.) in the lower part of the phytal zone (ca. 15 m depth). Above the mean sea level in the splash zone, the rocks are covered by the blue-green alga Calothrix scopulorum, able to fix atmospheric nitrogen (HÜBEL and HÜBEL, 1974). The physical structure of this zone has also developed according to the maximum power principle (most efficient use of available energy). The area of greatest physical noise, the splash zone is populated by the short, nearly crustlike Calothrix browsed upon by snails like the freshwater species Theodoxus fluviatilis. The mean sea level is dominated by filamentous algae (Cladophora, Stictyosiphon, Ceramium) swaying with the waves and forming a large "inner surface" due to the mode of growth. This inner surface provides microniches for tiny plants like sessile diatoms, microscopic animals like rotifers and turbellarians and numerous young of crustaceans from the zones deeper down which here find a nursery with abundant and suitable food for their first month (A.M. JANSSON, 1967).

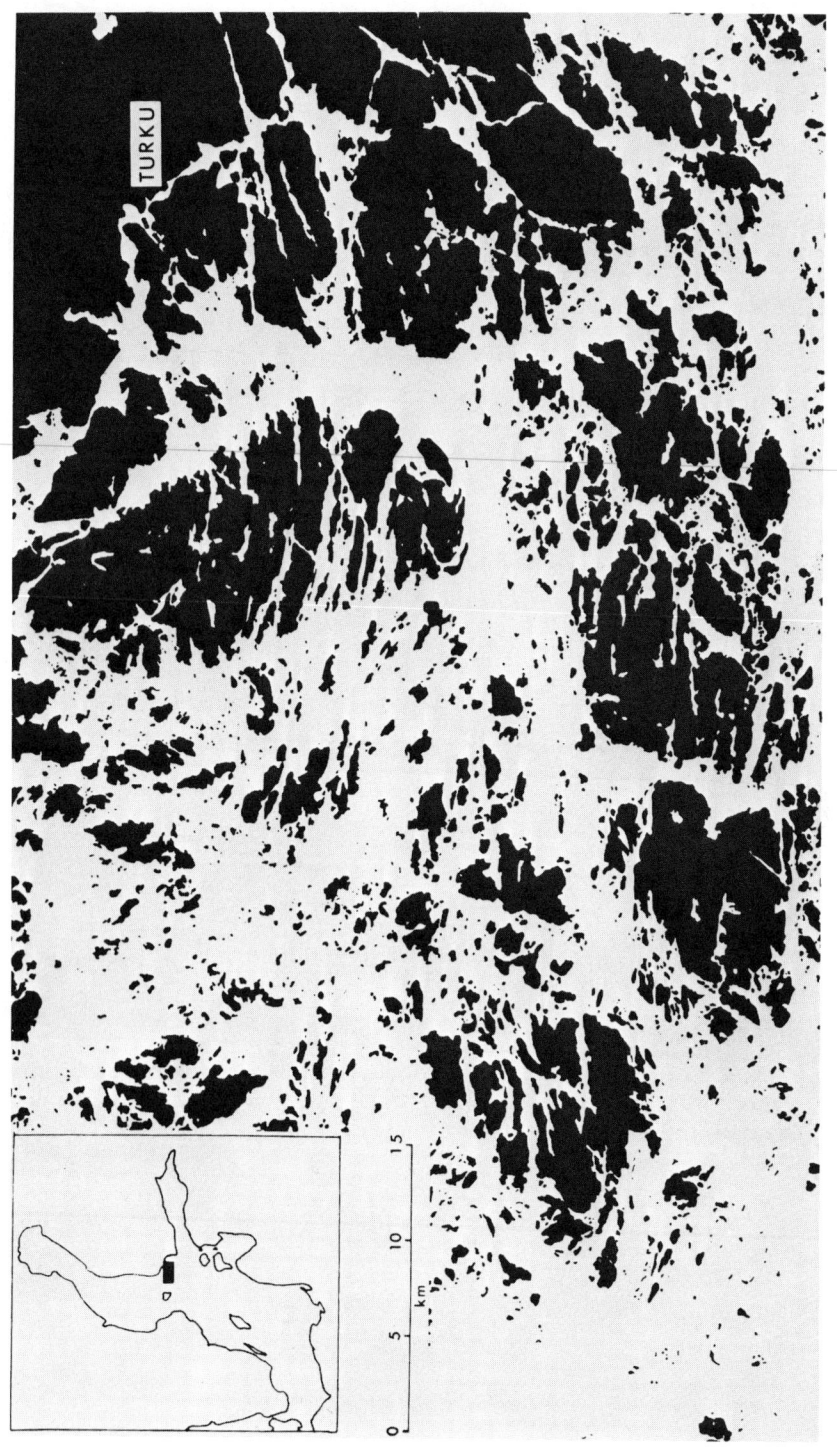

Figure 5. The Archipelago Sea off the city of Turku - a submerged rocky landscape (from a folder of The Archipelago Research Institute, University of Turku, Finland).

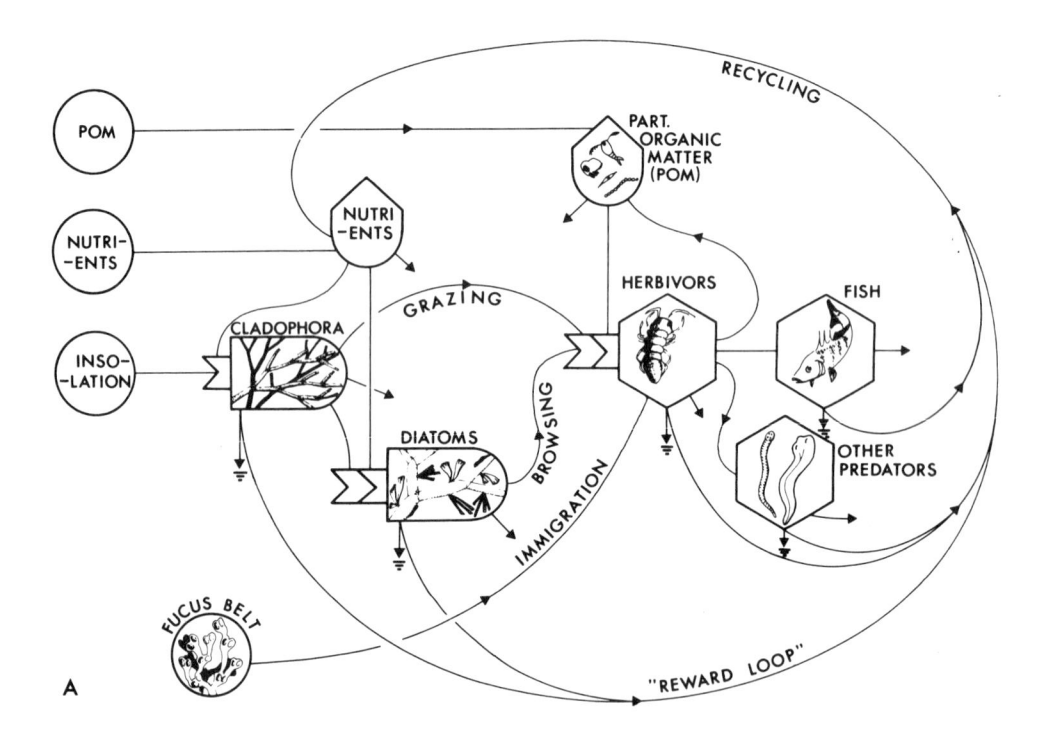

Figure 6. Model of the green algal belt (Cladophora belt) in the northern Baltic proper (from A.M. JANSSON, 1974).

A. Structure of model.

B. Simulation graphs from a perturbation where the nutrient input was increased 30%. Note the tendency of a second maximum of both Cladophora and diatoms at the end of the period.

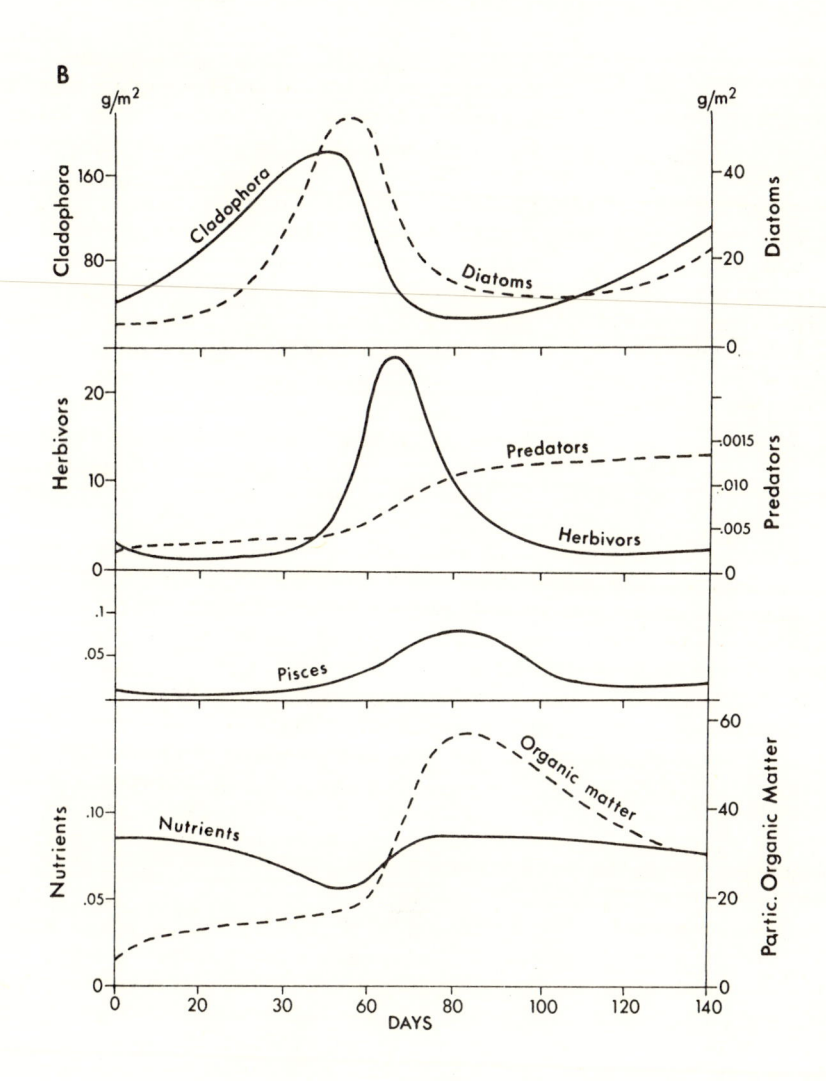

THE BALTIC - A SYSTEMS ANALYSIS OF A SEMI-ENCLOSED SEA

The Fucus belt below is dominated by the palmated, upright structure of the Fucus alga, exposing its surface both to light and to the continuous flow of nutrients in the turbulent water. Numerous filamentous epiphytes increase the number of species in this, for Baltic circumstances, highly diversified community of crustaceans, snails, bivalves and fish (SEGERSTRALE, 1944, HAAGE, 1975).

The red algal belt has a rather spare vegetation. On the hard bottoms below the dense vegetation there is quite apparently hard competition between the sessile plants and animals. The blue mussel, Mytilus edulis, is very successful and covers large areas. As a filter feeder it greatly benefits from the turbulent water, consuming the organic particles from the more shallow areas, turning them into flesh and respired nutrients. The fish fauna in the hard bottom phytal is the most diversified in the Baltic. It also shows the curious mixture of fresh and saltwater species. Gobiids (Pomatoschistus spp.) lump-suckers (Cyclopterus lumpus), pipefish (Syngnathids) and viviparous blenny (Zoarces vivparus) mingle with sticklebacks (Gasterosteus aculeatus), perch (Perca fluviatilis) and pike (Esox lucius). Their life in the clear oligotrophic water is dominated by the problem of concentrating food and the top carnivores in the fish fauna are therefore fast swimming hunters, ready to take the hook of the sports fisherman.

Within the phytal community, positive feedback mechanisms must have evolved to keep the order and timing of this complex system. The blue-green belt of the splash zone fixes nitrogen which is made available to the green belt below by the browsing of the snails and natural mortality of the alga. The annual green algae start growing in early spring when the ice has gone. They are stimulated by the increasing light and by the fresh supply of nutrients from below, where the perennial Fucus belt, through respiration and part decomposition during the dark winter period has accumulated waste products. As a "reward" the green algal belt rapidly builds up the well equipped "nursery", ready to receive the new generation of crustaceans from the Fucus belt in June. They return after a month to the now flourishing Fucus system before the green algae start to decompose. As an example of the dynamics of an algal belt, Fig. 6 shows a model of the green alga subsystem with the main functions represented: fixation of solar energy, nursery for fish, regeneration of nutrients. It was given values on storages and flows taken from thorough field investigations and laboratory experiments translated to 7 differential equations (one for each state variable) and simulated on a computer. The model showed a realistic behaviour when simulated. A 30% increase in input of nutrients gave two maxima of the green alga as they occur in eutrophicated areas (A.M. JANSSON, 1974).

In shallow areas with less turbulence, the freshwater influence is stronger and the vegetation is dominated by freshwater plants such as reed (Phragmites sp.) water millfoil (Myriophyllum sp.) and pond weed (Potamogeton spp.). The dense reed belts show a self-generating mechanism. The densely growing, vertical stems act as a coarse mechanical filter. Floating debris such as algae and fragments of driftwood remain and decompose in the reed belt. The numerous stems offer a large surface for epiphytes such as the green algae (Cladophora sp. and Enteromorpha spp., which when decomposing increase the amount of organic matter at the bottom. The reeds greedily take up and holds the nutrients released during organic decomposition (BJORK, 1967) and the belt grows. In former days when agriculture was more dominant, the reed belts were grazed by cattle, which brought back the nutrients to land through the dung. Nowadays, both an increasing waste flow from overpopulated areas and the absence of cattle have made the reed belts spread, to the annoyance of people using the coast as a recreation area.

The fauna in the reed belt consists of freshwater species where insects such as larvae of caddis flies and dragonflies and water bugs dominate. In the nutrient-rich water the concentration of energy is a small problem and the fish fauna consists to a large extent of sluggish species such as rudd, bream and tench.

In sandy areas, especially in the southern Baltic, large areas are covered by the eelgrass, Zostera marina. Associated with it are clams, mussels and snails (GOTHBERG and RONDELL, 1973, LAPPALAINEN, 1973). This association is an example of one of nature's innumerable positive feedback mechanisms. The animals help to keep water and leaves clear and clean, the plant is stimulated by the resulting increase in insolation and more substrate is offered to the animals.

Within a coastal area there is competition between systems. An increased organic input will transform the former hard bottoms into soft bottoms and algae will be substituted by higher plants such as reed. Competition for space is also strong within the algal systems but has been solved through succession of species. The green algae with high light optima dominate the upper zone during summer but are replaced by the less demanding red algae during autumn and winter.

Quantitative estimates of biomasses in the phytal subsystems are scarce in the Baltic. The most accurate ones exist for an investigation area of 160 km^2 in the Landsort region (A.M. JANSSON and KAUTSKY in print). Here the hard bottom, covering a total area of ca.27 km^2 carry a total biomass of plants of 2163 tons and of animals, of 8618 tons (dry weight with shells). Of these, the blue mussels accounted for 7797 tons. In the southern Baltic where sand and gravel is common, their biomass has been estimated at 3.4 million tons (fresh weight without shells) which is ca. 70%

of the total animal biomass in this area (DEMEL and MULICKI, 1954). According to NIXON, OVIATT, ROGERS and TAYLOR (1971) their main function in the system is as nutrient regenerators. They import organic material, metabolise it and export nutrients.

Production figures are less accurate. The turnover time in summer of the Fucus subsystem has been estimated by community respiration measurements as 70 days (JANSSON and WULFF, in print). Although a true comparison with data for other areas is difficult to make, the Baltic seaweeds must be regarded as far less productive than e.g. the Nova Scotia algae as summarized by MANN (1973).

4.2.2. The pelagic subsystem

The pelagic zone in the Baltic is dominated by the typical temperate heavy pulse of solar energy, which together with the physical characteristics of the area, has created a dynamic pattern of the energy transformation. The great physical "noise" in the system has checked the building of large size structures and for the solar energy fixing organisms, the phytoplankton, the microscopic size has been compensated for by a high turnover rate. Evolving in a physical system, a succession of different forms or strategies can be seen during the Baltic year similar to that described by PATTEN (1963) for the east north-American coast. In the beginning of the year when light is short and the water is dense and cold, the primary production is carried out by forms like diatoms which are heavy, have low light optima, long generation times, storage capacity, no mobility but floating devices. During summer with luxuriant light, high temperatures and less dense water, green algae and other small forms which are light, have high light optima, short generation times and mobility are more efficient producers. When the water is cooling off towards the end of the year these are replaced by diatoms and dinoflagellates. In this way, the system maximizes its use of available energy by using the most efficient "worker" for each situation.

During early spring, more energy is pumped into the system as solar radiation increases. The break-up of the ice is accompanied by a vertical mixing of the homogeneous water, upwelling is favoured and the spring bloom starts. Diatoms like Skeletonema costatum and Thalassisira baltica flourish, dinoflagellates like Gonyaulax catenata rapidly increase and the consumption of nutrients is vigorous. Nitrate might be depleted within four days and the producers appear to consume nutrients in excess (HOBRO, LARSSON and WULFF, in print). Primary production may attain values of over 1 $gC \cdot m^{-2} \cdot d^{-1}$ during this period (HOBRO and NYQVIST, 1972). The first herbivores to respond to this intensive fixation of energy are the unicellular ciliates which reproduce fast enough to be able to respond to such rapid changes (HOBRO, LARSSON and

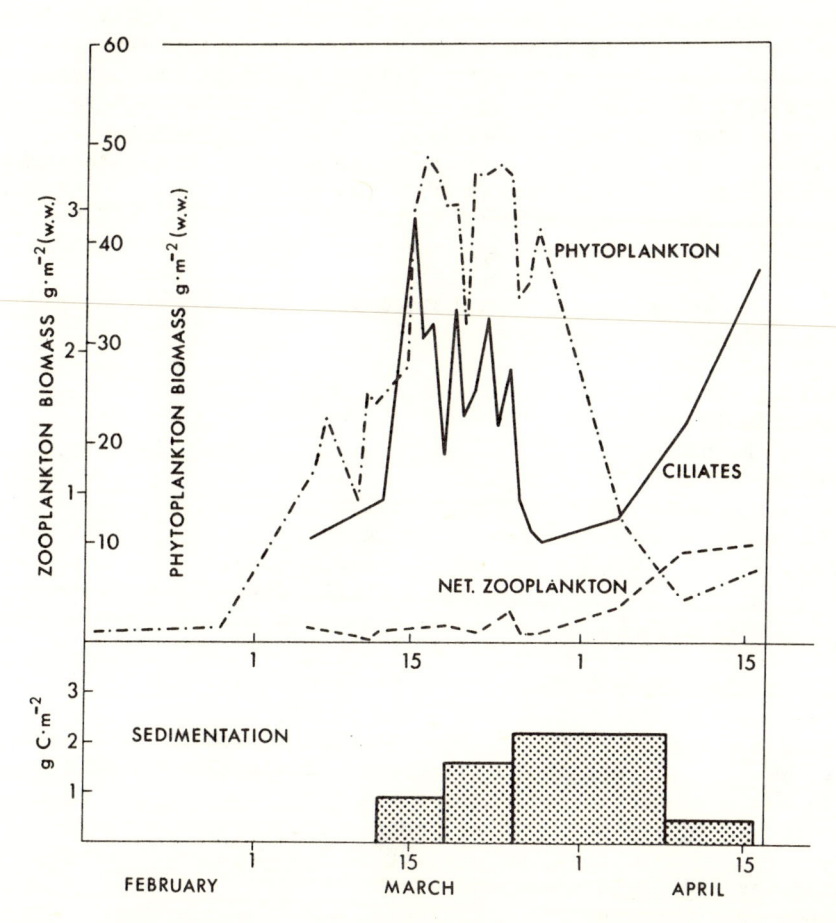

Figure 7. The dynamics of a spring bloom in the Asko-Landsort area (from HOBRO, LARSSON and WULFF, in print). The ciliates rapidly respond to the increase in phytoplankton and bacteria (not shown here). Sedimentation is fast during the maximum and decline of bloom and net zooplankton has a time lag.

WULFF, in print). They graze heavily both on the pelagic bacteria which now have a turnover time of only a few hours (HAGSTROM and LARSSON, in print) and on the smaller phytoplankton, saving energy for the system by making the necessary concentration of food for the bigger zooplankton (Fig. 7). These usually have minimum biomass during this time of the year (ACKEFORS, 1975) but at the approximate time of the bloom, the overwintering benthic eggs of some species like Acartia hatch, filling the water with nauplii which rapidly develop into copepodites. The rotifers have a built-in mechanism for fast reproduction and Synchaeta spp. increases rapidly towards the end of the bloom. With few grazers being present, a large proportion of the primary produced organic matter sinks out of the pelagic system to the bottom (KAISER and SCHULTZ, 1975). The heavy diatoms especially have been shown to settle during the bloom and more than one third of the fixed carbon is injected to the soft bottom system during this period (HOBRO, LARSSON and WULFF, in print).

The spring bloom makes use of most nutrients in the water and exports a considerable amount of them back to the bottom. The summer stage of the pelagic system consists of a low phytoplankton biomass where smaller forms such as dinoflagellates and monads dominate (BAGGE and NIEMI, 1971). Carbon fixation can still be intensive but the algal biomass is checked by intense grazing (KAISER and SCHULTZ, 1975). Microplankton increase in importance, apparently very much influenced by temperature (CISZEWSKI, 1975). The larger zooplankton now show medium biomasses and more species such as the typical copepods Temora longicornis and Pseudocalanus minutus elongatus (ACKEFORS, 1975) are important.

The warm and sunlit water would be favourable for phytoplankton growth were it not for the small amounts of nutrients. The system has a tool even for those situations, however, in the nitrogen fixing algae. During recent years blooms of blue-greens have become a typical Baltic feature (HORSTMANN, 1975) especially the species Nodularia spumigena which is able to use molecular nitrogen as a nutrient resource (HÜBEL and HÜBEL, 1974). During calm days in July-August, the water may be covered by large, dense patches or strings of this alga, accumulating at the surface by means of gas vacuoles. Efforts to quantify this bloom which may last for a couple of weeks have been made by using aerial photography and satellite pictures (NYQVIST, 1974). During 1973 a heavy bloom, extending along the Swedish Baltic coast from the Landsort area to Rugen (Fig. 8) was estimated to hold 1300-1700 tons of carbon. Rough calculations of the amount of fixed nitrogen of the same bloom has been made by ÖSTRÖM (1976) adopting BRATTBERG'S (1974) values for nitrogen fixation in the Stockholm Archipelago. Like the measurements of HÜBEL and HÜBEL (in print) they indicate an input of nitrogen from the atmosphere approaching the total input from rivers and human discharges. When declining,

THE BALTIC - A SYSTEMS ANALYSIS OF A SEMI-ENCLOSED SEA

Figure 8. Satellite picture from an ERTS-1 passage over the Baltic at 90 km height on July 3, 1973 at 0930-0936 hours. The dark streaks in the water are blooms of the blue-green alga Nodularia spumigena, able to fix atmospheric nitrogen. A rough calculation of the bloom shown in the picture amounts to 1300-1700 tons of C (NYQVIST, 1974).

the blue-green bloom yields nutrients, and an autumn bloom, mainly of diatoms, green alga and dinoflagellates captures energy for the pelagic system. Zooplankton now reaches its maximum and pelagic fish like herring and sprat are able to store fat reserves (ANEER, 1975, ELWERTOWSKI and MACIEJCZYK, 1964, cited in ACKEFORS, 1975). Triggered by temperature the autumn-spawning herring spawns in places where the concentration of copepod nauplii is highest (OJAVEER and SIMM, 1975). The autumn is thus a period of intensive development and growth to accumulate energy reserves to sustain the long and dark winter period.

During the winter, phytoplankton biomass is low, and for the total system respiration exceeds production. Zooplankton is scarce, the dominating genus being _Acartia_. The plankton-feeding fish, sprat and herring, having benefited from a good food supply in autumn, now have problems concentrating their food and have to draw energy from the soft bottom system. Reports show that herring during this period feed on oppossum shrimps, _Mysis_ spp., and the soft bottom amphipods, _Pontoporeia_ spp. (ANEER, 1975). Late autumn and winter is also a stormy period and in coastal areas a resuspension of sediments and "cleaning" of the archipelagos from decomposing algae takes place. The burning of all these energy reserves, from the fat of the herring to the resuspended organic matter, results in an accumulation of nutrients at the bottom, especially when ice has covered most of the coastal areas. Turbulence is checked, the particulate organic matter sinks out of the pelagic system to the soft bottoms and the water becomes cold and clear.

The described conditions, co-ordinated with the three main plankton bloom periods is mostly valid for the northern Baltic proper. Far south in the Belts, up to nine different blooms have been registered (VON BODUNGEN, 1975, SMETACEK, 1975), probably due to the changing hydrographical conditions where new energy reserves are pumped in every now and then. Far north in the Bothnian Sea and Bothnian Bay the spring and autumn blooms have merged and only one dominating phytoplankton maximum is discernable, a typical polar feature due to the shortness of the light season (LASSIG and NIEMI, 1975).

The total annual primary production in the Baltic proper can be estimated to at least 100 $gC \cdot m^{-2}$ which is of the same order of magnitude as the primary production in the shallow coastal waters of the oceans (RYTHER, 1970). The secondary production of zooplankton in the Baltic proper is calculated to 5 $gC \cdot m^{-2}$ (ACKEFORS, 1975). The relation between these two figures is a lot more intricate than a simple calculation of transfer efficiency would indicate. There are numerous feedback mechanisms working in the pelagic system. Of the organic matter sinking to the bottom consisting of dead plankton and fecal pellets, around 50% will be decomposed by bacteria before it reaches the winter water layer at 30-70 m depth ZSOLNAY (1973). NYQVIST (1974) estimated the amount of particulate organic matter reaching the 45 m level in the offshore area of Landsort to 38 $gC \cdot m^{-2} \cdot yr^{-1}$. There are few estimates of the number of aerobic bacteria in the pelagic water but they seem to be of the order of 10^4-10^5 cells ml^{-1} (ANKAR, 1972, HAGSTROM and LARSSON, in print). The nutrients released during their activity are used again by the primary producers. A fair amount of the phytoplankton/zooplankton activity results in dissolved organic matter. HOBRO, LARSSON and WULFF (in print) concluded that a large proportion of the initial net primary production was released as dissolved organic matter from the system, comprising the combined release or phytoplankton, zooplankton and bacteria.

The tight loops between the different parts of the pelagic system and the disputability of the common approach of breaking up a dynamic system in order to measure its efficiency more easily has clearly been shown by MCKELLAR and HOBRO (1976). They studied the relationships of phytoplankton/zooplankton in field experiments with plastic enclosures in the Asko-Landsort area. When enriched with zooplankton, primary production increased 10-fold in spite of the lower biomass of the phytoplankton due to grazing. The stimulatory effect could be traced both to higher productivity/chlorophyll ratios and to faster phosphate cycling. This is a perfect example of the efficient coupling between the parts of the system which has up to now been allowed to evolve below a relatively steady input of energy. It also questions the validity of using the traditional ^{14}C-technique for the evaluation of the productivity of the pelagic system. Values of ecological rates should whenever possible be assessed from studies where the important feedback mechanisms are represented. The importance of the positive feedback-loops for the stability of the pelagic system has also been shown in computer simulation experiments by SJOBERG (in print).

A dynamic energy flow model of the pelagic shallow water system in the Kiel Bight has been simulated by PROBST (1975) based on the results of the Kiel University research project "SFB/95" (HEMPEL, 1975). The model incorporates the main plankton and chemical components of the area, supposed to be uniformly distributed within the area and with curve-fitted rates for the nutrient flux from the sediment (Fig. 9). The behaviour of the model, exemplified here by simulation graphs for the biotic components, gave a realistic picture of the field dynamics.

4.2.3. The soft bottom system

Whereas both the phytal and the pelagic systems to a large extent may be regarded as producer systems running on solar energy and

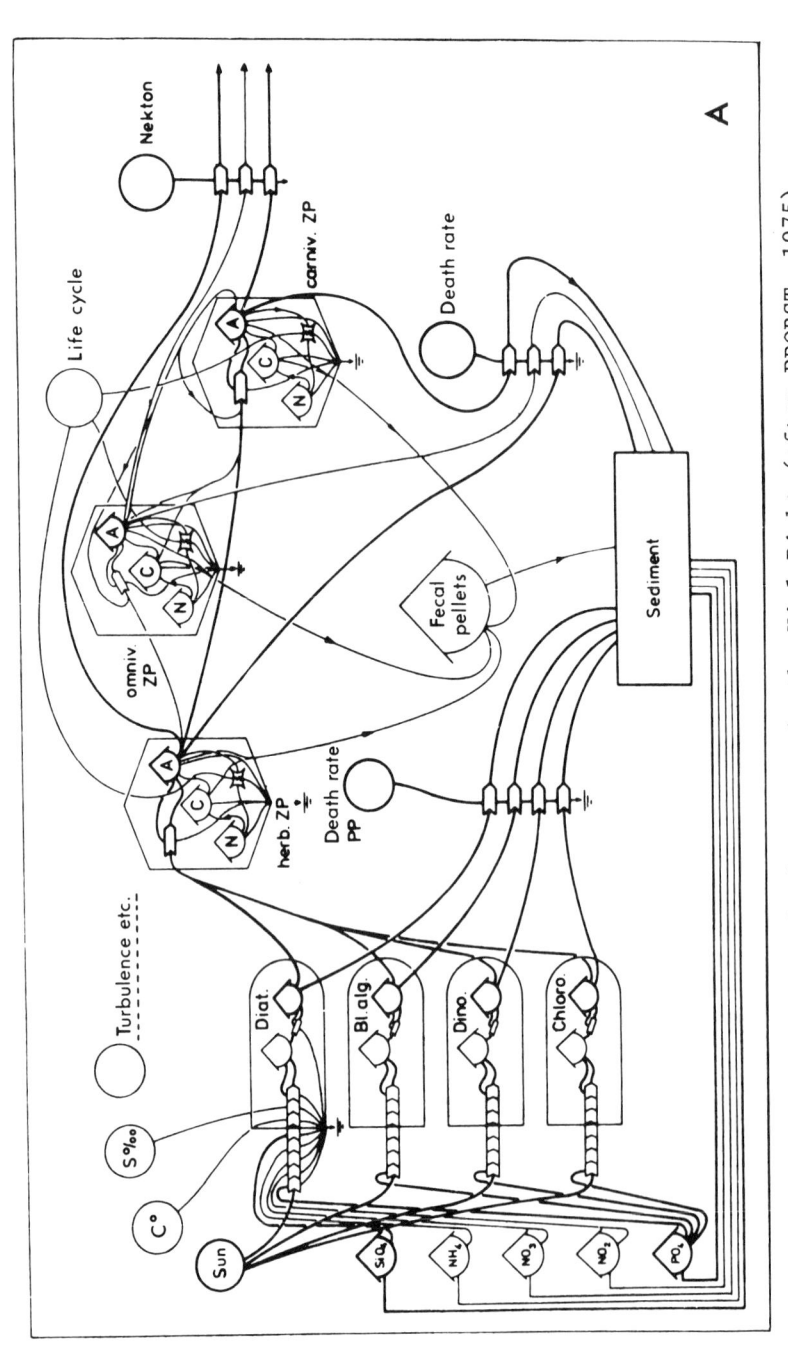

Figure 9. Energy model of the pelagic system in the Kiel Bight (after PROBST, 1975).
A. Model structure showing the uptake of recycled nutrients of diatoms (Diat.) blue-green (Bl. alg.), dinoflagellates (Dino.) and green algae (Chloro.). The algae are grazed by zooplankton (ZP), which are divided into herbivorous (herb.), omnivorous (omniv.) and carnivorous (carniv.) forms. Each population, forced by the life cycle pattern is represented by mauplii (N), copepodites (C) and adults (A). Through death rate pattern and production of fecal pellets of zooplankton, organic material is transferred to the sediment pool of the sea floor. Zooplankton populations are checked by the feeding of

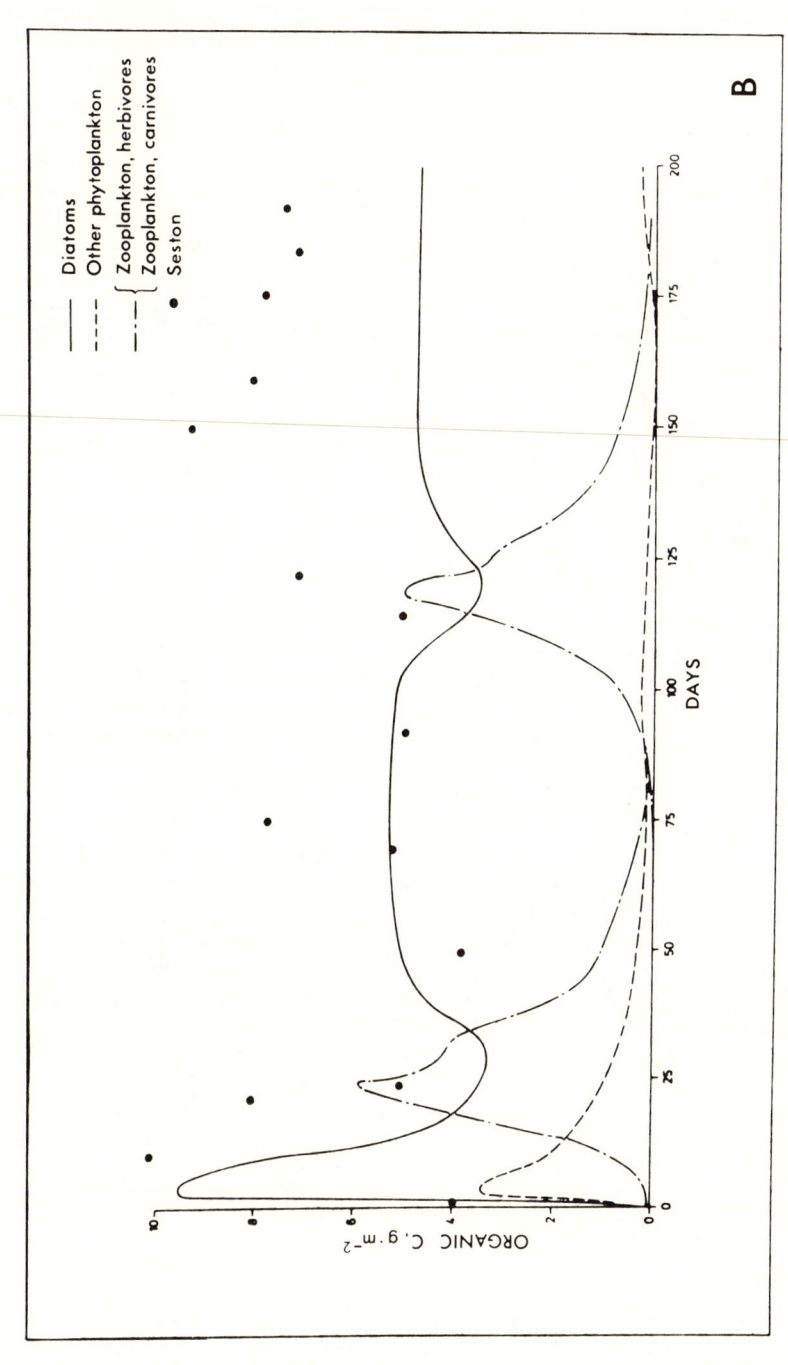

B. Simulation graphs of the biotic components compared to the pooled values for seston carbon from the field measurements of BODUNGEN (1975).

exporting organic matter, the soft bottom system is a typical
consumer system living on imports of already synthesized organic
material. From the pelagic zone an intermittent rain of dead
plankton and faecal pellets falls towards the bottom. Nearly
half of it was decomposed in the surface layer and recycled, the
rest needing more time for bacterial breakdown or consisting of
heavy particles, rapidly sinking to the bottom. As was stated
above, the spring bloom of heavy diatoms constitutes the great
injection of energy when the soft bottom receives a large part of
its necessary potential energy resources. A similar contribution
is made by the phytal system during early spring when the rocky,
shallow bottoms are covered with a thick layer of sessile diatoms.
Few herbivores except for some ciliates and rotifers are present
at that time and the diatom carpets are lifted off the substratum
by their production of gas bubbles and distributed to the soft
bottoms by the physical transport mechanisms. The annual filamentous algae go through several growth peaks and declining phases
during the light season and algal material is transported from the
shallow coast by currents.

A big reservoir of organic matter is therefore built up in the
Baltic soft bottoms. In some areas, erosion and activity of
animals will keep even pace or surpass deposition, in other places,
e.g. sheltered areas, the organic storage will increase. On the
whole, however, the Baltic can be regarded as a detritus based
system. The mud, though looking so homogeneous, contains numerous
food niches (Fig. 10) and nature has at its disposal many special
tools for utilizing these potential energy resources. Many of
these consist of bacteria, including carbohydrate hydrolysers,
nitrifiers and sulphate reducers. The system usually has a
vertical structure with an aerobic top player, a redox cline (sensu
HALLBERG, 1973) and below, an anoxic layer where especially the
hydrogen sulphide producing bacteria dominate. The low molecular
end products of the decomposition processes in this layer such as
H_2S, CH_4, and NH_3, diffuse upwards and, reaching the aerobic top
layer, they are oxidized by the aerobic bacteria, mostly chemoautotrophes. These obtain the necessary carbon by fixing external CO_2
which thus enters the detritus chain of the sediment (FENCHEL,
1969). In the gradient between oxygen and H_2S the white sulphur
bacterium Beggiatoa is often flourishing, moving up and down in
trying to keep to the optimal zone. It can be seen by the naked
eye, making white patches like cobwebs in places where the organic
load is heavy and responding to a temperature increase by increased
growth (ANKAR and JANSSON, 1973).

If light is present at the sediment surface, not only photosynthesis
and nitrogen fixation is carried out by algal carpets, but also
photoautotrophic purple bacteria fix the solar energy by using
H_2S (in oxic environments, pure sulphur) as hydrogen donor instead

Figure 10. Microflora and microfauna of a soft bottom in Oresund. Filamentous blue-green algae and spheric bacteria are active in the sulphur-cycle, while an elongated ciliate (above centre), a kidney-shaped ostracod (right of centre) and a curved nematode (lower left corner) prepare the organic material for bacteria. The boat-shaped organisms in lower margin are diatoms (producers). (From FENCHEL, 1969).

of H_2O (JORGENSEN and FENCHEL, 1974, BAGANDER, in print). This process is therefore called photoreduction. The three main processes which dominate a sunlit sediment surface, phytosynthesis, chemosynthesis and photoreduction are coupled together by "chemical food-chains" constituting another example of the maximum power principle in nature. Desulphovibrio bacteria in the anoxic zone use organic matter (and SO_3) producing H_2S. This is used by Beggiatoa bacteria with oxygen from the aerobic zone to form pure sulphur which is reduced by the purple sulphur bacteria in the oxidized zone. This is an example of how energy and matter is moved up and down in the sediment.

During oxic conditions phosphate is normally consumed by the sediment, the main limitation being the amount of cations (SCHIPPEL, HALIBERG and ODEN, 1973). During anoxic conditions, even at the sediment surface large amount of phosphate are given off to the bottom water. Bell jar studies of highly organic sediments have given figures of about 5 mg PO_4-P.m^{-2} day^{-1} (SCHIPPEL, HALIBERG and ODEN, 1973). Nitrogen in the form of NH^+_4 is now also released, the amounts depending upon the quality of organic matter (ENGWALL, in print). The anoxic periods of the deeper soft bottoms thus result in an export from the subsystem of not only hydrogen sulphide but also large amounts of low molecular, inorganic compounds, as in a large sewage tank: organics in - inorganics out.

But the soft bottom system also produces living organic material which is oxidized environments is an important step in the final decomposition of the sedimented organic matter. With regard to the size relationships which are a fairly good measure of the metabolic activity, the unicellular ciliates are a link between bacteria and meiofauna but they may play a less important role as concentrators of energy in this already concentrated "soup" of organics than in the "diluted" pelagic water. The meiofauna varies in numbers and biomass depending on the type of sediment. The harpacticoid copepods dominate the sand and gravel bottoms while the importance of the nematodes increase with decreasing particle size (SCHEIBEL and NOODT, 1975). Surprisingly high numbers of meiofauna for the Baltic have been found by ELMGREN (1973) who counted 5 x 10^6 ind.m^{-2} in a muddy sediment at 47 m depth. Towards the north the meiofauna decreases in number, several important groups having their northernmost limit in the Bothnian Sea (ELMGREN, ROSENBERG, ANDERSIN, EVANS, KANGAS, LASSIG, LEPPAKOSKI and VARMO, in print).

The macrofauna is poorer in the Baltic than in fully marine areas. The number of species decreases from south to north and for a rough characterization the summary of ZENKEVICH (1963) can still be used though the impoverished zone has now extended (Fig. 11). The southern areas are dominated by bivalve associations where Cyprina

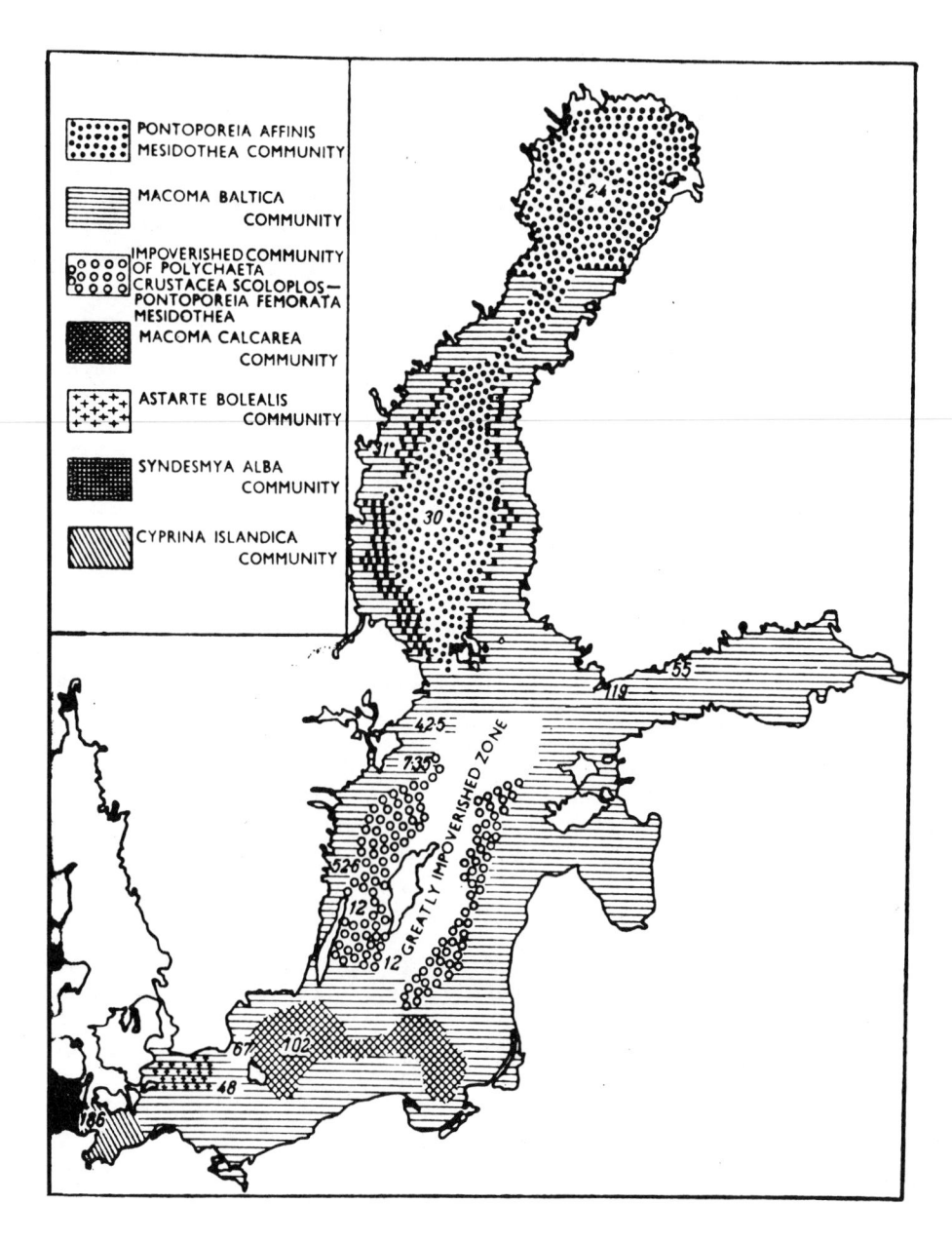

THE BALTIC - A SYSTEMS ANALYSIS OF A SEMI-ENCLOSED SEA

Figure 11. Faunal associations in the Baltic (from ZENKEVITCH, 1963).

islandica, Abra alba, Astarte borealis and Macoma calcarea greatly
contribute to the biomass, with polychaetes and crustaceans next
in order. Further north, the crustaceans gain importance mainly
through the isopod Saduria (Mesidothea) entomon and the amphipods
Pontoporeia affinis and Pontoporeia femorata. Saduria especially
dominates in the Bothnian Sea (HAAHTELA, 1975) and is to a great
extent living off P. affinis. In the northern Baltic proper, P.
affinis is by far the most productive macrofauna species (CEDERWALL,
in print) which by its burrowing and feeding activities in the
sediment probably controls both populations of bacteria and of
other animal species such as Macoma baltica (SEGERSTRALE, 1962) and
the oxygenation of the sediment.

The further concentration of the potential energy into fish is
hampered both by the critical oxygen conditions at depths greater
than 50 m and by the few species of fish. Compared to the North
Sea, the Baltic, with the exception of the western part and some
shallow areas, is a less favourable area for the feeding of fish
(HEMPEL and NELLEN, 1974). Both cod and flounder grow less and
the former is also limited to salinities above 10‰ for spawning.
One of the main feeding areas for cod, the Bornholm Sea, has
nowadays critical oxygen conditions. Further north where Saduria
entomon offers an important food contribution, spawning is impossible due to the low salinity and the fish has to return to the
south after feeding here, apparently a fairly new migration pattern.
Fourhorned sculpin and eel-pout are common on the coastal slope and
feed mainly on Pontoporeia and Harmothoe, the eel-pout also on
Mytilus and Macoma (ANEER, 1975).

A summary of the benthic ecosystem in the northern Baltic, in the
form of an energy flow diagram (Fig. 12) clearly shows not only
the great importance of bacteria and ciliates but also the dominance
of mud-eating animals (detrivores). ANKAR and ELMGREN (1976)
calculated that the minimum energy requirements for this system is
in the order of 40 gC·m^{-2}·yr^{-1}, less than half (100-150 kJ·m^{-2}·yr^{-1})
is left for the fish when respiration, consumption and mortality =
within the system have taken their share. For the Kiel Bight,
ARNTZ and BRUNSWIG (1975) estimated that at least half of "the
minimum production" of the potential fish food is consumed by the
bottom fish.

5. The Baltic as a total system

The physical system, the biological components and the bio-geo-
chemical cycles have been merged by evolution into one huge ecosystem whose characteristics might be difficult to discern without
broad generalizations. In the roughly two-layered system, with a
coastline long in proportion to the area, the coastal areas play
an important role for the total system in several respects. The
critical vertical transport of water in the heavily stratified

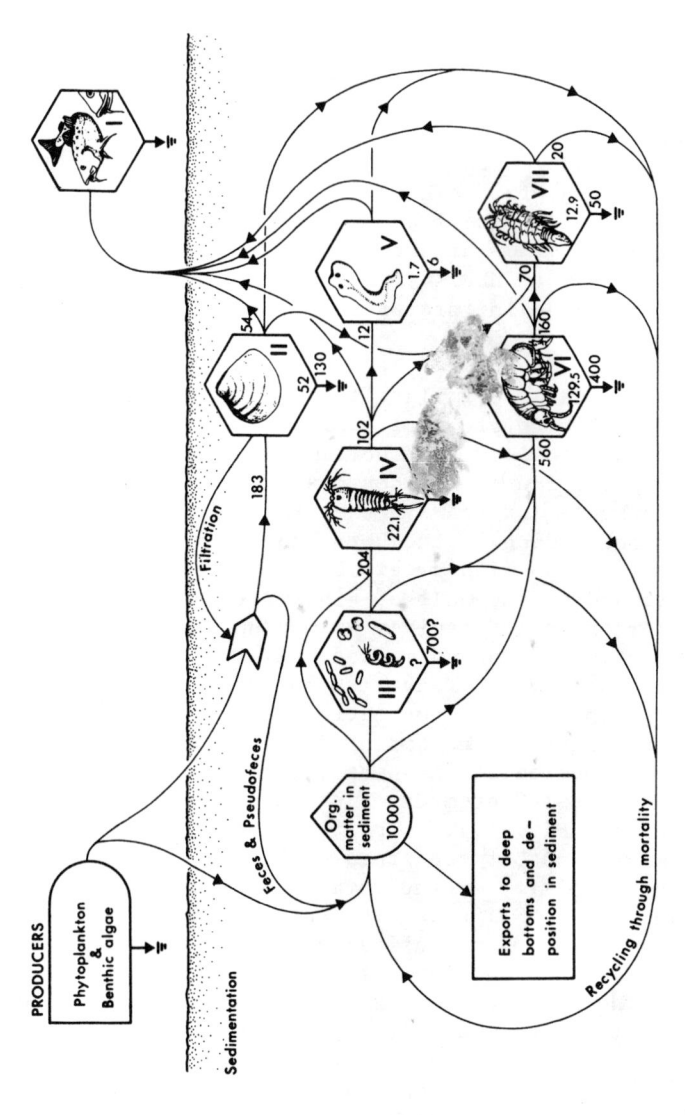

Figure 12. Rough energy flow model of the benthic ecosystem of the Askö-Landsort area. I Fish, II Planktivores, III Bacteria and ciliates, IV Meiofauna, detritovores, V Meiofauna, carnivores, VI Macrofauna, detritovores, VII Macrofauna, carnivores. Biomass measured, production estimated, respiration guessed at. Storages in kJ·m^{-2}·yr^{-1}. Figures to the left of the hexagons indicate assimilation (R + P). Feces and organic excretion are considered never to have left the organic pool in the sediment (from ANKAR and ELMGREN, 1976).

water is mechanically furthered by the coast where upwelling or
boundary mixing may transport large amounts of nutrients from the
bottom layers to the euphotic zone (WALIN, 1972). There is, so
far, little against a theory that many offshore plankton blooms
started initially at the coast as water-packages of high nutrients
and sparse phytoplankton which were then transported offshore
during culminating growth, conspicuous first far from land. The
resulting sedimentation would then be a positive feedback-loop to
the deeper offshore soft bottoms which once generated the nutrients.

Although the pattern of the blue-green bloom in Fig. 8 certainly
to a great extent is wind generated, it is tempting to correlate
the greater concentrations in the Landsort area and the Hano Bight
with those areas as upwelling centres. The area between Gothland
and the Swedish mainland shows higher primary production values
than the surrounding area (ACKEFORS, in print) which agrees well
with the reported upwelling in this area both during summer and
winter (SVANSSON, 1975).

The upwelling water may also stimulate the coastal vegetation though
the necessary nutrients here are more easily available through land
runoff and closeness to shallow soft bottoms in the well-mixed
water. In early spring the production maximum of both seaweeds
and phytoplankton travels like a wave over the Baltic starting in
the south in March and ending in the Bothnian Bay in July with
increased duration in areas with great stream discharge as in the
northernmost Baltic and in the Gulf of Finland (LASSIG and NIEMI,
1975). The shallow areas are rapidly warmed up, the water is
filled with larvae of plankton and macrofauna and the diurnal
pulse of production is heavy. Many fish now spawn, fresh water
species like pike and perch and some marine species like herring,
turbot and flounder arrive to the spring feast to spawn or feed.
The adults return to the sea again where later in the year the zooplankton maximum in August-September offers a second big meal before
the hard winter period (Fig. 13). The fish thus act as living
flows of energy, migrating between areas of high production and
contributing to the stability of the total system by cutting down
the peaks (ODUM and COPELAND 1969. The soft bottom subsystem
receives its big injection of potential energy after the spring
bloom in the pelagic system and the organic input is successively
channelled through the food-web. A second, smaller and in space
more restricted input of organic matter occurs during late autumn
when the storms tidy up in the coastal areas and decomposing annual
macrophytes are flushed away to deeper areas. The cooling of the
water, more rapid in the coastal areas also causes a downwelling
in areas with bottom topographical channelling, transporting both
nutrients and organic matter to deeper layers (SHAFFER, 1975).

During the winter a large part of the algal fauna migrates deeper
down, such as molluscs (SKOOG, 1971) and crustaceans whereas some

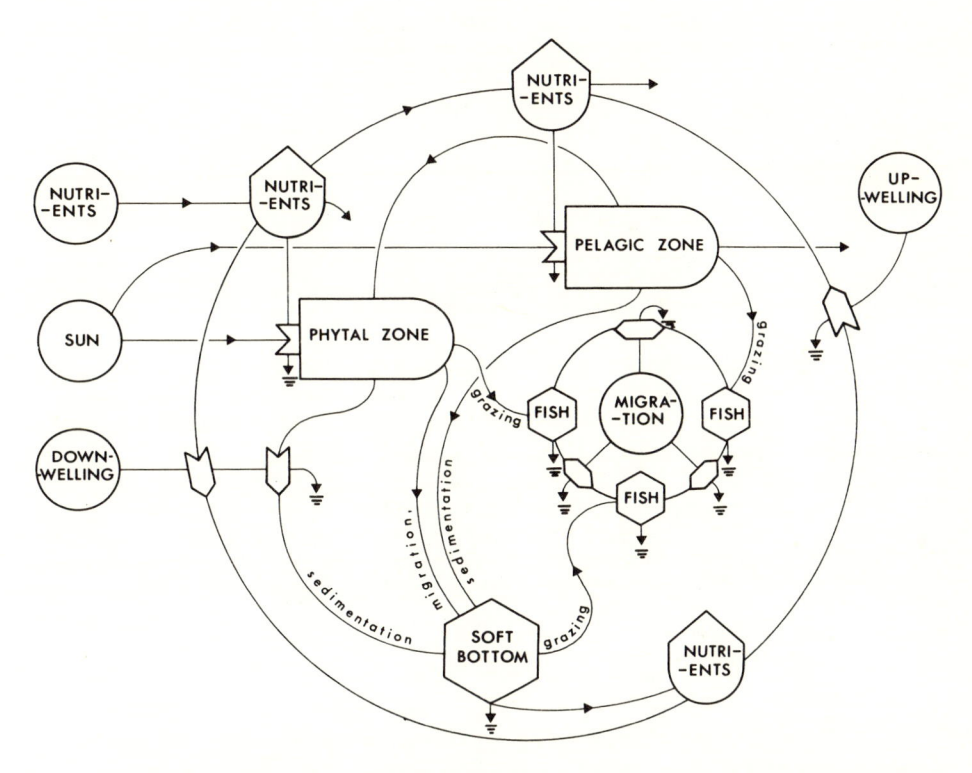

THE BALTIC - A SYSTEMS ANALYSIS OF A SEMI-ENCLOSED SEA 163

Figure 13. Diagram showing the fish as living flows of energy and matter, moving between the main parts of the total system, taking food where it is in excess. Driven by the migration pattern the pelagic feeding might be changed for benthic feeding during winter (e.g. herring) but the early warming up of the shallow phytal region makes a migration to, and spawning in these areas favourable.

cold-stenotherm animals from the soft bottom area such as the four-horned sculpin (WESTIN, 1970) move up the coastal slope utilizing this extra submerging food resource for spawning. During a short period there thus exists a "short-circuit" between the phytal and the deeper soft bottom subsystems. In the dark part of the year the soft bottoms in the shallow regions play the same role as the bottoms below the halocline - as nutrient regenerators, building up a storage which is transported to the euphotic zone at the start of the next season during the spring circulation.

6. Recent changes in the Baltic ecosystem

The Baltic is currently being driven out of a previously steady state by long-term processes which can be summarized as eutrophication and oceanization (ZMUDZINSKI, in print) and serious conditions have also arisen through the accumulation of hazardous substances.

6.1. The eutrophication of the Baltic

By structure the Baltic is a oligotrophic system, which means a low production, a slow mineral cycling and a poor water exchange. In the clear water the problem for the organisms of concentrating their food is great. Processes of animal migration such as those of plankton and fish have been evolved, and the fish fauna is dominated by fast hunters like salmon, pike and perch. During an increase of mineral supply, for example from sewage discharge, this system is replaced by a eutrophic type where primary production is high and those species not adapted to put energy into fast swimming are favoured. Pike and perch are replaced by bream and roach, fishes less desirable for man.

This is now happening in the coastal areas of the Baltic. The seaweeds, especially the brown algae such as Fucus are forced seawards (PEUSSA and RAVANKO, 1975) the green algae Cladophora and Enteromorpha increase and the normal vertical stratification of the seaweeds is changed (LINDGREN, 1975). The water becomes turbid by the enhanced primary production (e.g. WIKTOR and PLINSKI, 1975), organic deposition increases and the reed belts invade the archipelagos. To some extent the normal land elevation after the ice-age is responsible but the main factor is the increased discharge of nutrients from land. According to AHL and ODEN (1972) for rivers discharge, and ENGWALL (1972) for urban and industrial outlets, in total ca. 80% of the discharged phosphorus and ca. 50% of the nitrogen in Sweden is caused by man's activities. Whether phosphorus or nitrogen is the limiting factor for Baltic primary production has been much debated. Most authors have pointed to nitrogen (e.g. SEN GUPTA, 1968, FRANCKE and NEHRING, 1971, SCHULTZ and KAISER, 1973), a few have shown the intricate play between these nutrients, trace metals and season (RINNE and

TARKIAINEN, 1975, MALEWICZ, 1975) and some doubt that nitrogen can be a limiting factor as it is usually present in one form or another in the Baltic water (FONSELIUS, in print). In the offshore area the blooms of the blue-green alga Nodularia spumigena seem to have increased during the last few years and experiments have shown that this alga is stimulated by increasing phosphate concentrations in the water (HORSTMANN, 1975).

The serious discussion of the alarming state of the Baltic proper was started in the late sixties on the basis of the stagnation of the bottom water when FONSELIUS (1969), summing up the long series of hydrographical measurements of the Fishery Board of Sweden showed the declining oxygen conditions below the halocline (Fig. 14). The stagnation culminated in 1968-69 when a large part of the Gothland Basin and the Gulf of Finland contained hydrogen sulphide (Fig. 15). Huge amounts of phosphate were released from the sediments and when an injection of North Sea water at the end of 1968 started to flow through the Straits of Darss and reached the northern Baltic proper a year later, the phosphate-rich layers were mixed into the surface layer west of Gothland (Fig. 16). This is the area where both upwelling and high production values have been recognized. The stagnation and hydrogen sulphide formation soon took over again and periods of anoxic and oxic conditions in the water below the halocline characterized the following years.

The Baltic surface water is thus fertilized both from land and from the sea bottom with the same effect - an increase of primary production, an increase of sedimenting organic material and an increase of oxygen consumption of the bottom sediments. DYBERN (1970) has calculated the amount of BOD_5 and phosphorous in sewage discharged from the different countries and arrived at a total of 392 800 and 21 460 tons respectively, the most affected areas being the Bothnian coasts of Sweden and the Gulf of Finland due to the concentration of paper mills. From calculations of the carbon budget of the Baltic Sea, FONSELIUS (1972) estimated that as little as an extra 10% of the carbon annually produced in, or discharged into the Baltic, would, if it reached the deep water, suffice to use up all the oxygen transported there in a year. Although the stagnation of the Baltic is a physical phenomenon, there should therefore be little doubt that man's activities can affect the conditions of the bottom water. Comparing the secular changes in oxygen with those of salinity and temperature MATTHAUS (1973) states that "the increase in the oxygen deficit in the deep water can scarcely be attributed to hydrographical factors".

Through the injection of nutrients the stagnation process has been shown to stimulate offshore primary production (SCHULTZ and KAISER, 1973) and to increase zooplankton populations (SEGERSTRALE, 1965)

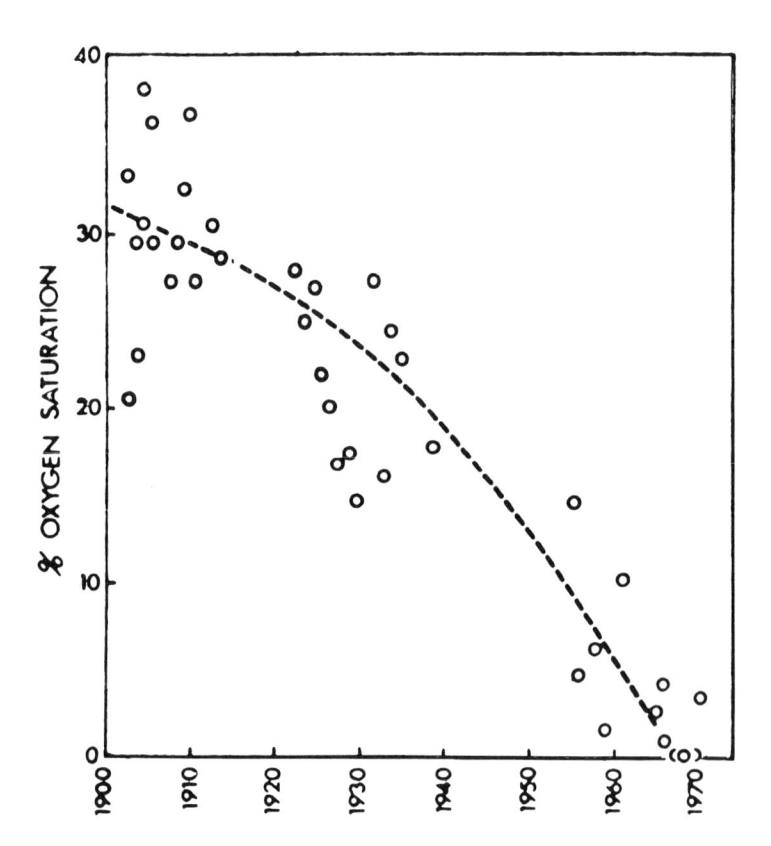

Figure 14. The successive decrease of the oxygen content in the bottom water in the northern Central Basin since 1900 (from FONSELIUS, 1969).

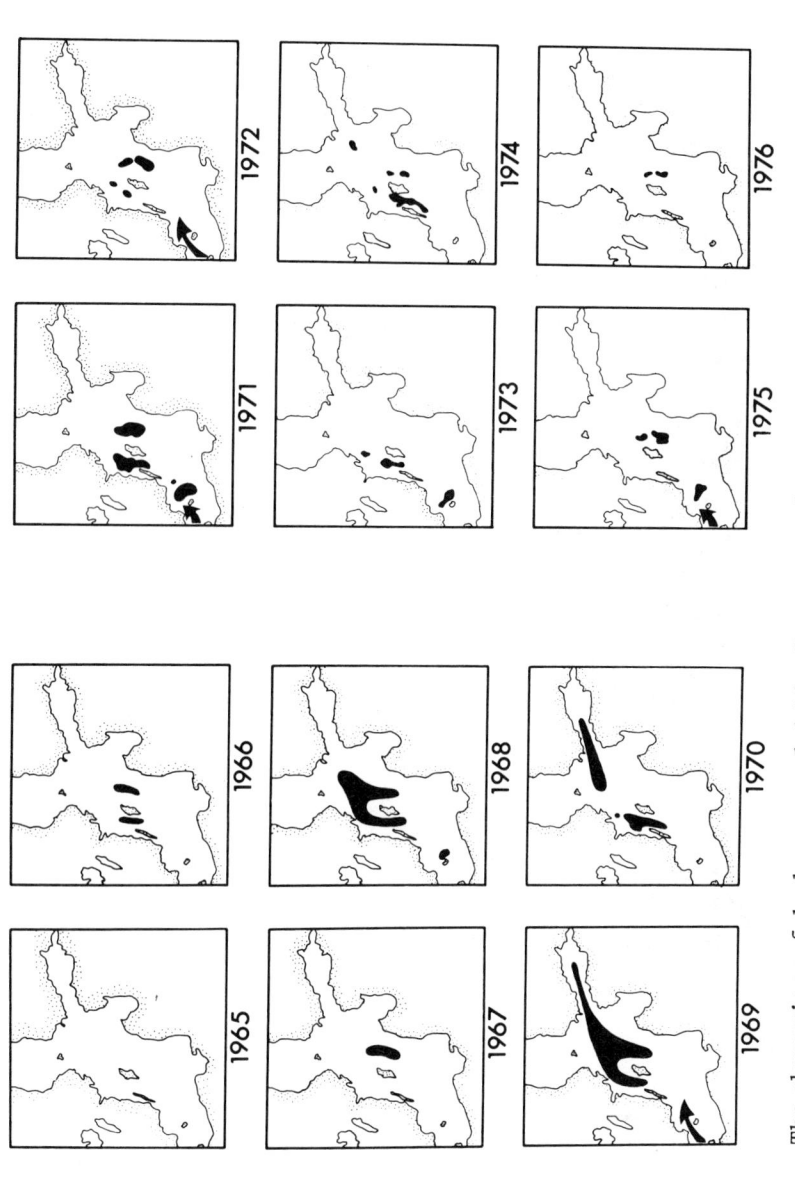

THE BALTIC - A SYSTEMS ANALYSIS OF A SEMI-ENCLOSED SEA 167

Figure 15. The dynamics of hydrogen sulphide formation (black) in the Baltic proper. The great stagnation culminating in 1969 is broken by a saltwater inflow (arrow) through the straits in 1968/69. The following periods are characterized by shorter stagnation - inflow periods (compiled from various sources such as FONSELIUS, 1969, Meddelanden fran Havsfiskelaboratoriet, Lysekil).

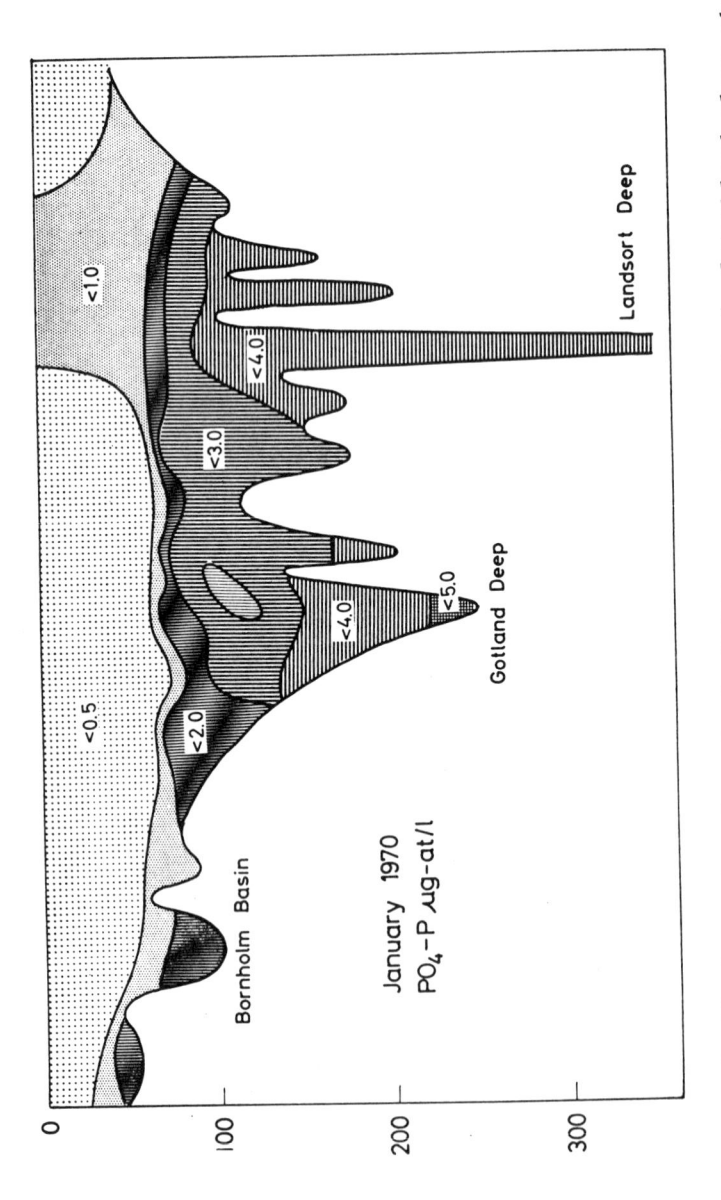

Figure 16. Longitudinal section through the Baltic around the island of Gothland along the line: Arkona Basin – Bornholm Basin – Gothland Basin – Northern Central Basin – Landsort Deep – Karlso Deep, showing the distribution of phosphate-phosphorus in January 1970 (from FONSELIUS, 1972).

and stock of herring (RECHLIN, 1969) and sprat (SCHULTZ, 1970).
The bottom fauna below the halocline has greatly suffered, however,
The formerly productive and important feeding area for fish,
Bornholm Basin now switches between oxic and anoxic conditions with
sparse and fluctuating macrofauna (TULKKI, 1965, HAGBERG, 1972) as
does also the Gothland Basin (SCHULTZ and KAISER, 1973) and the
Landsort area (CEDERWALL, in print). The hatching of bottom fish
like cod (DEMENTYEVA, 1972, LINDBLOM, 1973) and offshore flounder
LINDBLOM op. cit.) has also grown critical due to the low oxygen
concentrations in the main spawning areas.

6.2. The oceanization process

The increase in frequency of the injections of North Sea water has
increased the deep water body of the Baltic proper by ca. 200 km^3
and raised the primary halocline from its situation at 80 m depth
at the beginning of the century to the present 60 m level (FONSELIUS
1969). At the same time, the stability of the halocline has
increased (FONSELIUS, 1969) though it varies with the saltwater
inflows (VOIPIO and MALKKI, 1972). In the northern Baltic proper,
the salinity of the deep water has increased by roughly 1‰ and the
temperature by around 1°C (MATTHAUS, 1972) since the beginning of
this century.

The salinity increase can also be traced in the distribution of the
marine organisms. The common jellyfish (Aurelia aurita) has
extended further north (LINDQVIST, 1962) and autumnal occurrence in
the Asko-Landsort area of medium sized Cyanea capillata, previously
found only in small specimens occasionally in deep plankton hauls,
is now a regular feature.

The recolonization of the Bornholm Basin after the great stagnation
in the late sixties also shows clear features of oceanization in
species distribution, zoogeographical elements and feeding types
(LEPPAKOSKI, 1975). Most of the Arctic relicts have been replaced
by Atlantic-boreal and cosmopolitan species and the previous
dominance of suspension feeders has been exchanged for non-selective
deposit feeders.

6.3. Accumulation of hazardous substances

The concern about chemical pollution arrived early in Sweden and
in the Baltic, a first analysis of DDT in eggs of guillemots led
to an extensive mapping of the DDT and PCB content in marine
organisms (JENSEN, JOHNELS, OLSSON and OTTERLIND, 1969, 1972).
It turned out that the total DDT (DDT and DDE) and PCB contents in
herring, salmon and seal (ringed, common and grey) were up to ten
times higher than in corresponding populations in the North Sea
(Fig. 17). The white-tailed eagle from the Archipelago of Stockholm showed a 100 times higher concentration of total DDT and PCB

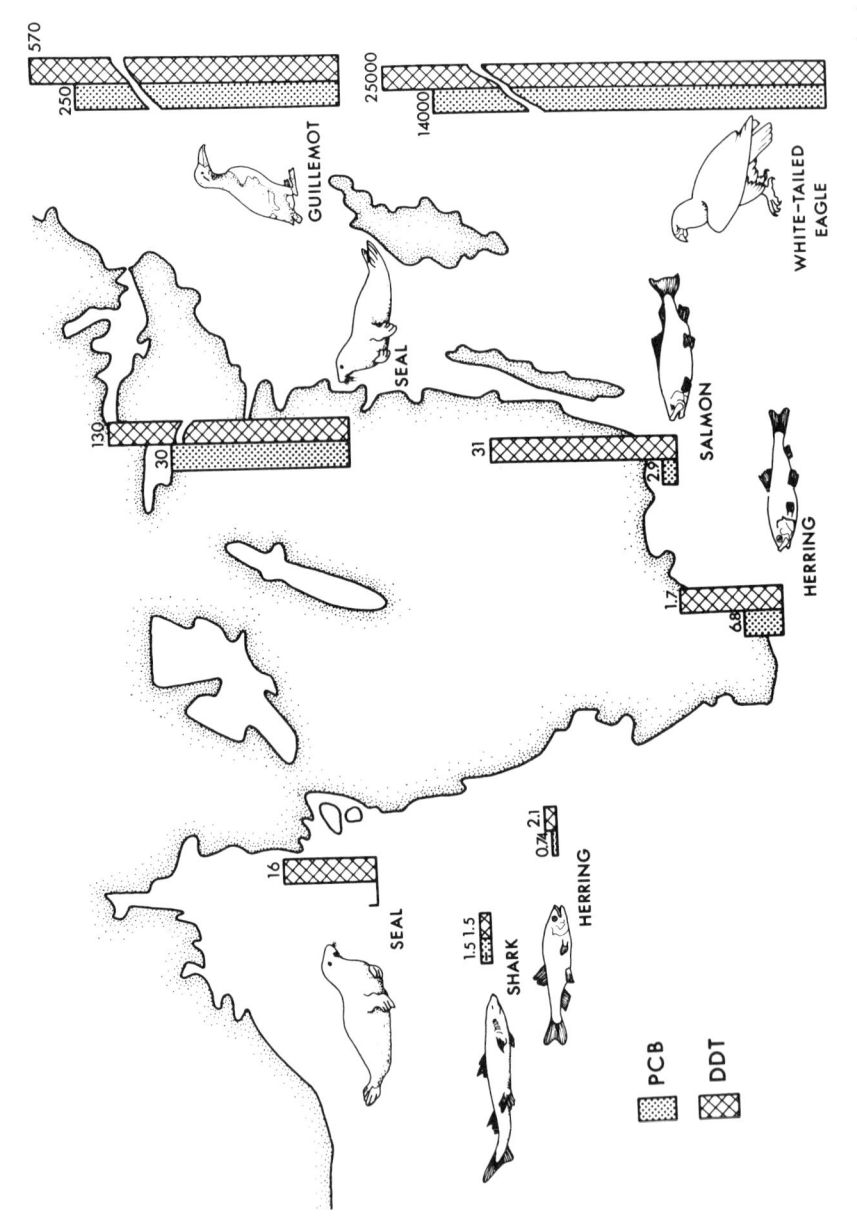

Figure 17. Amounts of DDT (sum of all DDT derivatives) and PCB in organisms from the Baltic and the Atlantic. The concentrations are expressed as mg per kg fat (from JENSEN, JOHNELS, OLSSON and OTTERLIND, 1969).

than populations from northern Sweden. Eggs from guillemots contained five to ten times higher concentrations than populations in the North Sea and the Atlantic.

Both DDT and PCB are found in higher concentrations in the eastern Bornholm Basin than in the western and the concentrations decrease northwards to the Bothnian Basin. The mercury content in aquatic organisms has also shown high values in Baltic animals, especially in pike (ACKEFORS, 1971). After the Swedish mercury ban in 1966, the levels have decreased and the mercury content in feathers of guillemots are now down to levels of the previous century (OLSSON, pers. comm.). This is a good example of how man really is able to change the dangerous trend of pollution by drastic but necessary methods. DDT concentrations can also be expected to decrease as these compounds are slowly broken down in the system. Sweden now has an almost total DDT ban. PCB is far more resistant and the present amounts will move up the trophic levels of the system and like DDT concentrate in fatty substances, affecting the metabolism of the organism. The greatly increased number of sterile seals in the Baltic has now been proved to be caused by the presence of PCB (BELLE, OLSSON and JENSEN, in print).

The closed nature of the Baltic system, the long residence time of the water and the low temperature which effectively slows down the decomposition processes make the Baltic a huge sink for substances from land runoff and the atmosphere. Many of the chlorinated hydrocarbons are supposed to enter as atmospheric outfall from the industries in the Baltic countries and on the European continent. They are absorbed to organic material and are apparently accumulated in large amounts in the bottom sediment (ODEN and EKSTEDT, 1976), which constitutes a pool for the further distribution in the system. The same mechanism seems to work for heavy metals such as Hg, Pb, Zn, Cu, Cr, Cd, which show a strong correlation with the amount of organic matter (ANDRULEWICZ, in print).

6.4. Oil pollution

As the Baltic is intensively used as a transportation area, the risk of oil spills is great both from the handling of the oil and from wreckage of oil tankers. An oil spill in an archipelago would affect a long coastline with potential effects on large and productive areas. A detailed study of an oil spill in the Archipelago Sea in 1969 fought with peat and fire had a great effect on the population of nesting eider ducks of which ca. 30% died. On the phytal and soft bottom systems, however, few effects were traced a year after (LEPPAKOSKI, 1973). The presence of ice and the early time of the year, before the algal belts had fully developed and migrating waterfowl arrived constituted unusually lucky circumstances.

Effects even of greatly diluted mixtures of oil and emulsifiers on Baltic organisms have been demonstrated, both as increased mortality and deformation of herring larvae (LINDEN, 1974) and as decreased fecundity and growth of important littoral crustaceans (<u>Gammarus oceanicus</u>, LINDEN, in print).

7. What systems are next?

In the foregoing, much evidence has been given of how the Baltic system, with the characteristic features Man has long appreciated is rapidly changing. This is partly due to changing meteorological conditions (which may also be caused by Man), but part of it can be directly traced to industrial and domestic wastes. We still have a possibility of deciding, within certain limits, what types of Baltic systems we want to live with, using the resources on a long-term basis. This "ecological engineering" requires a sound knowledge of the behaviour of the Baltic, of the sensitivity of its various parts and of how they are connected to form a functioning whole. The advantages of modelling as the best tool yet known for understanding and exploring the tolerance limits of a complex system like the Baltic cannot be too highly stressed (JANSSON, 1976).

An example of the complex interplay between biological, chemical and physical variables is shown in Fig. 18, which is a rough total model of the Baltic by SJOBERG, WULFF and WAHLSTRÖM (1973). In the surface water, the pelagic primary producers build organic material using sunlight and nutrients as energy sources. These come partly from land runoff and precipitation, partly as a positive feedback from the decomposition of organic material. Of the non-decomposed part, some settles on the shallow bottoms. Some fall through the halocline, building up an organic pool in the deep water, which if oxygen is present, is decomposed by aerobic bacteria to nutrients, which are partly bound in the sediment. If oxygen is used up by the decomposition processes in the sediment, anaerobic bacteria take over, forming low molecular gases such as hydrogen sulphide. In the model, the same compartment was used both for oxygen and H_2S, the latter symbolized by "negative oxygen". During these anoxic periods nutrients are released from the sediment. Through seasonal mixing or seawater inflows, these are mixed into the surface water, stimulating the primary producers to further growth. The organic pool will increase, the sedimentation is speeded up and the bacterial populations in the deep water will build up larger populations, quickly depleting the small amounts of oxygen. The anaerobic state will soon be reached and bottom nutrients will increase. The cycle goes into its second loop.

When the model was run using recent values, mainly from FONSELIUS (1969, 1972) and BOLIN (1972), a steady state was reached after ca. 150 years with an anaerobic deep water. Perturbations of

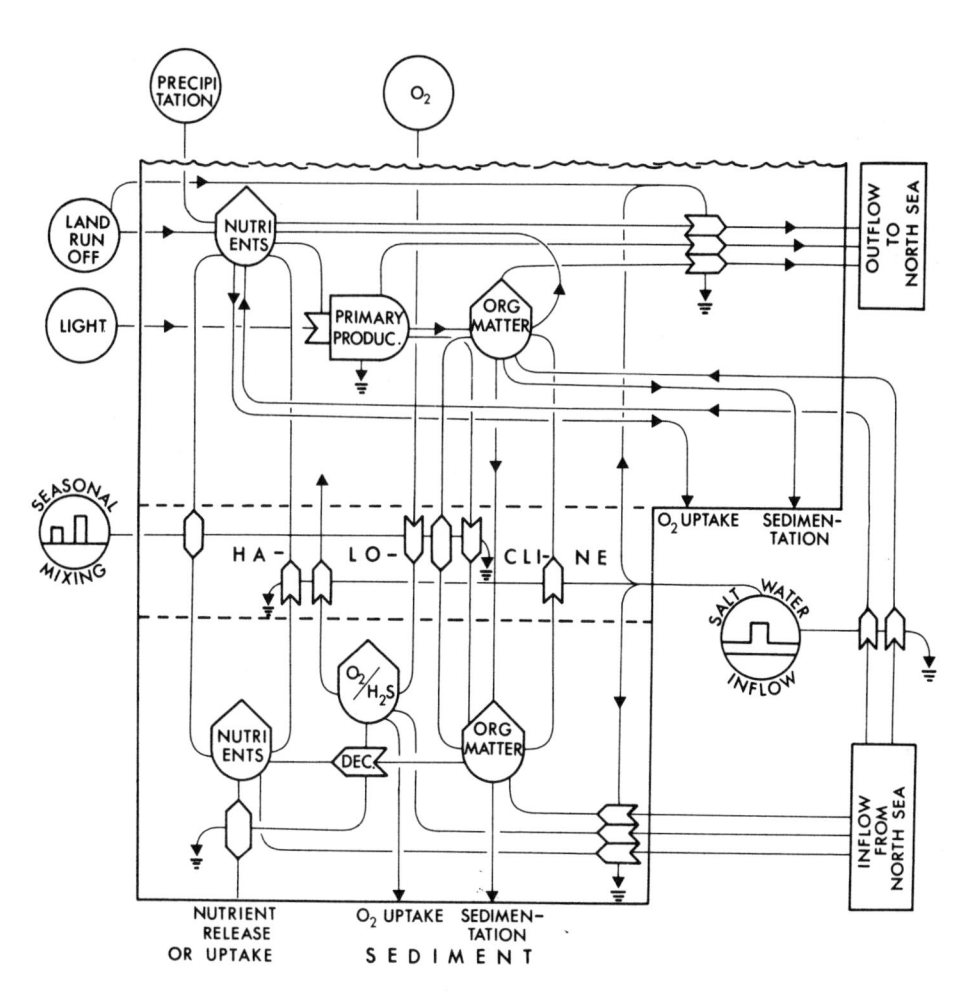

Figure 18. Energy flow model of the Baltic showing the surface water, the halocline, the bottom water and the sea floor with its main forcing functions and state variables. (Simplified from SJOBERG, WULFF and WAHLSTROM, 1972).

this steady state showed the exchange with the North Sea and sedimentation to be critical processes. The most interesting feature, however, was the effect of changing the nutrient release rate per unit surface of anaerobic sediment. If it was doubled, a new steady state was reached and the anoxic bottom areas increased three times with a new release rate six times higher.

This shows very clearly the amplification function of the biological system and the necessity of integrated models where biology, chemistry and physics are coupled in an adequate way. The understanding and co-operation especially between physical and biological oceanographers has to be strengthened in order to achieve a more adequate information of the dynamics of the total system as a basis for a better management of the potential resources of the sea.

REFERENCES

ACKEFORS, H. (1975). Production studies of zooplankton in relation to the primary production in the Baltic proper. Merentutkimuslaitoksen Julkaisu/Havsforskningsinstitutets Skrift No. 239, 123-130.

ACKEFORS, H. (1971). Mercury pollution in Sweden with special reference to conditions in the water habitat. Proceedings of the Royal Society of London B, 177, 365-387.

ACKEFORS, H. (in print). Primär och sekundärproduktion i Östersjön.

AHL, T. and S. ODEN, (1972). River discharges of total nitrogen total phosphorus and organic matter into the Baltic Sea from Sweden. Ambio Special Report, 1, 51-56.

ANDRULEWICZ, E. (in print). Relation between organic matter and metal content of sediments. Prace Morskiego Instytut Rybacki.

ANEER, G. (1975). Composition of food of the Baltic herring (Clupea harengus v. membras L.), fourhorn sculpin (Myoxocephalus quadricornis L.) and eel-pout (Zoarces viviparus L.) from deep soft bottom trawling in the Asko-Landsort area during two consecutive years. Merentutkimuslaitoksen Julkaisu/Havsforskningsinstitutets Skrift, No.239. 146-154.

ANKAR, S. and R. ELMGREN, (1976). The benthic macro- and meiofauna of the Asko-Landsort area (Northern Baltic Proper). A stratified random sampling survey. Contributions from the Asko Laboratory, Stockholm, Sweden, 11, 115 pp.

ANKAR, S. and B-O. JANSSON, (1973). Effects of an unusual natural temperature increase on a Baltic soft-bottom community. Marine

Biology, 18, 9-18.

ANKAR, S. (1972). Intensive studies on the micro- and macro-benthos of a soft bottom and a survey of the abundance of bacteria in other parts of the ecosystem. Report to the Swedish Environmental Protection Board, Contract 1, 10/35 and 7-85/70.

ARNTZ, W.E. and D. BRUNSWIG, (1975). An approach to estimating the production of macrobenthos and demersal fish in a western Baltic Abra alba community. Merentutkimuslaitoksen Julkaisu/ Havsforskningsinstitutets Skrift No. 239, 195-205.

BAGGE, P and A. NIEMI, (1971). Dynamics of phytoplankton primary production and biomass in Loviisa Archipelago (Gulf of Finland). Merentutkimuslautoksen Julkaisu/Havsforskningsinstitutets Skrift No. 239, 19-41.

BJORK, S. (1967). Ecologic investigations of Phragmites communis. Studies in theoretic and applied limnology. Folia Limnologica Scandinavica 14, 248 pp.

BODUNGEN v., B., (1975). Der Jahresgang der Nagrsalze und der Primarproducktion des Planktons in der Kieler Bucht. Dissertation Christian-Albrechts-Universitat, Kiel, 166 pp.

BOJE, S., (1974). Intersuchungen zum Energie-und Stoffumsatz des sublitoralen Meeresbodens in der Kiel Bucht. Dissertation, Christian-Albrechts-Universitat, Kiel, 77 pp.

BOLIN, B., (1972). Some preliminary views on the problem of modelling the Baltic for ecological purposes. Oikos Supplementum 15, 7-8.

BRATTBERG, G. (1975). Kravefixering och primarproduktion i St ckholms Skargard 1972 och 1973. Report to the Swedish Environmental Protection Board, Contract No. 7-74/74.

BAGANDER, L.E. (in print). Chemical dynamics of Baltic sediments -bacterial sulphate reduction. Ambio Special Report No. 4.

CEDERWALL, H., (in print). Production of Pontoporeia affinis in the northern Baltic proper. Prace Morskiego Instytut Rybacki.

CISZWSKI, P., (1975). Effects of climatic conditions on microzooplankton biomass in the southern Baltic. Merentutkimuslaitoksen Julkaisu/Havsforskiningsinstitutets Skrift No. 239, 137-138.

DAHL, E., (1973). Ecological range of Baltic and North Sea species. Oikos Supplementum, 15, 85-90.

DEMEL, K. and Z. MULICKI, (1954). Quantitative investigations on the biological bottom productivity of the South Baltic. Prace Institutu Rybackiego w Gdyni, 7, 25-126. (in Polish).

DIETRICH, G. and F. SCHOTT, (1974). Wasserhaushalt und Stromungen In: Meereskunde der Ostsee, L. Magaard and Reinheimer, editors, Springer-Verlag, 33-41.

DIETRICH, G., (1950). Die naturlichen Regionen von Nord- und Ostsee auf hydrographischer Grundlage. Kieler Meeresforschungen 17, 35-69.

DYBERN, B-I. (1970). Report of the ICES Working Group on Pollution of the Baltic Sea, Co-operative Research Report, Series, A, B, 15.

DYMENTYEVA, T.F. (1972). On the causes responsible for changes in fish production of the Baltic Sea. Ambio Special Report, 1, 63-66.

EHLIN, U., I. MATTISSON, and G. ZACHRISSON, (1974). Computer based calculations of volumes of the Baltic area. In: 9th Conference of the Baltic Oceanographers, Kiel, April 1974, Paper No. 7.

ELMGREN, R., R. ROSENBERG, A-B. ANDERSIN, S. EVANS, P. KANGAS, J. LASSIG, E. LEPPAKOSKI, and R. VARMO, (in print). Benthic macro- and meiofauna in the Gulf of Bothnia (Northern Baltic). Prace Morskiego Instytut Rybacki.

ELMGREN, R., (1975). Benthic meiofauna as indicator of oxygen conditions in the Northern Baltic Proper. Merentutkimulsaitoksen Julkaisu/Havsforskningsinstitutes Skrift No. 239, 265-271.

ELWERTOWSKI, J. and J. MACIEJCZYK, (1964). Application of mathematical analysis to the study of the influence of physiological and ecological factors on the change of fat contents in the adult sprats of the Bay of Gdansk. Prace Morskiego Instytut Rybacki Gydni, 13, 7-18.

ENGVALL, A-G. (in print). Nitrogen exchange at the sediment-water interface. Ambio Special Report No.4.

ENGVALL, R. (1972). Swedish municipal and industrial discharges into the Baltic. Ambio Special Report No.1, 41-49.

FALKENMARK, M. and Z. MIKULSKI, (1975). The Baltic Sea - semi-enclosed sea as seen by the hydrologist. Nordic Hydrology, 6, 115-136.

FALKENMARK, M and Z. MIKULSKI (1974). Hydrology of the Baltic. Water balance of the Baltic Sea, a Regional Co-operation Project of the Baltic Countries, Project Document No.1.

FENCHEL, T. (1969). The ecology of marine microbenthos IV. Structure and function of the benthic ecosystem, its chemical and physical factors and the microfauna communities with special reference to the ciliated Protozoa. Ophelia, 6, 1-182.

FONSELIUS, S. (1972). On biogenic elements and organic matter in the Baltic, Ambio Special Report 1, 29-36.

FONSELIUS, S., (1970). On the stagnation and recent turnover of the water in the Baltic. Tellus, 22, 533-544.

FONSELIUS, S. (1969). Hydrography of the Baltic deep basins III. Fishery Board of Sweden, Series Hydrography, 23, 97 pp.

FRANCKE, E. and D. NEHRING, (1973). Physical and chemical variations in the eastern part of the Gotland Basin in 1969/70. Oikos Supplementum 15, 14-20.

GÖTHBERG, A. and B. RÖNDELL, (1973). Ekologiska studier i Zosterasamhället i norra Östersjön. Information fran Söttvattenslaboratoriet Drottningholm, 11, 37 pp. (In Swedish).

HAAGE, P. (1975). Quantitative investigations of the Baltic Fucus belt macrofauna. 2. Quantitative seasonal fluctuations. Contributions from the Asko Laboratory, University of Stockholm, Sweden, 9, 88 pp.

HAATELA, I. (1975). The distribution and size of Mesidothea entomon (Crustacea, Isopoda) in the Northern Baltic area with reference to its role in the diet of cod. Merentutkimuslaitoksen Julkaisu/Havsforskningsinstitutets Skrift No. 239, 222-228.

HAGBERG, A. (1972). Undersökningar över bottenfauna. Meddelands fran Havsfiskelaboratoriet, Lysekil, 117, 8-9 (in Swedish).

HAGSTROM, A. and U. LARSSON, (in prep.). On the relationships between phytoplankton and bacteria in primary production studies.

HALLBERG, R.O. (1973). The microbiological C-N-S cycles in sediments and their effects on the ecology of the sediment-water interface. Oikos Supplementum 15, 51-62.

HELLE, E., M. OLSSON, and S. JENSEN (in print). PCB levels correlated with pathological changes in seal uteri. Ambio Special Report No. 5, 5-6.

HEMPEL, G. (1975). An interdisciplinary marine project at the University of Kiel "Sonderforschungsbereich 95". Merentutkimuslaitoksen Julkaisu/Havsforskningsinstitutets Skrift No. 239, 162-166.

HEMPEL, G. and W. NELLEN (1974). Fische der Ostsee. In: Meereskunde der Ostsee, L. Magaard and G. Reinheimer, editors, Springer-Verlag, 215-232.

HOBRO, R. and B. NYQVIST, (1972). Pelagical studies in the Landsort area during 1970-1971. Report to the Swedish Environmental Protection Board, Contract No.1 10/35 and 7 85/70.

HOBRO, R., U. LARSSON, and F. WULFF, (in print). Dynamics of a phytoplankton spring bloom in a coastal area of the Northern Baltic. Prace Morskiego Instytut Rybacki.

HORSTMANN, U., (1975). Eutrophication and mass production of bluegreen algae in the Baltic. Merentutkimuslaitoksen Julkaisu/ Havsforskningsinstitutets Skrift No. 239, 83-90.

HÜBEL, H. and M. HÜBEL (1974). In-situ Messungen der diurnalen Stickstoff-Fixierung an Mikrobenthos der Ostseeküste. Archiv für Hydrobiologie, Supplement 46, Algological Studies 10, 39-54.

HÜBEL H. and M. HÜBEL (in print). Nitrogen fixation in coastal waters of the Baltic. Prace Morskiego Instytut Rybacki.

JANSSON, B-O. (1972). Ecosystem approach to the Baltic problem. Bulletins from the Ecological Research Committee/NFR, Stockholm Sweden, 16, 82 pp.

JANSSON, B-O. (1976). Modelling of Baltic ecosystems. Ambio Special Report No. 4.

JANSSON, B-O., and F. WULFF (in print). Baltic ecosystem modelling. In: Ecosystems model in theory and practice, C.A.S. Hall, editor, Wiley-Interscience.

JANSSON, A.M. (1967). The food-web of the Cladophora-belt fauna. Heloglander wissenschaftliche Meeresuntersuchungen, 15, 574-588.

JANSSON, A.M. (1974). Community structure, modelling and simulation of the Cladophora ecosystem in the Baltic Sea. Contributions from the Asko Laboratory, University of Stockholm, Sweden, No. 5, 130 pp.

JANSSON, A.M. and N. KAUTSKY (in print). Quantitative survey of hard bottom communities in a Baltic archipelago. 11th European Symposium on Marine Biology, 1976.

JENSEN, S., A. JOHNELS, M. OLSSON and G. OTTERLIND, (1969). DDT and PCB in marine environment. Fauna och Flora 64, 142-148. (in Swedish).

JENSEN, S., A. JOHNELS, M. OLSSON and G. OTTERLIND (1972). DDT and PCB in herring and cod from the Baltic, the Kattegat and the Skagerrak. Ambio Special Report No. 1, 71-85.

JERNELOV, A. and R. ROSENBERG (1976). Stress tolerance of ecosystems. Environmental Conservation, 3, 43-46.

JORGENSEN, B.B. and T. FENCHEL (1974). The sulphur cycle of a marine sediment model system. Marine Biology, 24, 189-201.

KAISER, W. and S. SCHULTZ (1975). On primary production in the Baltic. Merentutkimuslaitoksen Julkaisu/Havsforskningsinstitutets Skrift No. 239, 29-33.

KULLENBERG, G.E.B. (1974). Some observations of the vertical mixing in the Baltic. In: 9th Conference of the Baltic Oceanographers, Kiel, April, 1974.

LASSIG, J. and A. NIEMI (1975). Parameters of production in the Baltic measured during cruises with R/V Aranda in June and July 1970 and 1971. Merentutkimuslaitoksen Julkaisu/Havsforskningsinstitutets Skrift No. 239, 34-40.

LAPPALAINEN, A. (1973). Biotic fluctuations in a _Zostera marina_ community. Oikos Supplementum 15, 74-80.

LINDBLOM, R. (1973). Abundance and horizontal distribution of pelagic fish effs and larvae in the Baltic Sea. Meddelande fran Havsfiskelaboratoriet, Lysekil, 140, 33 pp.

LINDEN, O. (1974). Effects of oil spill dispersants on the early development of Baltic herring. Annales Zoologici Fennici, 11, 141-148.

LINDEN, O. (in print). Lethal and sublethal effects of oil pollution on various life stages of the amphipod _Gammarus oceanicus_ Prace Morskiego Instytut Rybacki.

LINDGREN, L. (1975). Algal zonation on rocky shores outside Helsinki as a basis for pollution monitoring, Merentutkimuslaitoksen Julkaisu/Havsforskningsinstitutets Skrift No. 239, 334-347.

LINDQVIST, A. (1962). Sjokalven i Bottenhavet. Ostkusten, 4, 9. (in Swedish).

LEPPAKOSKI, E. (1973). Effects of an oil spill in the Northern Baltic. Marine Pollution Bulletin, 4, 93-94.

LEPPAKOSKI, E. (1975). Macrobenthic fauna as indicator of oceanization in the Southern Baltic. Merentutkimuslaitoksen Julkaisu/Havsforskningsinstitutets Skrift No. 239, 280-288.

MALEWICZ, B. (1975). Some factors limiting primary production in the coastal waters of the Southern Baltic. Merentutkimuslaitoksen Julkaisu/Havsforskningsinstitutets Skrift No. 239, 67-71.

MANN, K.H. (1973). Seaweeds: their productivity and strategy for growth. Science, 182, 975-981.

MATTHÄUS, W. (1972). Zur Hydrographie der Gotland See I. Säkulare Variationen von Salzgehalt und Temperatur. Beiträge zur Meereskunde, 29, 35-51.

MATTHAUS, W. (1972). Secular changes in oxygen conditions in the deep water of the Gotland Basin. Oikos Supplementum 15, 9-13.

McKELLAR, H. and R. HOBRO. (1976). Phytoplankton-zooplankton relationships in 100-liter plastic bags. Contributions from the Asko Laboratory, University of Stockholm, Sweden No. 13, 83 pp.

NIXON, S.W., C.A. OVIATT, C. ROGERS and K. TAYLOR (1971). Mass and metabolism of a mussel bed. Oecologia (Berlin), 8, 21-30.

NYQVIST, B. (1974). Österjön blommar. Forskning och Framsteg, 6, 1-2. (In Swedish).

ODEN, S. and J. EKSTEDT (1976). PCB and DDT in Baltic sediments. Ambio Special Report, 4, 127-131.

ODUM, H.T. and B.J. COPELAND (1969). A functional classification of the coastal ecological systems. In: Coastal ecological systems of the United States, H.T. Odum, E.A. McMahan and B.J. Copeland, editors. A report to the Federal Water Pollution Control Administration, Contract RFP 68-128, 1878 pp.

OJAVEER, E. and M. SIMM (1975). Effect of zooplankton abundance and temperature on time and place of reproduction of Baltic herring groups. Merentutkimuslaitoksen Julkaisu/Havsforskningsinstitutets Skrift No. 239, 139-145.

PATTEN, B.C. (1963). Optimum diversity structure of a summer community. Science, 140, 894-898.

PEUSSA, M. and O. RAVANKO (1975). Benthic macroalgae indicating changes in the Turka sea area. Merentutkimuslaitoksen Julkaisu/ Havsforskningsinstitutets Skrift No. 239, 339-343.

PROBST, B. (1975). Ein Modell zur Darstellung des pelagischen Kreislaufes in einem marinen Flachwasserökosystem der westlichen Ostsee. Dissertation, Christian-Albrechts-Universität zu Kiel, 53 pp.

REMANE, A. and C. SCHILIEPER (1958). Die Biologie des Brackwassers. E. Schweizerbartsche Verlagsbuchhandlung, 348 pp.

RINNE, I. and E. TARKIAINEN (1975). Chemical factors affecting algal growth off Helsinki. Merentutkimulsaitoksen Julkaisu/ Havsforskningsinstitutets Skrift No. 239, 91-99.

SCHEIBEL, W. and W. NOODT (1975). Population densities and characteristics of meiobenthos in different substrates in the Kiel Bay. Merentutkimuslaitoksen Julkaisu/Havsforskningsinstitutets Skrift No. 239, 173-178.

SCHIPPEL, F., R.O. HALLBERG and S. ODEN (1973). Phosphate exchange at the sediment-water interface. Oikos Supplementum 15, 64-67.

SCHULTZ, S. and W. KAISER (1973). Biological effects of the salt water influx into the Gotland Basin in 1966/70. Oikos Supplementum 15, 21-27.

SCHULTZ, S. (1970). Der Lebensraum Ostsee. Ökologische Probleme in einem geschichteten Brackwassermeer. Biologische Rundschau, 8, 209-218.

SCHWENKE, H. (1974). Die Benthosvegetation. In: Meereskunde der Ostsee, L. Magaard and G. Rheinheimer, editors, Springer-Verlag, 269 pp.

SEGERSTRALE, S. (1944). Weitere Studien uber die Tierwelt der Fucus-Vegetation an der Süd-Küste Finlands. Commentationes Biologicae, 9, 28 pp.

SEGERSTRALE, S. (1957). Baltic Sea. In: Treatise on marine ecology and palaeo-ecology 1, Ecology, J.W. Hedgepeth, editor 751-800.

SEGERSTRALE, S. (1962). Investigations on Baltic populations of the bivalve Macoma baltica (L.). Part II. Commentationes Biologicae 24, 26 pp.

SEGERSTRALE, S. (1965). On the salinity conditions off the south coast of Finland since 1950, with comments of some remarkable hydrographical and biological phenomena in the Baltic area during this period. Commentationes Biologicae 28, 28 pp.

SEN GUPTA, R. (1968). Inorganic nitrogen compounds in ocean stagnation and nutrient resupply. Science, 24, 884-885.

SHAFFER, G. (1975). Baltic coastal dynamics project - the fall downwelling regime off Asko. Contributions from the Asko Laboratory, University of Stockholm, Sweden, 7, 61 pp.

SJOBERG, S. (in print). Are pelagic systems inherently unstable - a model study. Ecological modelling.

SJOBERG, S., F. WULFF and P. WAHLSTROM (1973). The use of computer simulations for systems ecological studies in the Baltic. Ambio Special Report 1, 217-222.

SMETACEK, V. (1975). Die Sukzession des Phytoplanktons in der westlichen Kieler Bucht. Dissertation, Christian-Albrechts-Universitat, Kiel, 151.

SOSKIN, I.M. (1963). Long term changes in the hydrological characteristics of the Baltic. Hydrometeorological Press, Leningrad, 160 pp (in Russian).

SVANSSON, A. (1972). The water exchange of the Baltic. Ambio Special Report 1, 15-19.

SVANNSON, A. (1975). Interaction between the coastal zone and the open sea. Merentutkimuslaitoksen Julkaisu/Havsforskningsinstitutets Skrift No. 239, 11-28.

VOIPIO, A. and P. MALKKI (1972). Variations of the vertical stability in the Northern Baltic. Merentutkimuslaitoksen Julkaisu/Havsforskningsinstitutets Skrift No. 23, 3-12.

WALIN, G. (1972). Some observations of temperature fluctuations in the coastal region of the Baltic. Tellus, 24, 187-198.

WELANDER, P. (1974). Two-layer exchange in an estuary basin with special reference to the Baltic Sea. Journal of Physical Oceanography, 4, 542-556.

WESTIN, L. (1970). The food ecology and the annual food cycle in the Baltic population of fourhorn sculpin, Myoxocephalus quadricornis (L.), Pisces. Institute of Freshwater Research, Brottningholm, 50, 168-210.

WIKTOR, K. and M. PLINSKI (1975). Changes in plankton resulting from eutrophication of a Baltic firth. Merentukimuslaitoksen Julkaisu/Havsforskningsinstitutets Skrift No. 239, 311-315.

WILMOT, W. (1974). A numerical model of the gravitational circulation in the Baltic. ICES Special Meeting 1974 on models of Water Circulation in the Baltic, 11, 40 pp.

SEKTZER, I.S. (1973). On the ground-water discharge to the Baltic Sea and methods for estimating it. Nordic Hydrology, 4, 2.

ZENKEWICH, L. (1963). Biology of the seas of the USSR. Wiley and Sons, 955 pp.

SMUDZINSKI, L. (in print). Long scale biological changes in the Baltic Sea. Prace Morskiego Instytut Rybacki.

ZSOLNAY, A. (1973). Distribution of labile and residual carbon in the Baltic Sea. Marine Biology, 21, 13-18.

OSTROM B. (1976). Fertilization of the Baltic by nitrogen fixation in the bluegreen alga *Nodularia Spumigena*. Remote sensing of environment 4, 305-310.

ACKNOWLEDGEMENTS

I thank Ragnar Elmgren, Stig Fonselius, Gary Shaffer and Fredrik Wulff for critical reading of the manuscript. Sig Fonselius also kindly put an unpublished manuscript at my disposal. Maureen Moir corrected the English and Bibi Mayrhofer made the illustrations.

The work was carried out within the research project: "Dynamics and energy flow in Baltic ecosystems", sponsored by the Swedish Natural Science Research Council, grant B 2969-013.

TACTICS OF FISH MOVEMENT IN RELATION TO MIGRATION STRATEGY AND WATER CIRCULATION

F.R. HARDEN JONES, M. GREER WALKER and G.P. ARNOLD

FISHERIES LABORATORY, LOWESTOFT, SUFFOLK

THE MIGRATION PATTERN

The title of this paper was misprinted in the programme - immigration appearing for migration - and correcting the mistake provides an opportunity to define some terms. A man <u>emigrates</u> when he quits one country and <u>immigrates</u> when he enters another to settle: the journey only needs a single ticket. We often use the word migration in the sense of moving from one place to another. But when it is applied to animals it has a special meaning; a migration is a coming and going with the season and the journey needs a return ticket.

Many fish migrate and their movements can be reduced to a simple triangular pattern shown in Figure 1 which links the movements of the young fish with those of the adults.

In general terms these movements are related to those of the water currents. Thus the young stages must usually drift passively from the spawning ground to the nursery ground. The spawning migration of the adult fish has often been thought of as an active movement against and directed by the prevailing current and the return of the spent fish - exhausted by the toils of spawning - as a passive movement with the current. These two movements - the first active and contranatant, the second passive and denatant - are the essential features of the traditional theory of fish migration. By way of comment we would remind you of the advice given to Dr Watson by Mr Sherlock Holmes: "It is a capital mistake to theorize before one has the data"[1].

[1] "A scandal in Bohemia".
Copyright (c) Controller HMSO, London 1978.

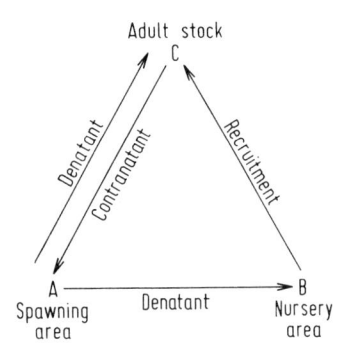

```
                    Adult stock
                         C
                       ↗ ↑ ↖
                      /  |  \
              Denatant Contranatant Recruitment
                    /    |    \
                   ↙     |     \
                 A ───────────→ B
              Spawning  Denatant  Nursery
                area              area
```

Figure 1 Diagram to show the migration pattern of young and adult fish.

The adult fish feed in one area, winter in a second, and spawn in a third. So their migratory movements can also be represented by a triangular pattern, the sequence of movements differing according to the time of the spawning season; spring spawners follow one sequence F, W, S, autumn spawners another F, S, W.

Characteristics of migration

There are four points that can be made about fish migrations which would, in the present climate of belief, gain general acceptance. We will consider them under the headings of distribution, straying, regularity, and homing.

Distribution

The distribution of a particular stock often appears to be contained within a regional hydrographic circulation. There are, for example, stocks of cod at Newfoundland, Labrador, West Greenland, Iceland, Faroe Bank, the Barents Sea, the Norway coast, the North Sea, and the Baltic. In general terms these sea areas can be identified with well defined hydrographic systems and here the key word is gyral. While there is, at times, significant mixing between cod at West Greenland and those at Iceland, and while there may be more than one stock of cod in the North Sea, the different stocks are, for the fisheries biologist, more or less independent units which can be described in terms of recruitment, growth, and mortality.

Straying

Some fish will stray beyond the boundaries of the hydrographic system within which their stock is contained. The loss of such fish is reflected in an apparent increase in natural mortality. A fish that strays - an emigré - is lost to the parent stock so far as reproduction is concerned.

Regularity

Migrating fish follow clearly defined routes within the area occupied by a particular stock and the movements are regular. This is well known to fishermen who profit by their knowledge. Our colleague Dr David Cushing has put a figure to this regularity in connexion with the spawning seasons of temperate species such as the salmons, herring, cod and plaice. He derived indices for the mean dates of peak spawning with standard deviations of about 7 days (Cushing, 1969).

Homing

The feeding, wintering and spawning areas occupies by a stock are usually stable over a period of at least several decades. The fish return to places formerly occupied: in other words they home.

The Grand Strategy

Migratory behaviour is one of several features in the life histories of fish which are ultimately directed towards reproduction. Here the Grand Strategy must be to ensure a sufficient number of viable offspring to maintain the population up to the limit, in terms of numbers or of weight, that can be fed. In this connexion migration can be regarded as a special adaptation towards abundance, the movements allowing a stock to exploit and to make the best of the resources of different areas during the different stages of its life history.

So it is no accident that most - but not all - of the species which form the basis of the Great Fisheries of northern waters are migratory; the fish are abundant because they are migratory, and the Fisheries are Great because they are based on species that are abundant. One might expect the migration patterns to be related to zonal differences in climate. For example, the production cycles of northern waters are more restricted, in space and time, when compared with those of southern waters. In temperate and arctic waters a spawning will not be successful unless the eggs are shed in the right place and at the right time. Well developed migratory patterns could be advantageous in tropical waters: they could be a condition of existence in higher latitudes.

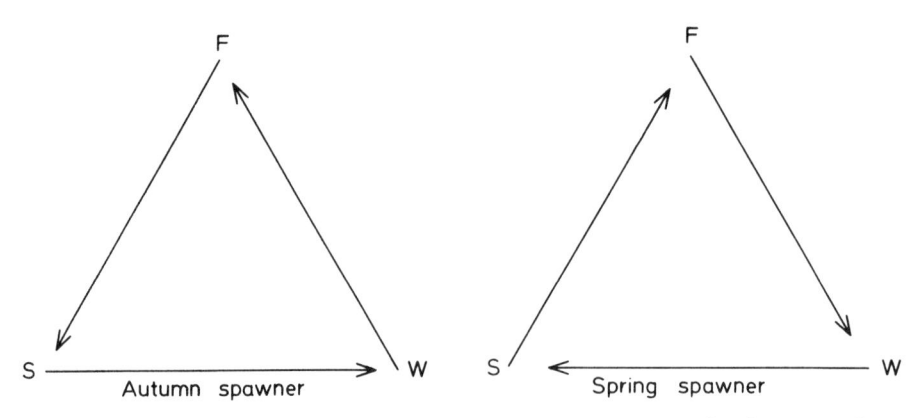

Figure 2 Diagram to show the migration pattern of adult spring or autumn spawners. F, S and W indicate feeding, spawning, and wintering areas respectively.

Table 1 Distance, in km, covered each year by herring and cod when completing their migration circuit

	Herring	Cod
Length of fish	30 cm	80 cm
North Sea	1600 km	1300 km
Northern waters	3000 km	2600 km

This raises an interesting question to which we have not found a satisfactory answer. To what extent do the commercial fisheries depend on migratory fish in tropical, temperate and arctic waters? We suspect that while migratory species account for a large proportion of the catch in temperate and arctic waters, non-migratory species - which are diverse and numerous on tropical shelves and reefs - may be relatively more important in lower latitudes.*
Are there counterparts of the non-migratory reef fish in northern waters? The only supposed non-migratory group that comes readily to mind are the sandeels which make up 15% by weight of the total North Sea catch.

*But upwelling areas in tropical waters, where the conditions parallel those in higher latitudes, may sustain fisheries largely dependent on migratory fish.

The tactics

We have referred to the Grand Strategy; that of maintaining the population up to the limit of the available food supply. Now for the tactics which serve that strategy and here we will stay close to the sense of the word as given in Fowler's Dictionary: tactics in the sense of the movement of bodies; and so in this connexion, the movement of fish. We shall give some account of what we have been finding about fish movements; where we are having some success; and how our findings fit in with what is known of water currents. As the story develops you will see that we have problems; and that there are areas where we need help.

Energy problems

The story starts with the simple migration pattern shown in Figure 2. A fish will complete this circuit once a year and the distance covered during the year - the length of the migration circuit - varies between stocks. Table 1 shows that the migration circuits of cod and herring in the North Atlantic are about twice as long as those of cod and herring in the North Sea but in both areas the fish complete the circuit in one year. It is true that Norwegian sea herring are somewhat larger than North Sea Herring, and that there may be bigger cod in the Barents Sea than ever came out of the North Sea. But nevertheless many fish of the same size appear to complete the longer as readily as the shorter migration; and between species, a 35 cm herring may cover the same distance during a year as a 90 cm cod.

Is there any relation between the size of the fish and the distance covered and speed maintained during migration? This is a difficult question to answer as conventional tagging experiments are not very satisfactory. One does not know how long the fish has been waiting at the recapture position, and the speed is necessarily a minimum speed. The best that can be done is to examine the distribution of recoveries within a relatively short time of release. We have analyzed the recoveries of a tagging experiment on cod, carried out by our former colleague Mr G C Trout.

Trout's tagging experiments with cod

The Arcto-Norwegian cod winter to the west and south of Bear Island before moving to their spawning area centred about the West Fjord, within the Lofoten Islands. Mature fish leave Bear Island in December and arrive at the Norway coast in January. In January 1959 cod were caught, tagged and released at Fuglöy Bank; the release position is shown in Figure 3. Fifty-four of the fish were recovered along the line of the Norwegian coast in the next 90 days. Figure 4 shows the relation between fish length and distance travelled in km, and speed in fish lengths per second.

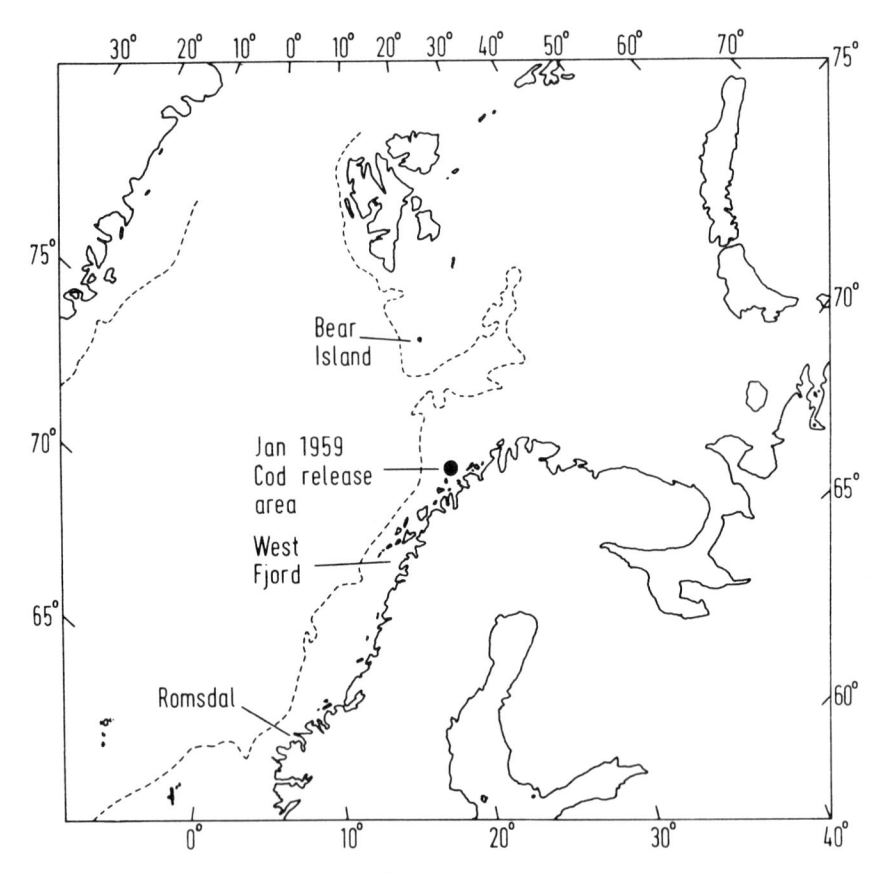

Figure 3 The position (Fuglöy Bank) at which cod were tagged and released in 1959 by Mr G C Trout.

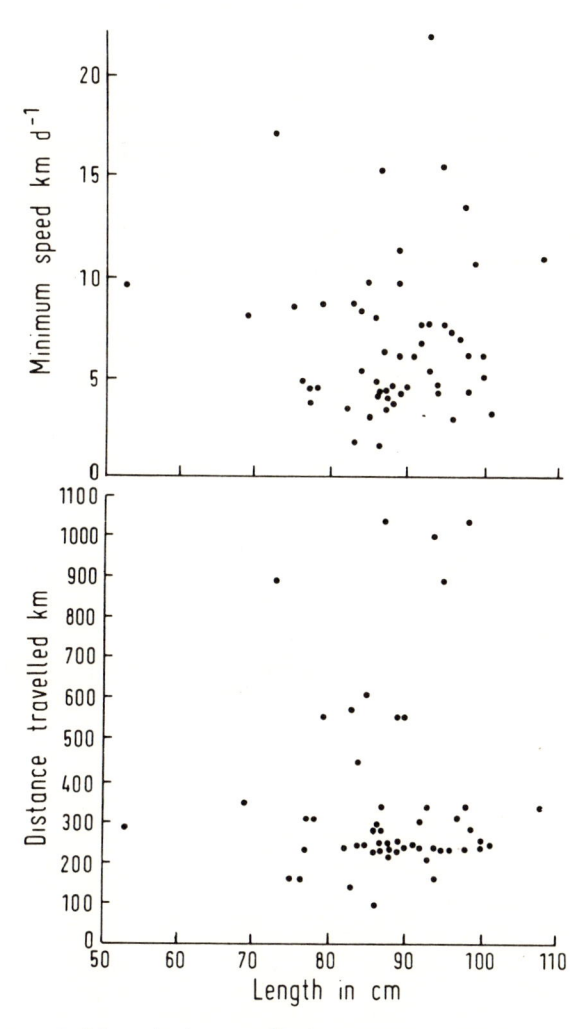

Figure 4 The relation between fish length, distance travelled and
minimum speed for 54 fish recaptured within 90 days of
release at Fuglöy Bank.

Table 2 Minimum speeds of fish on migration estimated from tagging experiments

Species	Mean Length cm	Speed km d^{-1}	Speed cm s^{-1}	Speed L s^{-1}	Author
Sole	30	7- 16	9- 18	0.28-0.53	Anon (1965)
Plaice	35	1- 7	2- 8	0.06-0.23	Bannister (unpubl)
Herring	25	4- 30	5- 35	0.20-1.40	Bolster (1955)
Mackerel	35	16- 23	19- 27	0.54-0.77	Bolster (1974)
Salmon (sockeye)	70	9- 22	10- 25	0.15-0.36	Harden Jones (1968)
Cod	80	6- 28	7- 32	0.09-0.40	Trout (unpubl)
Albacore	77	26- 44	30- 51	0.39-0.66	Clemens (1961)
Bluefin	250	93-185	108-214	0.43-0.86	Mather et al (in press)

There is no obvious relation between them. This is not what one might expect if the fish were actively swimming against the north-going Atlantic current; in a contranatant migration the bigger and more powerful fish should do better than the smaller fish.

The speeds at which cod travel are rather slow and this is generally true for fish on migration as can be seen from Table 2; even the large bluefin tuna and albacore which make transoceanic migrations in 60 days, are not travelling very quickly.

If the data in Table 2 are used to calculate the Reynolds number according to the equation: $R = 100\ V\ L$ where V = speed in cm s^{-1} L = length in cm values with the range 10^5-10^6 or less are obtained which are consistent with low drag coefficients: the fish have no great problems in hydrodynamics. But such calculations are hardly justified: the speeds given in Table 2 are ground speeds and the appropriate speed for the Reynolds number is the speed through the water.

The two speeds

The distinction between the two speeds is important. There could be certain advantages in travelling quickly over the ground. For example, if time were short or if there was something to be gained by arriving at the feeding, wintering or spawning area before the other fish. And for a small fish a quick journey might reduce the risk of being eaten on the way.[1]

[1] Small fish might try to reduce the risk of predation by travelling together: the fish convoy is the shoal (Brock and Riffenburgh, 1960).

But if there is something to be said for travelling fast over the ground, there is also something to be said for going slowly through the water. It is this: the force to be exerted against surface drag is proportional to the square of the fish's velocity through the water. An energy conscious fish would therefore swim slowly.

So it could be argued that when on migration the wise fish would make arrangements to move quickly over the ground while moving slowly through the water. Obviously the journey should be made economically; there is no advantage in expending more energy on swimming than can be harvested in the feeding area.

Very little is known how fish move when on migration, and we have already cast some doubts on the validity of the traditional contranatant hypothesis. But we are now starting to learn something about how fish move from one position to another. A variety of tracking techniques have been used and interest in this sort of work is world wide. At the Fisheries Laboratory, Lowestoft we have been studying the movements of plaice in the open sea.

Tracking work with plaice

We have combined two acoustic techniques into a reliable and effective tracking system which can be used from a research vessel at sea. In the late 1960s we were fortunate to have an opportunity to use the British Admirality Research Laboratory's 300 kHz sector scanning sonar (Cushing and Harden Jones, 1967) which was later installed on the Ministry of Agriculture, Fisheries and Food's Research Vessel "Clione". Details of the equipment and its use on RV Clione have already been described (Voglis and Cook, 1966; Mitson and Cook, 1971).

The receiving beam is scanned electronically over a 30° sector with a resolution of 0.3° in bearing and 7 cm in range and an effective range out to about 300 m. The stabilization system allows the system to scan horizontally (to give bearing in azimuth) or vertically (to give depth).

When used in conjunction with the acoustic transponding tags developed at the Fisheries Laboratory, Lowestoft (Mitson and Storeton West, 1971) it has been possible to follow individual fish in the open sea for periods up to 55 h and over distances up to 60 km.

In these experiments a tagged fish was released and followed by the research ship and kept under continual surveillance. During the track the ship was positioned regularly at 15 min intervals by

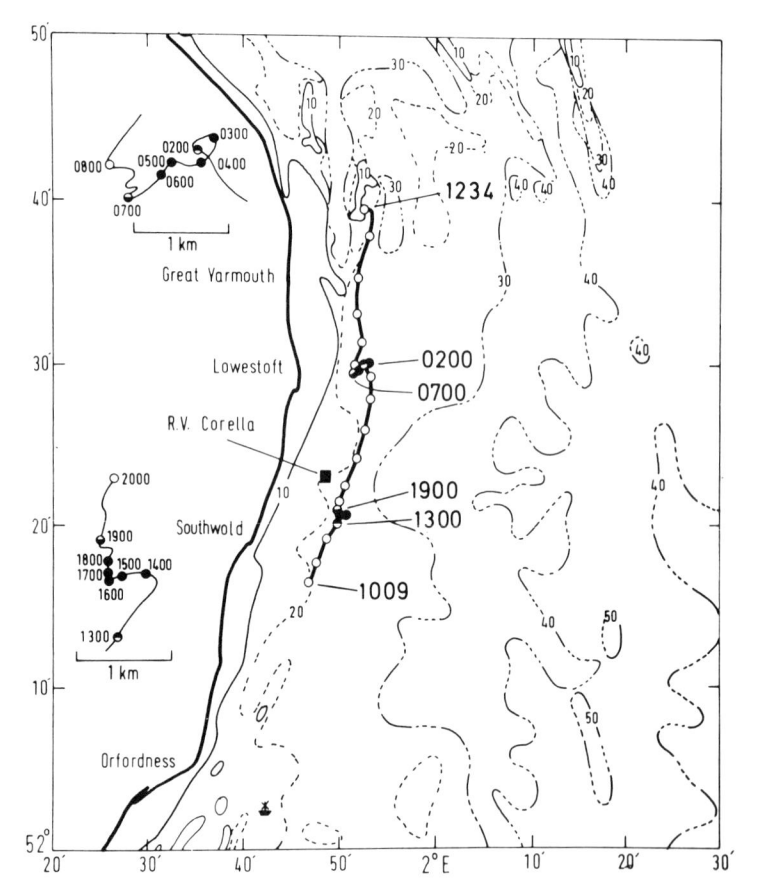

Figure 5 Track chart of plaice 7, 41 cm long, released at 1009 h
GMT 12 December 1971. Hourly positions of the fish
are indicated and the times of slack water are given

O north going tide
● south going tide
◒ low water slack
◓ high water slack

TACTICS OF FISH MOVEMENT

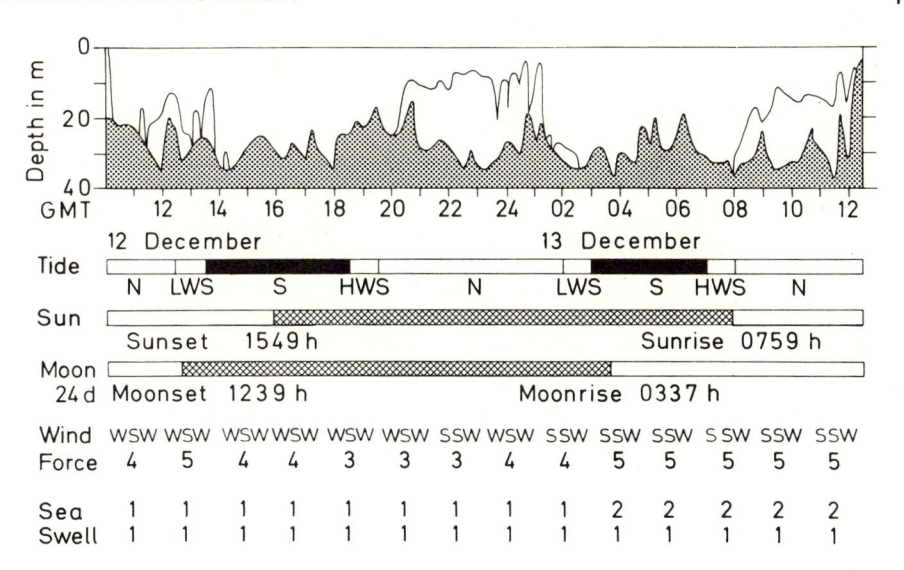

Figure 6 The depth of plaice 7 in relation to the direction of the tide and other environmental factors.

reference to the Decca Navigator System which was accurate to ± 50 m in the working area. The signal returned by the transponding tag was clearly recognizable on the B-scan of the sonar display and allowed the fish to be positioned relative to the ship to within ± 2.5 m in range and 1-2° in bearing, corresponding to about 2 m in depth[1].

We have tracked 12 plaice (Pleuronectes platessa) off the East Anglian coast and a detailed account of the work will be published soon (Greer Walker, Harden Jones and Arnold, 1977). Here we will look at the details of one fish - plaice 7 - released off Orfordness on 12 December 1971. The fish was tracked for 26 h during which it moved 43 km to the north.

The track chart (Figure 5) shows that the fish gained ground to the north during the northerly tide, the fish coming into mid-water after high water slack and returning to the bottom at low water slack (Figure 6). These vertical movements, which occur about every 12 h, can be described as semi-diurnal and are clearly

---

[1] A 16 mm film of the sonar display was shown during the meeting.

Table 3 Speed over the ground, direction of tide, and position in the water column. Data for Fish 7 rounded off and simplified

Time GMT	Speed over ground cm s^{-1}	Direction of tide	Position in water column
1000-1300	80	North	Midwater
1300-1900	9	South	On bottom
1900-0200	68	North	Midwater
0200-0700	8	South	On bottom
0700-1230	105	North	Midwater

related to the tidal cycle. The fish was abandoned at 1234 h on 13 December when it "went aground" in shallow water at the Cross Sand. Over the period of the track its ground speed was 39.4 km d^{-1}, or 46 cm s^{-1}, equivalent to 1.1 L s^{-1}. These values are very similar to those of migrating fish summarised in Table 2. The relation between the direction of the tide, the position of the fish in the water column and the distance moved over the ground are summarised in Table 3. It is clear that the speed of fish 7 was greatest when the tide was set in the direction of overall movement.

During this, and some other tracks, water current measurements were made from a second research ship or specially laid moored current meters which allowed us to estimate the speed of the fish through the water; it rarely exceeded 1 L s^{-1}. Fish 7 appeared to be swimming downstream with the tide, but three other fish consistently swam slowly through the water heading southeast as they were being carried northwards over the ground.

The semi-diurnal pattern of vertical migration, locked to the tidal cycle, appeared to be characteristic of almost all the plaice that moved appreciable distances during the period of surveillance. We have called this behaviour selective tidal stream transport, and it depends on the fish leaving the bottom on one slack water and returning to it on another. The vertical movements are usually very clear cut and we are not at all certain of the stimulus, or stimuli, to which the fish responds. But ascents are more closely related to slackwater than descents. Descents are usually - but not always - preceeded by excursions to the bottom. Our observations suggest that some form of 6 h biological clock, reinforced by contact with or sight of the bottom, are essential parts of the sensory chain.

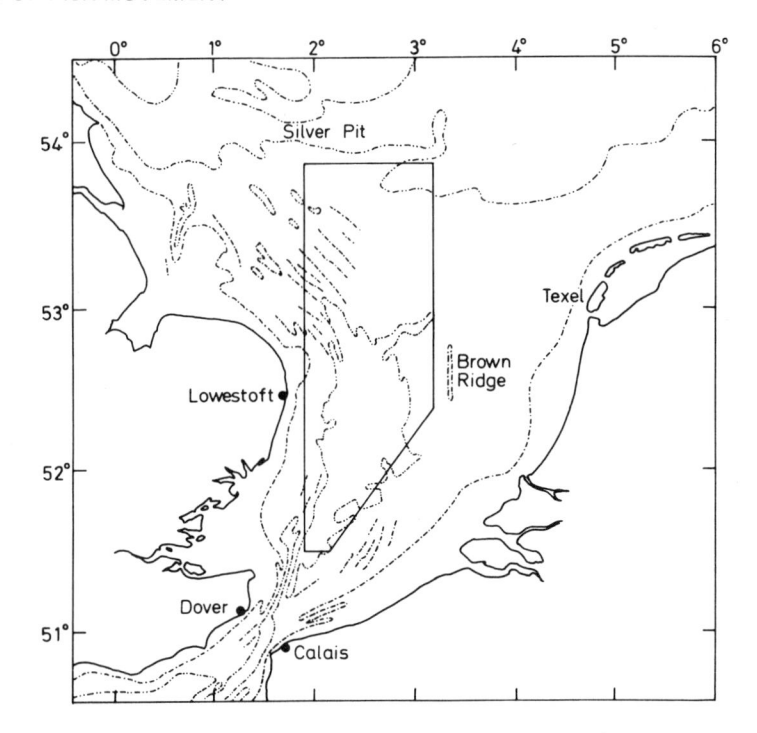

Figure 7 Plaice spawn in the Southern Bight within the area 51°
30-52°00N, 2°-3°E. In November and December ripening
plaice move from north of 53°N towards the spawning area
in the south. Midwater trawling to test the selective
tidal transport hypothesis was carried out with the area
indicated.

Migration by selective tidal stream transport

Similar behaviour has been observed in or inferred for other
animals, particularly those which have well defined on or off shore
movements, such as shrimps and elvers: there is nothing new in the
idea itself. Selective tidal stream transport would allow a fish
to travel quickly over the ground while moving slowly through the
water and stimulated by de Veen's (1967) observations on soles, we
wondered if this behaviour could provide a migratory mechanism for
plaice.

A stock of plaice spawns in the Southern Bight of the North
Sea with peak egg production in January. In November and December
the mature plaice move from their feeding area south of the Dogger
Bank into the Southern Bight and so towards their spawning grounds.
A chart of the area is shown in Figure 7.

Table 4 A comparison between midwater tows for plaice made on consecutive northerly and southerly tides in the Southern Bight of the North Sea in November-December 1974-75

Number of valid pairs of tows	Southerly tide tow greatest	Northerly tide tow greatest	Tows equal
41	33	6	2

Table 5 A comparison between midwater tows for plaice made on consecutive northerly and southerly tides in the Southern Bight of the North Sea in February 1976

Number of valid pairs of tows	Southerly tide tow greatest	Northerly tide tow greatest
8	1	7

Table 6 Catch of plaice in midwater on consecutive northerly and southerly tides in the Southern Bight of the North Sea, November 1974 - February 1976

Season	Northerly tide hours	catch n	catch rate	Southerly tide hours	catch n	catch rate	Ratio
Autumn	148	65	0.44	154	254	1.65	1.0 : 3.7
Winter	24	81	3.37	26	19	0.74	4.6 : 1.0

If the fish migrated using selective tidal stream transport, the mature plaice should come off the bottom into midwater on south-going tides and return to the bottom on north-going tides. We have tried to test this hypothesis by fishing along the line of the migration route with a midwater trawl and comparing the catch of plaice taken on tows made on consecutive northerly and on southerly tides. If the plaice migrate in this way, the catch of plaice taken in a tow made on a southerly tide should exceed that made on a northerly tide.

We used a paired haul technique, the research vessel towing an Engel midwater trawl for 3 h during the peak of one tide, followed, 3 h later, by a second 3 h haul during the peak of the next tide. The headline of the net was set so that the gear fished that part of the water column which the tracking work showed plaice to occupy when in midwater. The results, summarised in Table 4, are consistent with the tidal stream transport hypothesis.

Going one step further, we predicted that after spawning the fish would reverse their behaviour and return to their feeding grounds, on the northerly tides. Preliminary results obtained in February 1976 and summarised in Table 5 suggested that this is true: northerly tide tows were greater than southerly tide tows and the catch ratios were reversed. (Table 6).

A tidal streampath chart

Other tracking experiments (Greer Walker and Arnold, in preparation) have shown that cod behave similarly to plaice although the amplitude of the vertical movements appears to be less. Could selective tidal stream transport be a general mechanism for migration? When this idea was put forward at a meeting in Aberdeen in November 1975 (Harden Jones, 1977) a participant suggested that we should look at the tidal streams in the North Sea; is there any pattern which would fit in with the distribution and movements of different stocks of fish? We have tried to construct a tidal streampath chart for the North Sea and the Rostock Atlas 'Anon, 1968) provided the basic data. Chart 3 in the Atlas shows the direction of the tide at maximum flow. The first step in making the streampath chart was to mark the direction of flow at intervals along each isoline as shown in Figure 8. The lines were then joined up to produce the final streampath chart shown in Figure 9.

It is a curious chart and, from the point of view of fish migration very interesting. The pattern is different from the residual current chart for the North Sea and the most obvious feature is the split between the north-south streampaths and those running west to east across the German Bight. The chart certainly separates the Southern Bight and German Bight stocks of plaice, and links the Flamborough plaice with those of the Southern Bight and

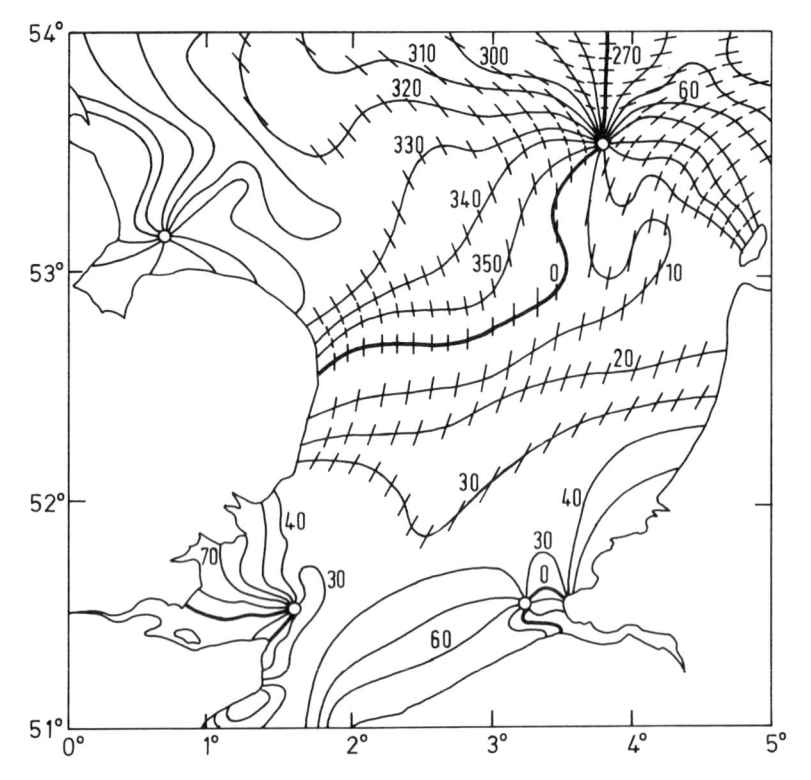

Figure 8 Lines joining points at which the maximum tidal flow runs in the same direction. From chart 3 in the Rostock Atlas (Anon, 1968). The numerals indicate the direction of maximum flow and the dashes on the isolines mark the lines of flow.

English Channel. We suspect that this chart could provide new insights into several problems: for example the streampaths suggest that it might be quite difficult to leave the Minch between the Butt of Lewis and Cape Wrath. It would be of interest to see similar charts for other areas; would Georges Bank be a suitable case for treatment?

There is another chart in the Rostock Atlas which is relevant to the tidal streampath hypothesis. Chart 5 gives an index of directivity for the tidal streams, based, at any station, on the value of the slowest tide expressed as a percentage of the fastest. Thus long and thin tidal ellipses, where the speed of the slowest tide is less than 10% of the fastest, are directional; and rounded ellipses, where the slowest tide may be as much as 60% of

TACTICS OF FISH MOVEMENT

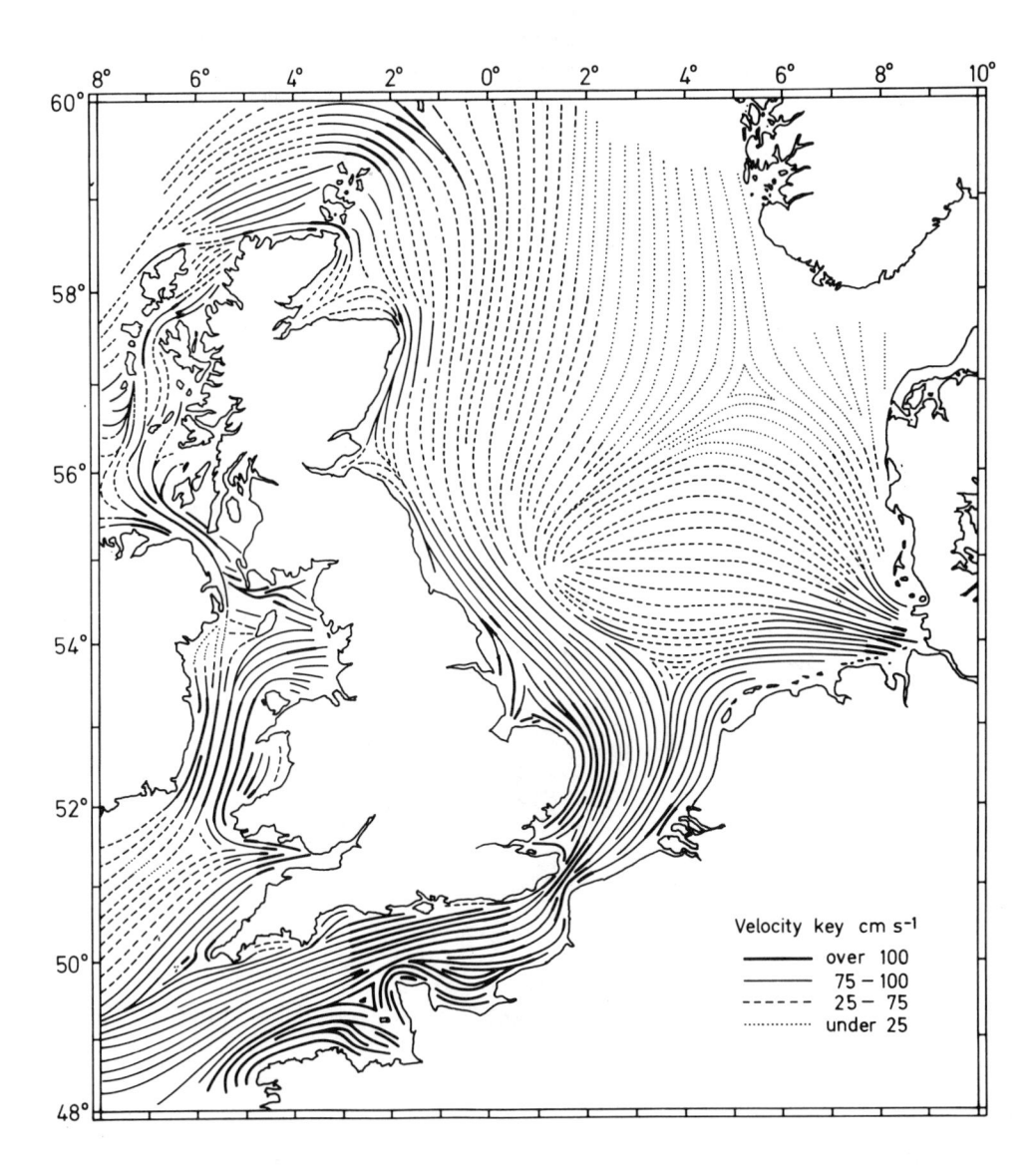

Figure 9 A tidal streampath chart for the North Sea and adjacent waters reconstructed from the data given in chart 3 of Rostock Atlas (Anon, 1968).

the fastest, are less so. Part of chart 5 from the Rostock Atlas is redrawn here in Figure 10 and we have selected two Admiralty Tidal Stations, N and P, as representative of the ellipses in areas of low and high directivities respectively. There are marked differences in the ellipses for the two stations (Figure 11) and the progressive vector diagrams in Figure 12 show that a fish using selective tidal transport will travel quicker in one regime than the other. It may be more than coincidence that the migration route of plaice moving between the Dogger and the southerly spawning grounds goes through the area where the tidal ellipses are directional.

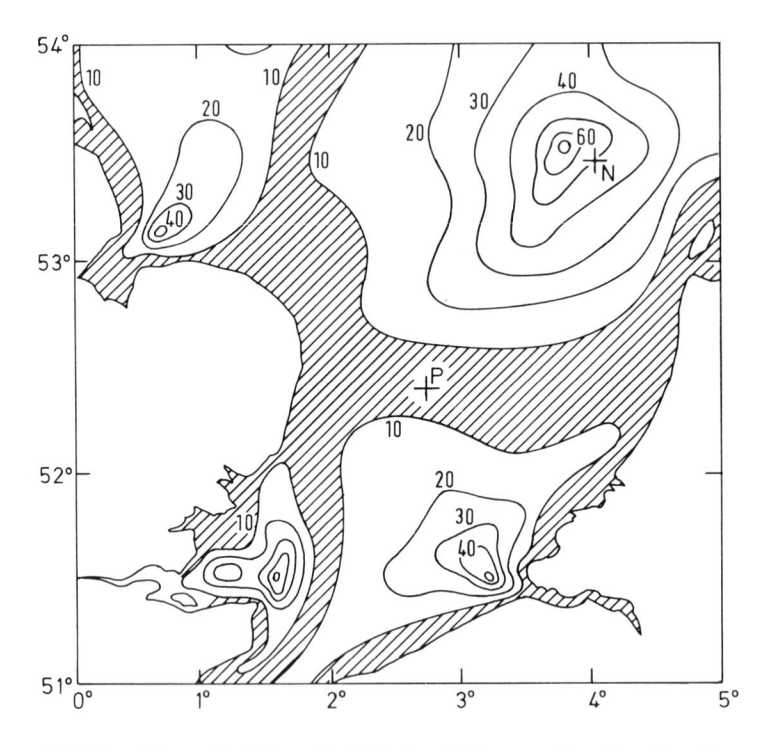

Figure 10 The directivity of tidal ellipses in the Southern Bight of the North Sea. Directivity is measured by expressing, at one station, the slowest tidal speed as a percentage of the fastest. The isolines join stations of equal directivity. Redrawn from chart 5 in the Rostock Atlas (Anon, 1968). The position of two tidal stations representative of areas of high and low directivity are indicated. Station positions from International Chart 2182A.

TACTICS OF FISH MOVEMENT

    Another point concerns the clue or clues that trigger off the vertical movements which connect the fish with the transporting tide. The rounded ellipses are probably associated with lower linear and angular accelerations or velocities than the long drawn out ellipses and the signal from the sea might not be strong enough to initiate the vertical movements. There could, perhaps, be "no-go" areas and, intuitively, we would expect to find them where the fish collect - on the feeding, wintering and spawning areas - rather than on the migration routes. But we are now several steps ahead of the facts and it is time to heed Mr Holmes' advice.

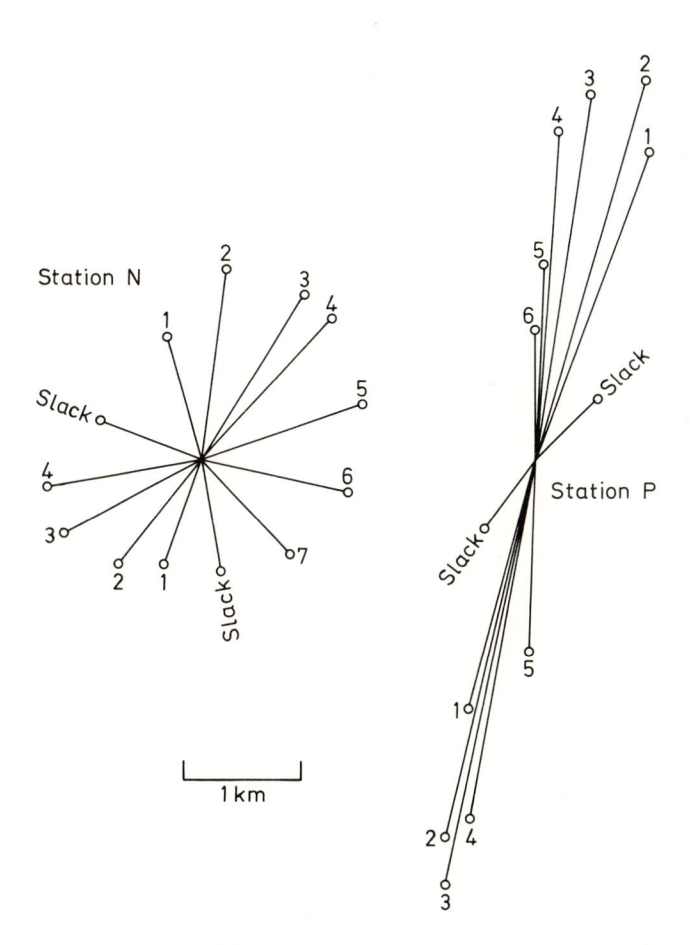

Figure 11 Ellipses at Tidal Stations N and P. Data from International Chart 2182A.

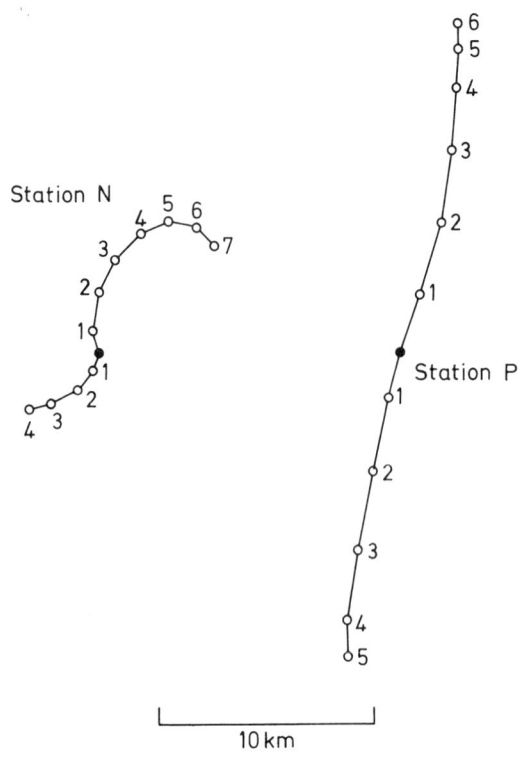

Figure 12 Progressive vector diagrams for fish using selective tidal transport at Tidal Stations N and P.

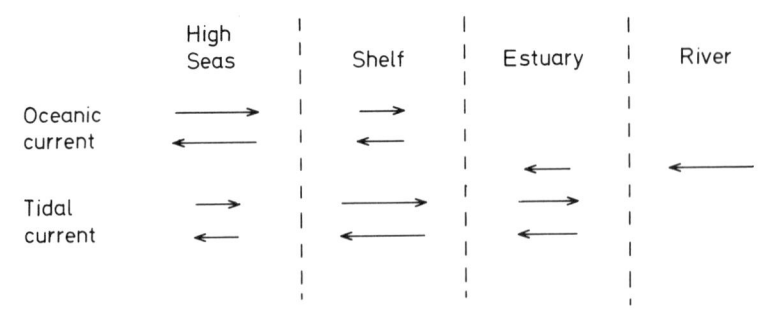

Figure 13 Current regimes in relation to fish migration. Oceanic currents are dominant on the high seas, tidal currents over the shelf. In estuaries the discharge reinforces the ebb tide and in the river there is only the downstream flow.

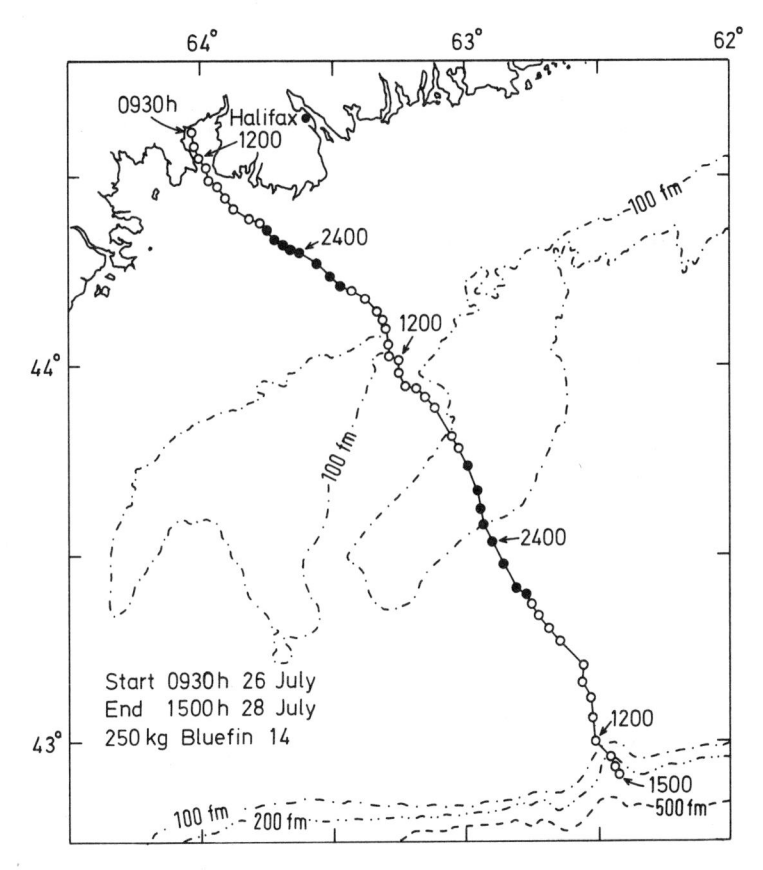

Figure 14 The track of a bluefin tuna tracked in the open sea.
          Hourly positions are indicated. O, day, ●, night.
          Redrawn from Lawson and Carey (1972). The fish was
          released at 0930 h in St. Margaret Bay, Nova Scotia.

Migration in other areas

Even if we have found a basis for migratory movements in certain areas, this cannot be the whole story. On the high seas, ocean currents predominate, while tidal influences are strongest on the shelves. The point is made in Figure 13 which shows that there are four distinct current regimes or environments. Ocean currents dominate the high seas; tidal currents are more important on the shelves; in estuaries the discharge supplements the ebb; while in the rivers there is only the downstream flow. One might expect the behaviour of a migratory fish to be different in each regime and that a fish which passed through all four - such as a salmon - should show marked behavioural changes at each interface.

We have tracked plaice and cod in areas dominated by tidal currents; and similar studies have been made by Stasko and others with salmon and Tesch with eels. While many of the salmon and eel tracks show features that are consistent with selective tidal stream transport, only the Lowestoft results include the depth data which make the results convincing. There has been some tracking work in the high seas regime, and we are thinking in particular of the work on bluefin tuna by Lawson and Carey (1972) from Woods Hole. One of their tracks is shown in Figure 14. This is a splendid track and the straightness of the course - by day and by night - and the regularity of the movement is rather different from the pattern observed with fish in shelf regimes.

We think that we are just at the beginning of something very interesting. To take a significant stride further we may need a new generation of telemetering tags, and at Lowestoft we have hopes of a device which will include a compass to let us know which way the fish is heading. But if we are to make new progress - as distinct from more observations - there is work to be done together and those to be involved must include oceanographers, biologists, and the gentlemen who work so skilfully with electronics. But it is up to the biologists to state the problems clearly before their colleagues in other disciplines can give real help.

References

Anon, (1965). Report of the working group on sole. Co-op. Res. Rep., (5) 126 pp.

Anon, (1968). "Atlas der Gezeitenströme für die Nordsee, den Kanal und die Irische See". Seehydrographischer Dienst der Deutschen Demokratischen Republik, 2nd edition. Rostock 58 pp.

Bolster, G.C., (1955). English tagging experiments. Rapp. Cons. int. Explor. Mer, 140: (2) 11-14.

Bolster, G.C., (1974). The mackerel in British waters. In Sea Fisheries Research, 101-16. Edited by F R Harden Jones, Elek Sci., Lond., 510 pp.

Brock, V. E., and Riffenburgh, R.H., (1960). Fish schooling: a possible factor in reducing predation. J. Cons. int. Explor. Mer, 25: 307-17.

Clemens, H. B., (1961). The migration, age, and growth of Pacific albacore (Thunnus germo), 1951-1958. Fish. Bull. Calif., (115) 128 pp.

Cushing, D.H., (1969). The regularity of the spawning season on some fishes. J. Cons. int. Explor. Mer, 33: 81-92.

Cushing, D.H., and Harden Jones, F.R., (1967). Sea trials with modulation sector scanning sonar. J. Cons. int. Explor. Mer, 30: 324-45.

Greer Walker, M., Harden Jones, F.R., and Arnold, G.P. (in press). Movements of plaice (Pleuronectes platessa L.) tracked in the open sea. J. Cons. int. Explor. Mer, 38:

Harden Jones, F.R., (1968). Fish migration. Arnold, Lond., 325 pp.

Harden Jones, F.R., (1977). "Performance and behaviour on migration". In "Fisheries Mathematics" edited by J.H. Steele. Academic Press, London and New York.

Lawson, K.D., and Carey, F.G., (1972). An acoustic telemetry system for transmitting body and water temperature from free swimming fish. Tech. Rep. Woods Hole, WHO1 - 71-67, 21 pp.

Mather, F.J., Mason, J.M., Schuck, H.A., and Jones A.C., (1977). Life history and fisheries of Atlantic bluefin tuna (Thunnus thynnus). Adv. Mar. Biol., 15: (in press).

Mitson, R.B., and Cook, J.C., (1971). Shipboard installation and trials of an electronic sector-scanning sonar. Radio electron. Engr., 41: 339-50.

Mitson, R.B., and Storeton-West, T.J., (1971). A transponding acoustic fish tag. Radio electron. Engr., 41: 483-89.

de Veen, J.F., (1967). On the phenomenon of soles (Solea solea L.) swimming at the surface. J. Cons. int. Explor. Mer, 31: 207-36.

Voglis, G.M., and Cook, J.C., (1966). Underwater applications of an advanced acoustic scanning equipment. Ultrasonics, 4: 1-9.

VARIABLE OCEAN STRUCTURE

WALTER H. MUNK

INSTITUTE OF GEOPHYSICS AND PLANETARY PHYSICS
SCRIPPS INSTITUTION OF OCEANOGRAPHY, UNIVERSITY OF
CALIFORNIA, SAN DIEGO, LA JOLLA, CALIFORNIA 92093

The most important development since the Tokyo meeting is a growing appreciation of mesoscale variability in the oceans; the "DC" (or zero frequency) circulation accounts for only 1% of the geostrophic kinetic energy; nearly all of the kinetic energy is contained in scale lengths of a few hundred kilometers and scale times of a few months.

Before 1950, oceanographers were almost exclusively occupied with the zero frequency ocean structure. The associated general circulation consisted of the windspun gyres in the upper oceans and a sluggish deep motion resulting from bottom water formation in the Weddell and Norwegian Seas. In 1958, John Swallow set about to confirm Stommel's deep circulation theory, using acoustically tracked, neutrally buoyant floats. Swallow expected a steady, broad drift of order one millimeter per second. Instead, he found much larger velocities that were not well correlated over distances of some ten kilometers and that showed significant time changes in a month or so. The ARIES measurements in 1959 confirmed these findings: superimposed on a mean current of order one centimeter per second were fluctuations of order ten centimeters per second, with small spatial and large temporal structure. An ocean with 1 ± 10 cm per second motion is far different from one with 10 ± 1 cm per second! Since then John and Mary Swallow have discovered the writings of Major James Rennell, a marine surveyor of the East India Company, who took 17 voyages between Halifax and Bermuda, and found the currents to be <u>casual</u>, so this realization goes back a long way.

As early as 1950 oceanographers working on the windspun ocean circulation had become uneasy about DC oceanography. Soviet oceanographers, under the influence of V.B. Stockmann, undertook a series of expanding experiments involving moored current meters: 18 days in the Black Sea in 1956, 14 days in the Atlantic in 1958, 60 days in the Arabian Sea in 1967. Even then, these time series were found to be too short to give a clear picture of the variability of the currents; and so in 1970 in the northeast tropical Atlantic, the POLYGON experiment under the leadership of Brekhovskikh and Kort involved moored current meters over a period of six months. The experiment showed the existence of 10 centimeters per second currents with correlation scales of 100 kilometers and time scales of months.

I am more familiar with the MODE expedition in the northwest Atlantic during March and June, 1973. This was a major undertaking involving 50 oceanographers from three countries, 15 institutions, 6 ships, and 2 aircraft. The results were generally in accord with the POLYGON measurements. Wunsch has written amusingly about this first major collective experiment for U.S. oceanographers: people had worked together at earlier times, but according to the "sandbox principle," that is, like children playing together in a sandbox, but each in fact building his own castle. But for MODE the efforts of all concerned had to be coupled from the very start, and even theoreticians had to make decisions in "real time". The results of MODE were consistent with POLYGON, but carried our understanding much further. Now the resources of the POLYGON and MODE experiments have been combined into POLYMODE. Among the problems to be studied is whether mesoscales constitute an "eddy zoo" of which the Gulf Stream rings are the most dramatic manifestation, somewhat akin to hurricanes.

It is not surprising that mesoscale features have now turned up everywhere. They were found by Bernstein and White in BT cross sections in the central north Pacific, and by Adrian Gill in ocean weather ship observations.

We may think of the mesoscale fluctuations as the weather in the sea. The 100-kilometer ocean correlation scale compares to 1,000 kilometers in the atmosphere, and the two-month ocean time scale to four days in the atmosphere. Space resolution is very hard to come by, and so the oceanographer is faced with a tougher job than the meteorologist. If we give up on monitoring and charting mesoscale eddies, it is as if the meteorologists had given up on storms and confined themselves to problems of climate. A pilot coming into London Airport would not find it very useful if he were furnished only with the mean September winds. In just the same way a submarine sonar officer does not find seasonal charts very useful, and he has learned to depend instead on his own local observations. A similar situation may hold for deep-sea fishermen. The suggestion is that the past lack of success in describing and predicting biological distributions, air-sea interactions, ASW conditions, etc., is the result of not having taken into account this dominant mesoscale variability.

Could the technical means be developed for monitoring ocean structure on the mesoscale? A major effort in this direction might well be worthwhile.

OCEAN VARIABILITY - THE INFLUENCE OF ATMOSPHERIC PROCESSES

R.W. STEWART

INSTITUTE OF OCEAN SCIENCES, 512 FEDERAL BUILDING

VICTORIA, B.C. V8W 1Y4

When the abstract for this talk was written in April, 1976 I commenced it with:

> "One can identify at least four processes by which the atmosphere may affect the ocean:
>
> (a) momentum transferred from the atmosphere to the ocean is one of the prime sources of oceanic motion;
>
> (b) heat and water vapour transfer from ocean to atmosphere affects both the temperature and the salinity of surface oceanic water;
>
> (c) atmospheric cloudiness affects the heat balance at the ocean surface and therefore the oceanic temperature;
>
> (d) wind-stirring of the ocean affects the distribution of heat and salt."

Since then I have become aware of very recent results both within my own Institute and elsewhere which show that another process may be very important:

> (e) The response of the ocean to fluctuations in the atmospheric surface pressure may not be prompt enough nor complete enough to prevent these fluctuations from being reflected in substantial pressure fluctuations in the ocean. The resulting pressure gradients must induce corresponding motions.

Since the behaviour of the atmosphere is subject to variation on a wide range of time scales, one can reasonably expect that variations in all of these parameters will be responsible for some of the variations in the oceans. However, with few exceptions there is a mismatch between the characteristic time and space scales in the ocean and those in the atmosphere, so that the atmospheric influence is usually not particularly direct.

Some obvious comparisons between the ocean and the atmosphere are worth examining, so that attention can be drawn to the appropriate conclusions:

1. <u>Mass</u> : The mass of the atmosphere is equivalent to about 10 meters of water, distributed over the whole surface of the earth. The mass of the ocean is thus some 300 times that of the atmosphere.

2. <u>Density</u> : The density of water is approximately 1,000 times that of air.

3. <u>Speed</u> : Atmospheric motions characteristically have speeds of about 10 m s^{-1}, with extremes in jets and storms of nearly 100 m s^{-1}. Oceanic motions are characteristically about 0.1 m s^{-1}, with extremes of 1 m s^{-1}. The speed ratio is therefore about 100 to 1.

4. <u>Thermal Character</u> : The specific heat of air is about a quarter that of water. Therefore the thermal capacity of the ocean is more than 1,000 times that of the atmosphere, or in different terms, the top 3 meters of the ocean have the thermal capacity of the entire atmosphere.

From these elementary facts a few conclusions which are equally elementary, but which are very important, can be drawn:

<u>Kinetic energy</u> ($\frac{1}{2}mv^2$): With a ratio of 1/300 in mass but of 100 in speed, the atmosphere has about 30 times as much kinetic energy as does the ocean.

<u>Momentum</u> (mv): The same ratios give the magnitude of momentum in the atmosphere to be about one-third that of the ocean.

<u>Available thermal energy</u>: The extreme range of temperature in the atmosphere is about $2\frac{1}{2}$ times that of the ocean. Superficially, with the numbers given above, it would appear that the available thermal energy in the ocean (in the sense of energy usable to drive a heat engine) would be about 400 times that of the atmosphere. However, in fact, most of the ocean is cold so that the actual ratio is more like 10 : 1. Nevertheless, the ratio of available thermal energy to kinetic energy in the ocean is several hundred

times the corresponding figure in the atmosphere.

When we examine these figures, we see that although it is commonplace to say that the ocean is largely governed by the atmosphere, to drive the ocean is by no means easy for the atmosphere. The only thing which the atmosphere has a lot of, relative to the ocean, is kinetic energy. However, it is exceedingly difficult for the atmosphere to transmit kinetic energy to the ocean. The energy lost by the atmosphere is given by τv_a, where τ is the surface stress between the atmosphere and the ocean, and v_a is the velocity of the atmosphere. The energy received by the ocean is τv_w, where v_w is the velocity of the water. Since the magnitude of v_a is of the order of 100 times that of v_w, the energy transfer is evidently very inefficient. Only in the case of wave generation, where the appropriate v_w is the phase velocity of the waves, is the transfer efficient. In this latter case, the inefficiency is in getting energy from the waves into some sort of mean velocity in the water.

Momentum is, of course, conserved in the transfer, and of all the mechanical influences of the atmosphere on the ocean, it is momentum transfer which is most often discussed and which is, undoubtedly, the most important.

Stress on the ocean induces transport in the ocean. For some important problems, notably that of the redistribution of heat in the upper ocean, the distribution of this transport with depth is very important. However, in this paper, we will treat only the transport without considering its distribution.

For short times, i.e. short compared with $1/f$, where f is the Coriolis parameter, the transport induced per unit area is given by $\tau t/\rho$, where ρ is the density and t is the duration of the stress. The transport is initially in the direction of the stress. However, if t is much larger than $1/f$, the effect of the rotation of the earth becomes dominant and the transport is at right-angles to the stress, with magnitude $\tau/\rho f$

In most cases, simple horizontal transport does not accomplish much, since the horizontal gradients of properties tend to be rather small. Usually, far more important are induced vertical motions which arise when the horizontal transport is non-uniform, resulting in convergences or divergences of the horizontal motion which must be compensated for by vertical flows. If L is the scale of the fluctuation in stress, the vertical <u>displacement</u> resulting will be

$$\tau t^2/\rho L$$

for times short compared with $1/f$, and $\tau t/\rho f L$

for times long compared with $1/f$.

It is worth examining the magnitude of these expressions in some typical situations. A fairly large value for τ/ρ is about $10^{-3} m^2 s^{-2}$. (A typical value for τ/ρ is about one-tenth that number). Observations in macroscopically uniform wind-fields over water indicate that fluctuations with scales of a few kilometres and duration of tens of minutes are fairly common. Such values produce vertical displacements of the order of 1 meter or less, using the expression $\tau t^2/\rho L$

Large scale meteorological disturbances with a space scale of 1,000 km, and a time scale of about a day will yield about the same vertical displacement with the same value of stress.

A 1-meter vertical displacement would usually be lost in the "noise" of oceanographic observations.

Therefore under "ordinary" conditions, fluctuations in wind stress do not produce dramatic results in the ocean. We must look to "extraordinary" conditions.

Such occur. For example, t might be very long. A particular unusual stress pattern might persist for as long as 10^7s, as did the unusual hot dry weather in Europe in 1976. Although in this case the anomaly in τ/ρ is unlikely to be anything like so high as $10^{-3} m^2 s^{-2}$, nevertheless anomalous vertical displacements of 10m or more might occur. Such displacements would have important effects both dynamically and thermodynamically in the ocean.

In very intense, compact storms like hurricanes, τ/ρ can be ten times larger and L ten times smaller than that assumed in the above example, so that displacements of 100 meters can result. Such displacements, being upwards, can completely eliminate the upper mixed layer of the ocean.

Near the equator, (f) is anomalously small. Further, L may in this case more accurately represent the scale of the variation of f rather than the scale of the variation of τ. Near the equator, then, quite ordinary fluctuations in τ produce substantial anomalous vertical displacements if they persist for more than a few days.

Another circumstance which produces a small value of L is the presence of a coast. It is becoming commonplace that variations in wind stress have a dramatic effect upon coastal upwelling - a phenomenon which is increasingly being recognized as being very variable both in time and in space.

Thus we see that except for phenomena of unusual intensity or unusual duration, we should look for the immediate effects of fluctuations in wind stress to be revealed in equatorial regions and close to coastlines.

Let us now look at the effect fluctuations in atmospheric pressure referred to in (e) above. It must be recognized that even the fluctuations of the pressure on the surface are enormous compared with the wind stress. A perfectly ordinary variation in pressure amounts to 1 kilopascal (10 millibars). To compare this with the wind stress, we should put it into kinematic terms, by dividing by the density. One kilopascal is thus equivalent to 1 $m^2 s^{-2}$ in water, which should be compared with the $10^{-3} m^2 s^{-2}$ used above for a large wind stress.

The transport corresponding to such a pressure fluctuation, assuming it to be geostrophic, is given by

$$PD / \rho f L$$

where P is the pressure fluctuation, D is the depth over which the pressure fluctuation is effective and L is once more the horizontal scale of the fluctuation. If this is to be compared with the Ekman transport $\tau / \rho f$ it can be seen that the relative importance of the two is given by

$$PD / \tau L$$

For ordinary values of pressure fluctuation and of stress, P/τ is about 10^4. Typically L is about 10^6 m. Measurements taken at the top of a seamount by members of the Institute of Ocean Sciences at Patricia Bay, Canada, revealed pressure fluctuations of the order of 1 kilopascal at a depth of 300 metres. Other observers have seen apparently-meteorologically-related pressure fluctuations occurring at much greater depths. It would therefore appear that transports induced by meteorological pressure fluctuations may be much larger than the Ekman transports. It should be noted that the transports are large because they are distributed over a very large depth. Typically the speeds are substantially lower than those associated with Ekman transport.

Another way of expressing the effects of pressure fluctuation is to note that the integrated transport (to compare with the integrated transport in a major ocean current like the Gulf Stream) is given by

$$PD / \rho f$$

For a pressure fluctuation of 1 kilopascal and a depth of 1 kilometre in mid latitudes this yields $10^7 m^3 s^{-1}$, or 10 Sverdrups. Such a value is far from negligible compared with the largest ocean currents in the world.

Knowledge of these meteorologically induced pressure fluctuations in the ocean has been too newly acquired for its full significance to have been explored by oceanographers. It may yet turn out that despite the very large transports induced, their ephemeral nature together with the small velocities involved may mean that they will not prove terribly important. However the possibility that these transports may interact with bottom topography to produce important effects cannot be neglected, nor can the possibility that there is some asymmetry in the oceanic response so that fluctuating atmospheric pressure produces a net average ocean pressure. It is clear that this phenomenon warrants considerable further exploration both observationally and theoretically.

This paper has been written with a rather negative tone, indicating that the atmosphere drives the ocean only with difficulty because of the mismatch between the ocean and the atmosphere.

The clear exception to the mismatch is the upper layer of the ocean. In this region, of the order of 100 metres deep, the influence of the atmosphere is very direct and, since it contains only a few per cent of the total mass and of the total heat capacity of the ocean, it is able to respond relatively promptly. A great deal of work has been carried out in recent years on getting an understanding of the nature of this response sufficiently detailed that one could reasonably hope to predict upper-layer behaviour, particularly when the processes are not strongly influenced by lateral advection.

Recent efforts take into account atmospheric cloudiness, wind mixing, momentum transfer, the exchange of heat and water vapour with the atmosphere, and the effect of the density structure of the water on the evolution of the upper layer. While it cannot be said that the work is finished, very substantial progress has been made in the last half-dozen years. The subject is dealt with elsewhere in this assembly and I will not expand upon it here.

Apart from the large annual cycle, the major part of meteorological variation occurs with time scales of a few days. As we have seen, variations on this time scale are unable to change the ocean much in the same period of time because of the serious mismatch. However, this is a different thing from saying that variations in the atmosphere do not influence the ocean. For longer periods, the ocean is affected by the sum of all of the meteorological events occurring within those periods. The behaviour of the ocean, then, is that of an integrator of the meteorological phenomena.

It is important to recognize that integration, although it does produce a sort of averaging, does not eliminate fluctuation. A series of events, each of which lasts only a few days, will nevertheless statistically yield variations with periods of months and years. It may very well be that most of the long-period variation is observed in the ocean is of this nature. The more or less random fluctuations of the weather, on a scale of days, do not quite cancel out. They produce equally random fluctuation in the ocean with much longer periods.

Of course, the interaction between the ocean and the atmosphere is not all one way. The effect of the ocean on the atmosphere is well documented if only in such everyday phenomena as the influence of oceanic heat on the climate of Western Europe.

If random fluctuations of the atmosphere generate random but slow fluctuations of the ocean, the ocean in influencing the atmosphere will operate in non-random fashion. That is, an anomaly in the ocean will systematically influence the atmosphere for as long as it persists. It is probable that this feedback mechanism, with the ocean randomly driven by the atmosphere but the atmosphere systematically driven by the ocean, plays an important role in the evolution of climate.

However, I should like to bring attention back to a phenomenon referred to above which offers the possibility of random influences of the ocean on the atmosphere. It was noted above that the available thermal energy in the ocean is substantially greater than that in the atmosphere, particularly when compared with the kinetic energy. In both the atmosphere and the ocean this available thermal energy can be extracted by baroclinic instability to deliver kinetic energy. Although the process is very inefficient, nevertheless an important contribution is made to the kinetic energy of both fluids - perhaps even more to the ocean than to the atmosphere.

There is almost always a large degree of indeterminacy in the detailed behaviour of any instability. It is likely that the baroclinic instability in the ocean, in transforming available thermal energy into kinetic energy (no matter how inefficiently) will produce largely unpredictable events in the ocean. Unfortunately, then it seems probable that the ocean cannot be relied upon to be an essentially predictable "fly-wheel" in the climate machine. To some degree, it probably has a mind of its own!

VARIABILITY OF ECOSYSTEMS

N.M. VORONINA

SHIRSHOV INSTITUTE OF OCEANOLOGY

ACADEMY OF SCIENCES OF THE USSR, MOSCOW

Investigations of the last decades have shown the picture of the distribution of oceanic flora and fauna and made it possible to recognise large-scale communities which, together with the water masses they inhabit, constitute single ecosystems.

The most complete scheme of pelagical communities was given by C.V. Beklemishev (1969), who described in the open Ocean 2 cold-water glacial-neretic, 3 subpolar, 5 central and 3 equatorial communities which occupy large-scale gyrals carrying waters of definite structure. The communities of ecotones (fig. 1). The divisions given by other authors are generally similar to those of Beklemishev, differing only in details.

The question arises: How homogeneous is an ecosystem? To what degree is the composition of species constant? How constant are the quantitative relationships between the components and so their vertical distribution and general abundance of organisms? Let us consider the possible causes of variations. Most of them are associated with the essential property of biotope of pelagical ecosystem - its mobility.

I. Variability along boundaries

A constant water exchange exists between adjacent gyrals, resulting in an exchange of their inhabitants. The latter depends entirely on the hydrographic situation. It has been demonstrated for the Southern Ocean, that at a strongly expressed convergent water movement along the Polar front, antarctic and subantarctic species are completely separated. Not a single organism can

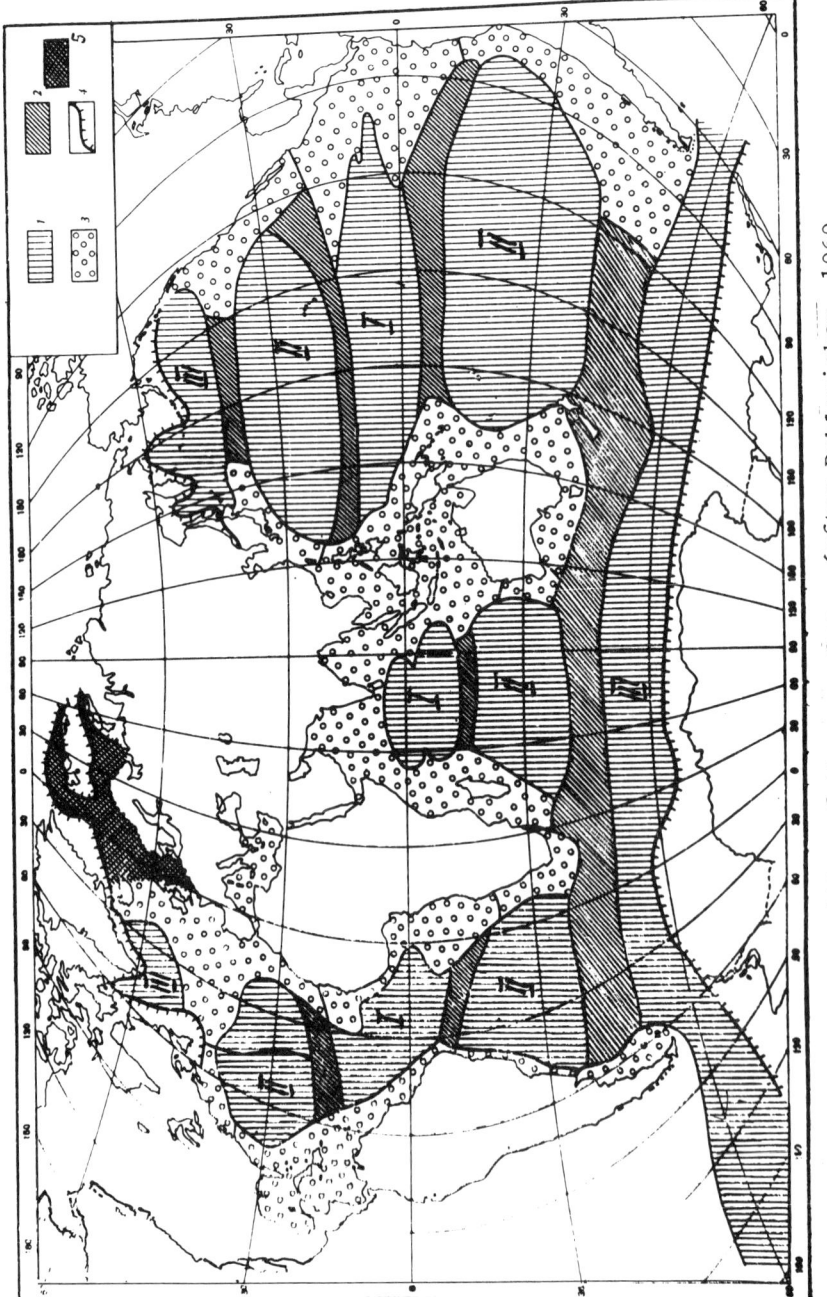

1. Pelagic communities of the open Ocean (after Beklemishev, 1969, with modifications). 1 – primary communities: I Equatorial, II Central, III Subpolar; 2 – secondary Transitional communities; 3 – gracial neritic communities; 4 – distant neritic communities; 5 – North Atlantic expatriation area.

penetrate into an "alien" water. But when the meridional circulation changes and water is carried across the front, a zone of mixed fauna is formed whose width may reach several hundred miles (Voronina, 1962).

Species that perform seasonal migrations are carried in winter with deep water. They may ascend in spring in an alien region. In this case the distance of expatriation depends on the rate of the meridional component of the deep water flow, which, as a rule, is not high, but may sometimes be considerable (Voronina, 1975). Thus the penetration of alien elements beyond biogeographical boundaries does not always indicate a mixing of waters.

The role of expatriants in the composition of tropical ecosystems of the Pacific ocean has been evaluated by Heinrich (1975). She showed that they spread within a considerable part of "alien" biotopes and contribute to the formation in each gyral of at least four areas (northern, southern, eastern and western) which differ in species composition and may be considered as dependent ecosystems of a lower rank (fig. 2). The immigrants account for 20% of the total number of species in the northern central community, and for 24% in the equatorial community. Thus some areas of adjacent ecosystems may have more similarity in their species composition, than some areas of one and the same ecosystem.

2. Variability associated with difference maturity

The upper layer is occupied by water of different age. In tropical upwellings the ascending water is "young", rich in nutrients, but poor in flora and fauna. In it a new community begins to develop. According to model computations, a maximum of phytoplankton is reached on the 15th day. Somewhat later, on the 20th - 30th days, a maximum of small zooplankton and after 10 more days, a maximum of large filter-feeders and predators is reached (Vinogradov et al., 1972). The developing community shifts with the current, and, as a result of asynchronic appearance of maxima of its constituents, the latter become disconnected in space. The more protracted the development of an individual, or the more remote from the producers is the place of a given group in the food chain, the further will the region of its abundance shift from that of the phytoplankton. Along the equator, in an east west direction, peaks of phyto-, zoo- and macroplankton alternately succeed each other (Voronina, 1964). Simultaneously, the diversity of plankton increases, its general abundance diminishes and predators become more numerous (Gueredrat et al., 1972). This complex of changes is evidence of the maturing of the community (Margaleff, 1963). Similar differences in maturity have been described for some regions of the Indian Ocean, differing in the character of vertical water movements (Timonin, 1972).

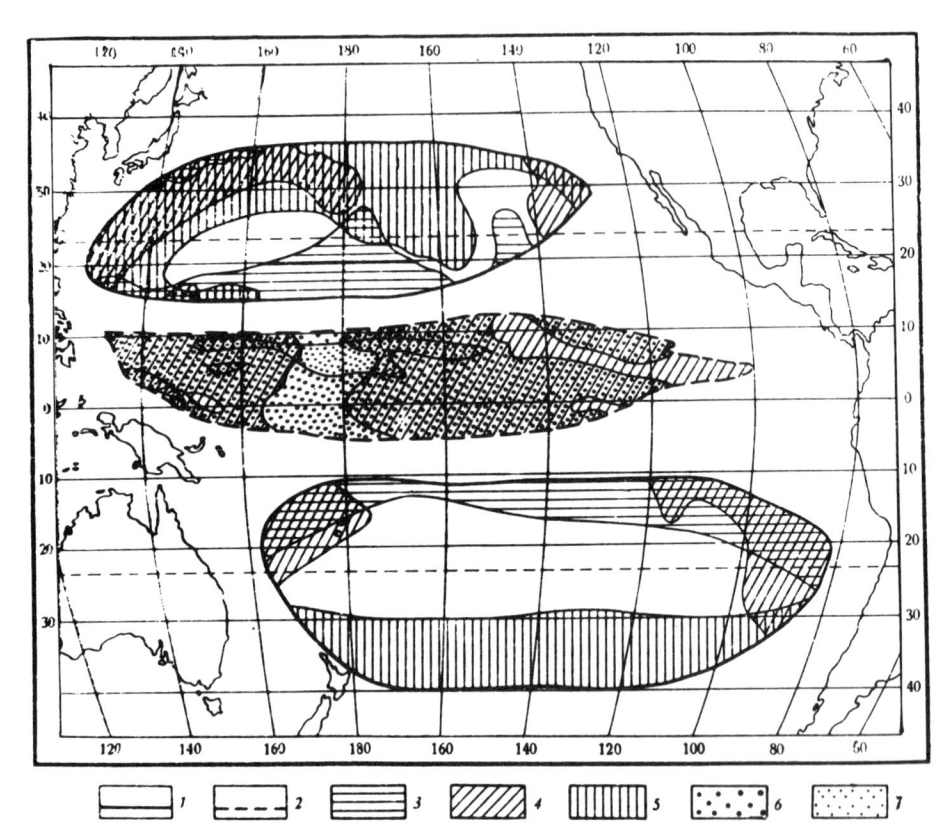

2. Distribution of the expatriation areas (after A.K. Heinrich, 1975). The boundaries of the communities: 1 - Central, 2 - Equatorial; areas of expatriation of species: 3 - Equatorial, 4 - distant neritic, 5 - transitional, 6 - central, 7 - areas of expatriation of both north and south central species.

3. Variability associated with local peculiarities of biotopes

The general abundance of plankton or its diverse constituents may be influenced by certain local peculiarities of the biotopes. Thus in the convergence zones where the water brought by meridional components of currents sinks, the main mass of organisms, inhabiting it remains within their inherent range of depths. This results in a constant enrichment of these zones with allochthonous material (King and Nida, 1957; Foxton, 1956). The rate of its daily increase in abundance, computed for an intercrossing of the Antarctic convergence, was found to amount to 3% of the total biomass (Voronina, 1968). Accordingly, on the map of biomass distribution, zones of convergences appear as narrow bands of rich waters.

Hydrobionts may be influenced by the character of vertical temperature distribution. Sharp gradients restrict the vertical movement in some species, preventing their penetration to the surface zone (Harder, 1968; Banse, 1959). Where such an obstacle comes in the way of seasonal migrants, areas may be formed with a greatly disturbed equilibrium between phytoplankton production and its consumption.

Some hydrographic features may have a selective effect on the populations. This may be illustrated on the example of _Euphausia superba_. In the Antarctic waters the reproduction of this abundant herbivorous species is confined to a relatively restricted area, mainly to the water circulation of the Weddell Sea. Here the relatively high upper boundary of dense bottom water impedes the sinking of the developing eggs of this species. In other regions, where this boundary or the bottom itself is in far greater depths (more than 1800 m), the hatched larvae are unable to cover the great distance separating them from the upper trophic layer (Voronina, 1974) and they perish. In such areas copepods dominate in plankton. So the position of the upper boundary of the bottom water may have an unexpected biological effect: it influences the quantitative relationships between the main inhabitants of the surface layer.

4. Variability, associated with seasonal character of production

The most important circumstance, determining the character of variability within ecosystems, is, I think, the presence or absence in their biotopes of seasonal changes which determine the development of primary production in time, namely its discontinuity or continuity. Accordingly, all the ecosystems can be divided into two types which differ in principle, show different trends of adaptations in animals, and have different annual cycles. These cycles are to a great extent responsible for spatial dissimilarities in the abundance of separate elements of communities in relatively homogeneous areas of the biotopes.

I shall attempt to characterize the peculiarities of these two types of ecosystems, using typical examples, restricting myself to the discussion of representatives of the better known lower levels of the trophic chain. Only the upper layers of the pelagial will be considered: horizons of photosynthesis plus the depths to which descend the direct consumers of phytoplankton.

Ecosystems with continuous primary production

In the open waters of tropical latitudes illumination is always sufficient in the surface layers, the critical depth is far greater than the depth of the mixed layer and the stratification is relatively stable. Therefore the process of primary production is uninterrupted. Fluctuations in its intensity depend mostly on the supply of nutrients and are rarely more than tenfold. If the existence of such biotopes is lasting, the possibility arises for the divergence of ecological niches. The specialization of animals is manifested, first of all, in their quantitative distribution. Sampling at narrow horizons (10 m) in the surface layer has revealed an extreme diversity of spatial associations of the maxima of different populations (Timonin, Voronina, 1975). According to our data in the equatorial Pacific seven diurnal patterns combined into three types can be recognized in the inhabitants of the upper 200 m layer (fig. 3).

I. Maximum concentrations occur within the upper mixed layer. In some species, as a rule, they are confined to the narrow surface horizon (of 0-10 m) (Acartia negligens, Clausocalanus mastigophorus). In others species the maximum is in its intermediate depths (10-40, less frequently 10-60 m) (Undinula darwini, Eucalanus attenuatus, Pterosagitta draco). In still others species the position of maximum changes irregularly within the mixed layer (Clausocalanus furcatus, Euchaeta marina, Sagitta regulatis etc.)

II. The maxima are very distinctly associated with the layer of maximum temperature gradients sinking with it to greater depths from east to west (Nannocalanus minor, Euchaeta longicornis).

III. The maximum concentrations are not confined to a definite water mass and may shift from the upper mixed layer to the thermocline and farther downward. The maxima of some species of this type are relatively constant in regard to the occupied depth (Eucalanus subtenuis, Acartia danae), while others may shift within a rather wide scale of depths (Neocalanus gracilis, Paracalanus parvus, Clausocalanus farrani, C. jobei, Sagitta enflata).

Furthermore, the diel changes in vertical distribution may greatly differ in different species and even age groups. From this point of view five patterns may be recognised for the

VARIABILITY OF ECOSYSTEMS

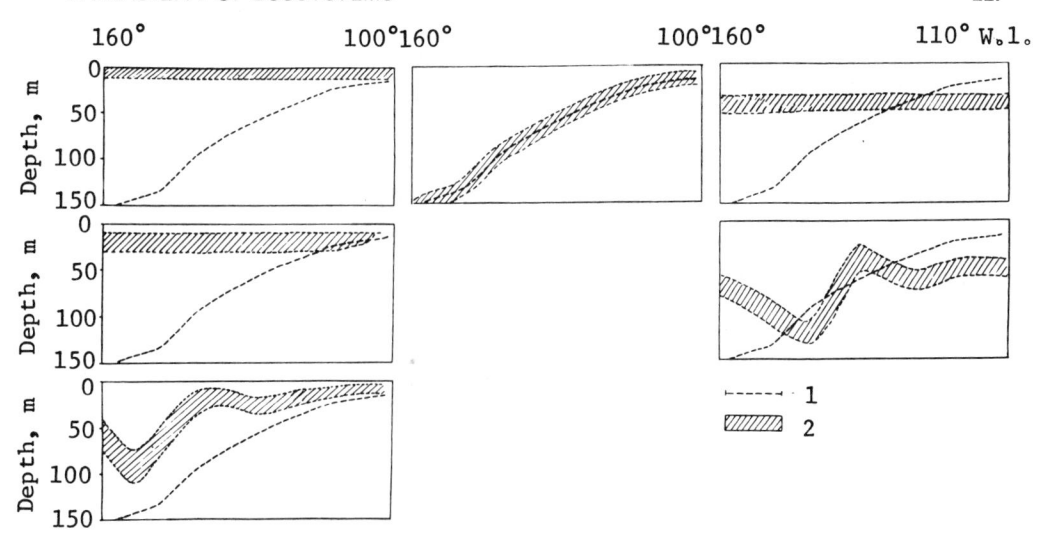

3. Types of spatial distribution of the layer of maximum concentrations in different species of tropical zooplankton according to materials from the Eastern Equatorial Pacific (based on A.G. Timonin and N.M. Voronina, 1975). 1 - layer of maximum temperature gradient, 2 - layer of maximum abundance. (See text).

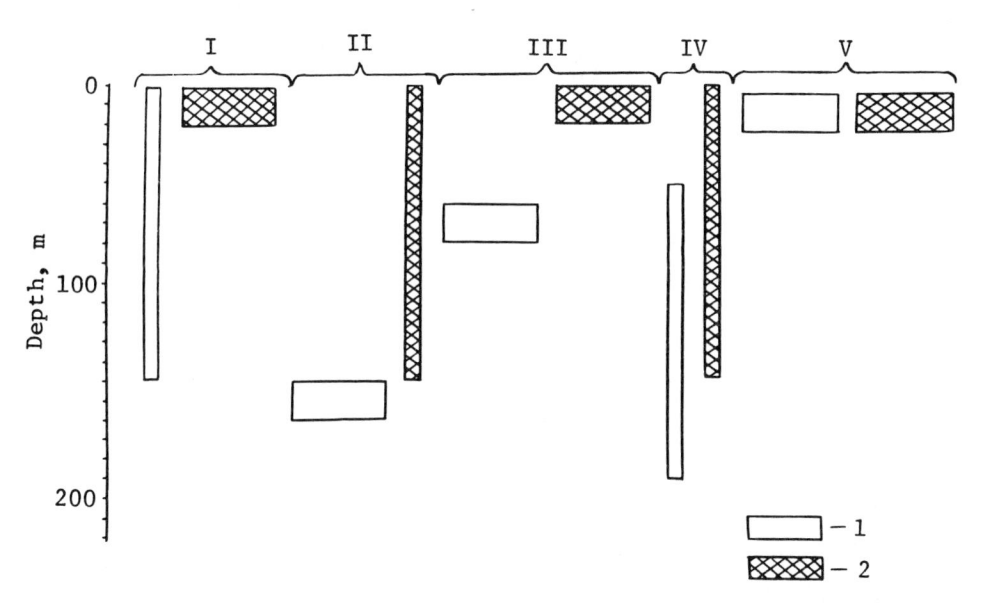

4. Types of diel variations in the distribution of different plankton species. Width of the rectangle is proportionale to the relative concentration of organisms (based on A.G. Timonin, 1975). 1 - day-time, 2 - night-time.

mesoplankton of the surface zone (Timonin, 1975) (fig. 4).

1. The populations spread in the daytime and concentrate in rather narrow layers in the night time.

2. The populations are concentrated during the day and spread during the night.

3. The populations form dense concentrations in deep layers in the daytime and in the upper layers in the night time.

4. The populations are dispersed within a relatively wide range of depths during the entire 24-hour period.

5. The populations are always confined to narrow surface layers.

In addition, representatives of interzonal plankton which in the daytime stays in great depths, ascend every night to the surface zone.

All this shows how very diverse are the means by which the populations of different species utilize the space of the upper trophic layer.

Contrary to what might be expected there is no high specialization in sources of food. The zooplankters are either predators or particle-grazers (Mullin, 1966): the latter, comprising the main mass of pelagic animals, have a very wide spectrum of food items including phytoplankton, small zooplankton detritus and bacteria (Paffenhoffer and Strikland, 1970; Pavlova et al., 1971, Poulet, 1976).

Actually, however, substantial differences may be observed in their diets. The size spectrum of filtered particles depends on the density of the filtering fringes and may differ in related species and even in their stages (Mullin and Brooks, 1967; Menach, 1974; Nival and Nival, 1976). Besides filtering the species can actively grasp the available food particles (Mullin, 1963; Gould, 1966; Richman and Rogers, 1969). This method is selective, its selectivity increases with increasing of food abundance (Perueva and Vilenkin, 1970). There are also some indications that filtration of even small particles is not passive, but may be modified depending on their concentrations. Filter-feeders are capable of shifting the grazing pressure from one size to another (Poulet, 1974). On the whole the feeding habits seem to be highly opportunistic and allow for the utilization of every kind of available food (Poulet, 1974; Piontovsky and Petipa, 1975). Therefore the ratio of items in the diet of a given species is subject to wide variations in space and in time, depending on the available resources, their abundance and the size scale (Petipa et at., 1975; Poulet, 1976).

How are these resources distributed? It is known that in the tropics there are layers in which phytoplankton and bacteria occur in concentrations dozens of times greater than in the immediately adjacent water (Vinogradov et al., 1970). From one to four enriched layers are recognized, in which the relative and absolute quantities vary from place to place. The frequency of occurrence of maxima at definite horizons differs in different areas (Semina, 1974). The ratios of definite components too are very unstable both horizontally and vertically. For example in Fig. 5 data are presented from some stations in the eastern equatorial Pacific, the same ones from which were obtained the data on zooplankton considered above.

Considering all these data, as well as great diversity of means by which animals of the lowest trophic level exploit the space of the productive layer, and allowing for possible differences in diet feeding rhythms, nutritive value of algae (Corner and Cowey, 1968) and rates of their assimilation, it is easy to understand that very dissimilar feeding conditions may be encountered by different filter-feeders at one and the same place. It seems that it is precisely the difference in the diet patterns of quantitative vertical distribution that determines the diversity of feeding habits of populations. The significance of this diversity for tropical plankton community has been pointed out by Mullin (1967). These partial divergences in the utilization of space and of food resources by the inhabitants of the same biotope makes possible the coexistence of a great number of species in the tropical zooplankton. The extensive overlapping of niches prevents the dominance of one or several species.

Owing to the constant availability of food in the tropics the reproduction of its consumers is continuous. That is why there is no predominance of ontogenetic stages in their populations and no great fluctuations in grazing pressure, such as are observed in temperate waters (Heinrich, 1963).

Not everywhere, however, are the food requirements of the species equally well satisfied (Vinogradov et al., 1975). Animals that do not store fat reserves live in a quasi-stationary equilibrium with their environment: under adverse conditions they perish, under favourable ones they grow rapidly and bring numerous offspring (Conover, 1968).

Owing to the complication of the vertical distribution of the community, any excess of food is always readily consumed at all layers. This combination of circumstances tends to maintain a stable relationship between primary production and its consumption, i.e. between the quantities of phyto- and zooplankton which is characteristic for tropical waters (Heinrich, 1971). As to the

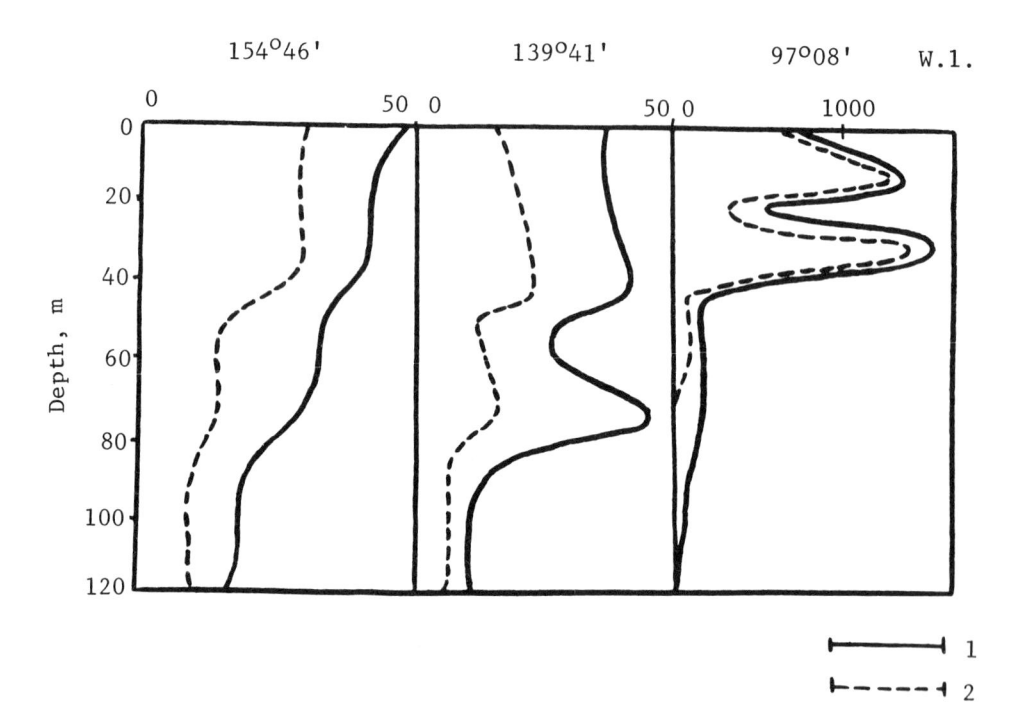

5. Vertical distribution of phytoplankton (mg/m^3) at three equatorial stations, 1 - total biomass, 2 - biomass of nannoplankton (after Yu. I. Sorokin, I.N. Sukhanova, G.V. Konovalova, E.B. Pavelyeva, 1975).

mesoscale variations in the relationship of separate species, they seem to depend on their local success or failure, that is on whether the layer most intensely exploited by the population during the breeding period meets rich or poor food supply. On these variations, in turn, depends the variability of distribution of the entire mass of zooplankton in adjacent areas (Timonin and Voronina, 1975).

Ecosystems with interrupted production

I shall take as an example the pelagial of the Antarctic where in the immense expanses of the open ocean uncomplicated by coastal influences the laws governing temporal processes, are most distinctly expressed.

The main feature of its biotope, like that of other biotopes in the temperate and polar latitudes, consists in sharp seasonal changes caused by the variations of solar radiation. The latter is responsible on the one hand, for the amount of light penetrating into the water and on the other hand for the processes of heat exchange which influence thermal and density stratification. Let us consider how these processes affect the general aspect of pelagic communities.

Light does not limit phytoplankton development at the surface from September to April in latitude 60°S, and from August to March in latitude 70°S. The concentrations of nutrients are high during the whole year (Hart, 1934; Hardy, Gunther, 1935; Clowes, 1938). In the final analysis the beginning of intensive phytoplankton growth depends on the establishment of the water structure (Gran, 1931), namely on the formation of the pycnocline which limits the vertical extension of the mixed layer to a less than critical depth. The formation of summer stratification depends on the melting of ice and the warming of water. In the northern Antarctic it appears in mid-October (Currie, 1954) and increasingly later the farther south. Accordingly, in the northern latitudes vernal phytoplankton bloom begins in October and reaches its maximum in December. These phenomena are delayed by one month in middle and by two months in southern latitudes (Hart, 1942).

The seasonal variations of phytoplankton distribution in the Antarctic consist generally in the formation of a vernal circumpolar belt of bloom in the northern latitudes and its gradual shift in a southerly direction. It takes about three months to cover the distance between the convergence and the continent. However substantial deviations from this scheme are possible, in view of local and year to year variations of glaciological and meteorological conditions, disturbing the latitudinal successiveness of biological processes. Seasonal fluctuations in abundance are of a very high amplitude: in extreme situations the concentrations of alga may vary by five orders of magnitude.

During the season firstly a regular succession is observed of dominant groups of phytoplankton according to their size and taxonomy (Hart, 1942; Steyaert, 1974) and, secondly, the surface maximum of abundance shifts to greater depths.

Such are the basic features of phytoplankton development. They determine the character of the cycles of its consumers. It is evident that only those species will prosper that are able to adjust themselves to using short-term periods of food abundance. Such are the copepods and the euphausiids. All representatives of the former and some species of the latter group are interzonal. During the long winter they stay away from the surface. Their sinking and spreading in depths from 250 to 1500 m has a double advantage: firstly it prevents a substantial part of the population from being carried away by meridional currents to the margins of their range or even beyond the boundary of their breeding zone, and secondly, it makes them less exposed to predation. This precisely is the biological purport of seasonal migrations. But they have also their drawbacks. We have already shown that the beginning of intensive phytoplankton development is subject to great variations and is rather unpredictable. This means that the animals ascending in spring to the surface may encounter in different places widely differing trophic conditions. Therefore they are unable to adjust themselves to a definite stage of the phytoplankton cycle. None of the species are specialized: firstly they do not differ in their diet (Voronina and Sukhanova, 1976) and, secondly, they are not spatially confined during the feeding period. They change their place and use the food reserves within the entire trophic layer irrespective of its hydrological partitions.

Most abundant in the Antarctic is the group of interzonal herbivorous copepods, which accounts for about 70% of the total biomass of mesoplankton. It is represented by three main species: Calanoides acutus, Calanus propinquus and Rhincalanus gigas. The cycles of these crustaceans follow practically the same course and are accompanied by the same reorganizations of the spatial structure of the populations. This process is shown in Fig. 6 where the vertical structure of the population is designated by lines, connecting the 50% levels of successive ontogenetic stages.

During the biological spring the overwintered individuals, represented in the main by older copepodite stages, concentrate within the narrow near-surface layer. Here they feed intensely, rapidly reach maturity and spawn. After hatching the populations reach their maximum numerical abundance. At first the young are confined to the narrow surface borizon, but with progressing development the older individuals sink to greater depth and the new generation gradually disperses. For a certain time the population occupies the whole euphotic layer, then the animals that have

VARIABILITY OF ECOSYSTEMS   233

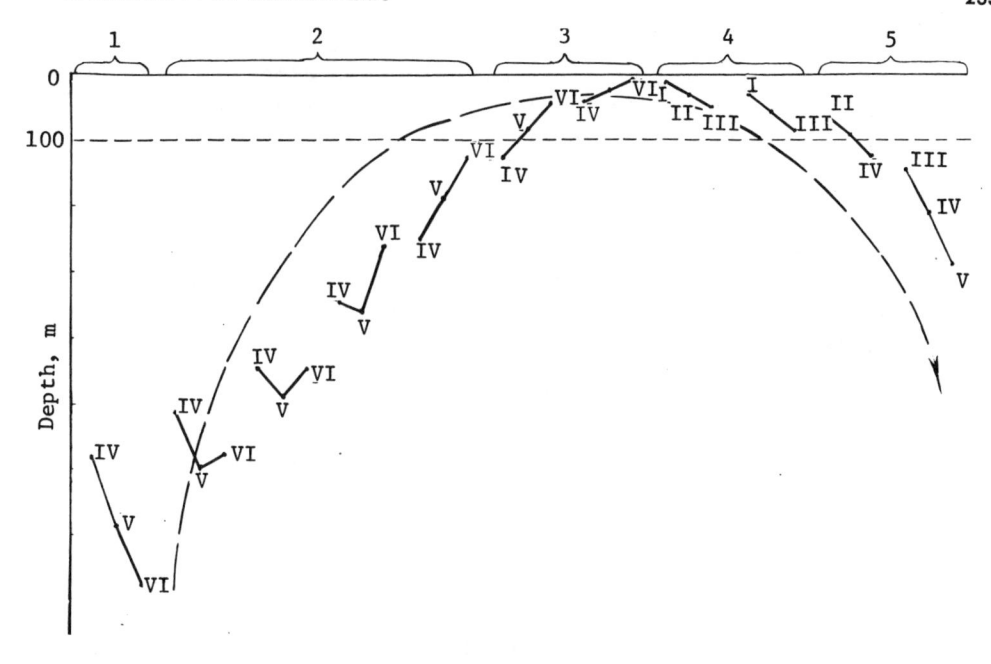

6. Seasonal changes in the vertical structure of populations of interzonal copepods (see text) (after N.M. Vononina, 1975).

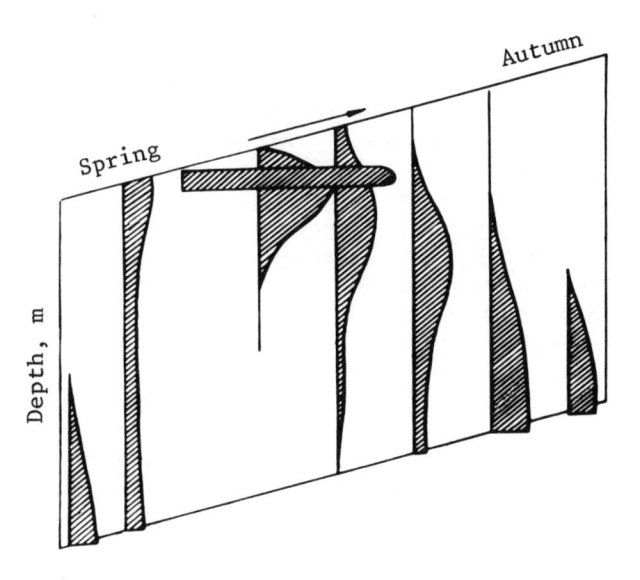

7. Seasonal changes in the vertical distribution of abundance of an interzonal species (after N.M. Voronina, 1972 a).

stored enough fat shift beyond its boundary. A distinct sequence is preserved in the vertical distribution of the cores of different stages: the younger are always positioned above the older ones. This order is also preserved in winter. (Over the greater part of the aquatory the wintering stock is represented by the copepodite stages IV, V and VI). During the vernal upward migration older individuals move faster, overtaking first the V and then the IV copepodites (Voronina, 1975). The migration is performed by a widely spread population but terminates in a dense concentration of all its animals at the surface. All copepods undergo this pattern of seasonal changes of the vertical structure of populations.

The pattern of vertical distribution of the total abundance of an interzonal species depends on the distribution of its dominant ages. Its seasonal changes are shown in Fig. 7.

Similar seasonal reconstructions of the vertical structure of populations are inherent also in euphausiids and some predatory plankters - copepods and chaetognaths.

Since individual body weight greatly increases with age the position of maximum biomass of an interzonal species within the surface layer depends in a greater measure on the age composition of its population than on its numerical abundance. In particular, during the summer period, it is associated with the dominance of copepodite stage IV. Later, although the weight of the animals is still increasing, they disperse in the water column, and no high concentrations are formed.

A very important feature, characteristic of the Antarctic plankton community, is the asynchronism of similar stages in the annual cycle of related species. This may be demonstrated by comparing the age composition of copepod populations at any stations chosen at random: everywhere Calanoides acutus occurs in a more advanced state than Calanus propinquus, while Rhincalanus gigas lags behind both of them (Fig. 8) (Voronina, 1966). This means that these populations do not coincide in time of ascent to the surface, reproduction and return to the depths. Asynchronism is also responsible for the divergence in numerical abundance and sequence of residence in the trophic layer i.e. for the seasonal variations in the vertical structure of the community. In Fig. 9 are schematically shown the seasonal changes of situation of the 50% levels of populations of these 3 Copepod species. Apparently different stages of the annual cycle differ in the sequence of cores of definite species along the vertical.

The cycles of zooplankters, like those of algae are characterized by the asynchronism of identical stages in different latitudinal zones. Ascent, reproduction and maxima of biomass occur earlier

VARIABILITY OF ECOSYSTEMS 235

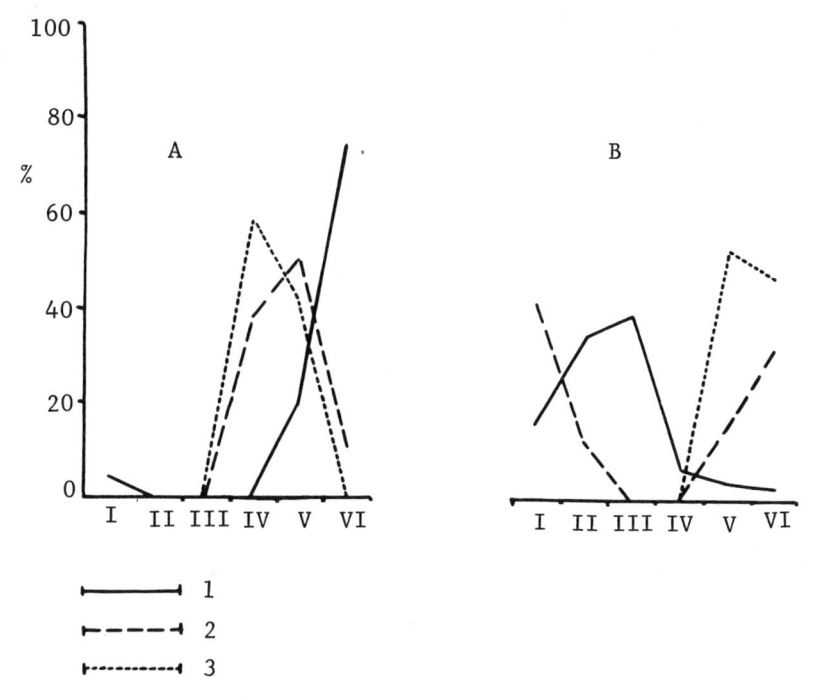

8. Age composition of copepod species at different stations:
   (1) Calanoides acutus, (2) Calanus propinquus, (3) Rhincalanus gigas.

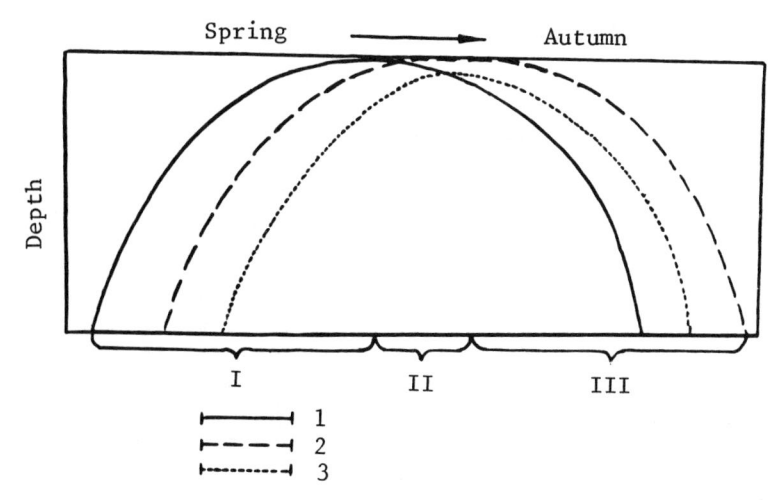

9. Seasonal variations in the vertical structure of the community (after N.M. Voronina, 1972 b). Simbols the same as in Fig. 8 indicate the position of 50% levels of different populations.

in northern latitudes and increasingly later the farther south (Voronina, 1969, 1975). Deviations from this successiveness are determined by hydrographic conditions on which the beginning of intensive bloom depends (Voronina, 1975).

The differences in the initiation of identical phases of the cycle at the convergence and near the continent as in phytoplankton, may reach up to three months. As a result of this asynchronism successive changes in age structure and abundance of populations may be observed along the meridians, similar to the temporal changes. Since the maxima of biomass of separate species alternate at any given point, they diverge along the meridian both in the horizontal and the vertical plane. On Fig. 10 the maxima of different species are shown by different shading for 3 meridional sections in the second part of the season. We see, that the summer maximum of C. acutus is always situated farther south and in greater depths than the maximum of C. propinquus while the maximum of R. gigas always occurs northward of both of them. So, there is little overlapping of zones of high biomass in species of similar biogeographical nature.

The cycles of separate populations determine, in their totality, the seasonal variations in the total quantity of plankton. For their description I used the data of the "Discovery" (Foxton, 1956). The Antarctic was divided into three latitudinal zones and data on zooplankton from 330 stations were averaged by months (Voronina, 1971). This method was successfully applied by Hart (1942) to data of plant pigments, but I could not make use of the zones recognized by Hart, because they were too fractioned for the zoological material available. All the data on pigments had to be recalculated in conformity with the new zones, but this did not alter the character of the curves.

The results obtained are shown in Fig. II. We see that there are two maxima of zooplankton during the year in the upper 100 m layer. The first, lesser maximum is associated with the vernal ascent of organisms to the euphotic layer, the second with the development of new generations. In the two northern zones the summer peak of zooplankton appears two months after the phytoplankton maximum. The cause of this delay has been repeatedly considered on the example of other communities (Cushing, 1959; Heinrich, 1971). As a matter of fact, effective consumption of phytoplankton begins only after the reproduction and growing up of new generations of zooplankters. Then a balance is reached which soon gives place to overexploitation - the basic cause of seasonal reduction of phytoplankton abundance. With decreasing grazing pressure a small autumn maximum is formed. This asynchronism of plant and animal maxima is reflected in their spatial distribution. In the horizontal plane the belt of bloom is always situated more southernly than the

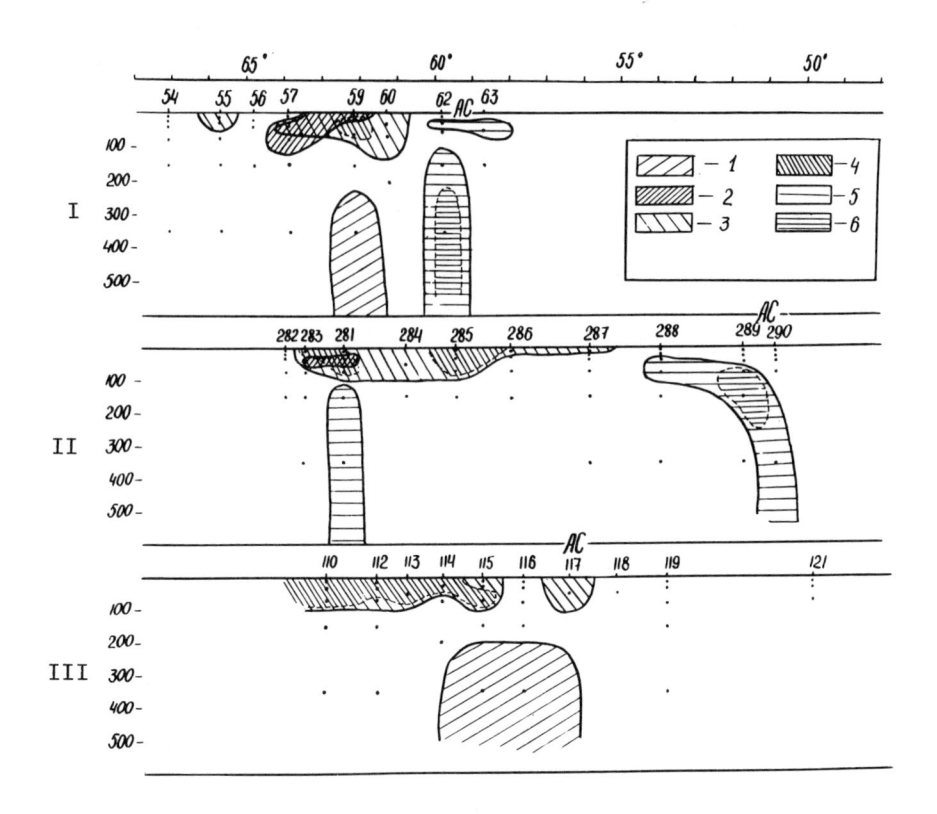

10. The position of biomass maxima of different copepod species
(after N.M. Voronina, 1970). I. Section south from New Zealand,
March-April 1956; II. Section along 98°E, April 1957; III.
Section from Mirny to Aden, May 1956. AC, Antarctic Convergence;
(1) C. acutus, 25 mg/m^3; (2) C. acutus, 50 mg/m^3; (3) C. propinquus,
25 mg/m^3; (4) C. propinquus, 50 mg/m^3; (5) R. gigas, 25 mg/m^3; (6)
R. gigas, 50 mg/m^3.

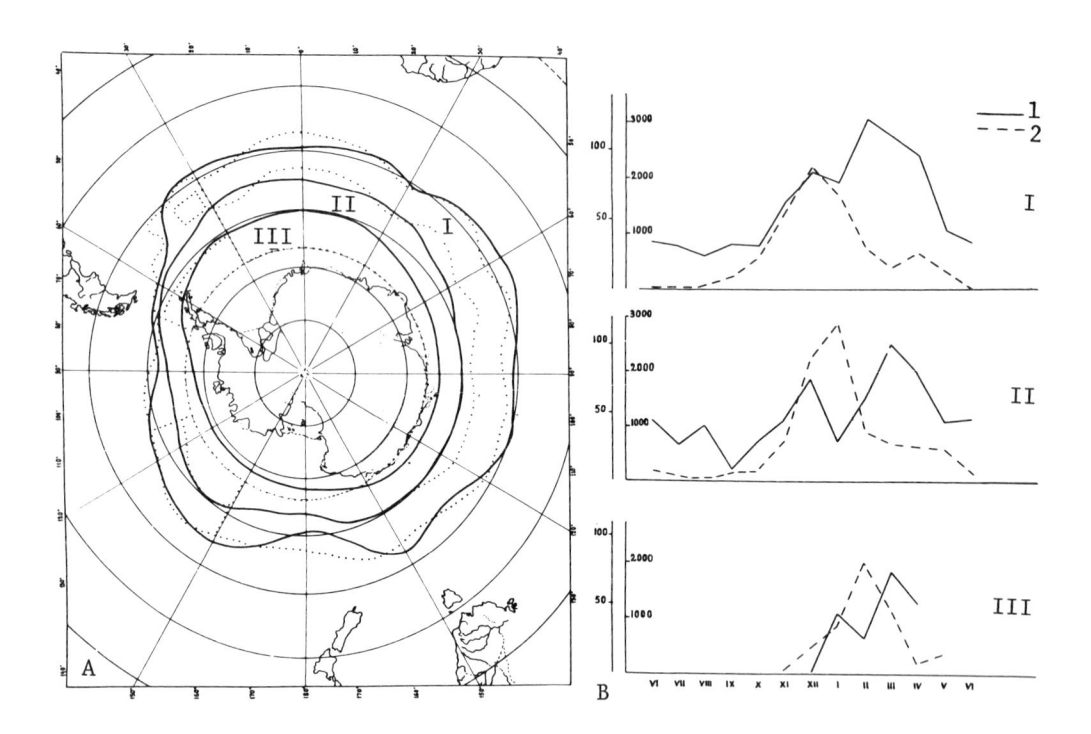

11. Seasonal changes in the quantities of phyto- and zooplankton in different latitudinal zones (after N.M. Voronina, 1971).
   A I-III - latitudinal zones; B 1 - zooplankton, 2 phytoplankton.

summer belt of zooplankton biomass adjoining it from the north.

Consequently, in systems with an oscillating regime of production all important structural features of the community: age composition of populations, vertical sequence of developmental stages, position of maxima of abundance, vertical sequence of different populations, dominance of species, and ratio of phyto- and zooplankton, undergo substantial changes during the annual cycle. It must be emphasized that most of the changes occur under practically constant environmental conditions. Aquatories with identical conditions may differ widely in the stages of a cycle which is initiated by the establishment of summer density stratification.

Summing up, we see that owing to scale-differences in the seasonal variability of biotopes, two types of regulation based on different principles have been developed in the communities. In the ecosystems where continuous production is possible, different methods are used for the simultaneous exploitation of the entire trophic layer by different populations, while in the ecosystems with seasonal fluctuations, each individual uses consecutively every layer of the trophic zone, but the time of exploitation diverges partially in separate species. These differences are in a considerable measure responsible for the different types of spatial variability within the ecosystems. In the first case the quantitative relationships between phyto- and zooplankton are stable and the variations in the ratio of the separate components of zooplankton are mainly accidental depending on the previous local hydrological situations, while in the second case all the structural characteristics of the community are subjected to continuous changes, and the situation observed depends primarly on the stage of the cycle.

Ecosystems with a continuous production process, the intensity of which widely varies in time, occupy an intermediate position. Such are, for instance, the western coastal regions with seasonal upwellings. Their communities include species inherent in both types considered above. Species of the first type are dominant during the oligotrophic, those of the second during the eutrophic period. Variability within such ecosystems presents an intricate combination of temporal and local factors.

Thus the causes of variability in the communities are manifold and result in considerable mixing along the boundaries and in substantial structural changes. But these circumstances should not cast doubts upon the reality of pelagic ecosystems, which has its roots in the existence of stable large scale gyrals and in the connection of geographical ranges with these gyrals.

REFERENCES

Banse K. (1969). Die Vertikalverteilung planktischer Copepoden in der Kieler Bucht. Berichte der Deutschen Komission für Meeresforschung, 15.

Beklemishev C.W. (1969). Ecology and biogeography of the open ocean. "Nauka", 291 pp.

Clowes, A. (1938). Phosphate and silicate in the Southern Ocean. "Discovery" Reports, 19, 1-120.

Conover R.J. (1968). Zooplankton - Life in a nutritionally dilute environment. American Zoologist, 8, 107-118.

Corner, E.D.S. and C.B. Cowey (1968). Biochemical studies on the production of marine zooplankton. Biological Reviews, 43, 393-426.

Currie, R.I. (1964). Environmental features in the ecology of Antarctic seas. Biologie Antarctique, Paris.

Cushing D.H. (1959). The seasonal variations in oceanic productions as a problem in population dynamics. Journal du Conseil, 24, 455-464.

Foxton P. (1956). The distribution of the standing crop of zooplankton in the Southern ocean. "Discovery" Reports, 28, 191-236.

Gauld D.T. (1966). The swimming and feeding of planktonic copepods. In: Some contemporary studies in marine science, H. Barnes, editor, 313-355.

Gran H.H. (1931). On the conditions for the production of plankton in the sea. Rapports et proces-verbaux des reunions, 75, 37-46.

Gueredrat J.A., R. Grandperrin, C. Roger (1972). Diversité spécifique dans le Pacifique équatorial: évolution de l'écosystème. Cahiers O.R.S.T.O.M. série Oceanographie 10, No. 1, 57-70.

Harder W. (1968). Reactions of plankton organisms to water stratification. Limnology and Oceanography, 13. 156-168.

Hardy C. and M.A. Gunther (1935). The Plankton of the South Georgia whaling grounds and adjacent waters 1926-1927. "Discovery" Reports, II, 1-156.

Hart T.J. (1934). On the phytoplankton of the South-west Atlantic and the Bellingshausen Sea. "Discovery" Reports, 8, 1-268.

Hart T.J. (1942). Phytoplankton periodicity in Antarctic surface waters. "Discovery" Reports, 21, 261-356.

Heinrich A.K. (1963). Age structure features of copepod populations in tropical regions of the Pacific. Okeanologiya, 3, 88-99.

Heinrich A.K. (1971). The productive cycles in the marine pelagic environment. In: Fundamentals of the biological productivity of the ocean and its exploration, C.W. Beklemishev, editor, "Nauka", pp. 43-63.

Heinrich A.K. (1975). The significance of the expatriated species in the structure of the Pacific tropical plankton communities. Okeanologiya, 15, 721-726.

Margalef R. (1963). Succession in marine populations. In: Advancing frontiers of plant science. Raghuvira, editor, 2, pp.137-188.

Mensah M.A. (1974). The reproduction and feeding of the marine copepod Calanoides carinatus (Kroyer) in Ghanian waters. Ghana Journal of Science, 14, 167-191.

Mullin, M.M. (1963). Some factors affecting the feeding of marine copepods of the genus Calanus. Limnology and Oceanography, 8, 239-250.

Mullin M.M. (1967). On the feeding behaviour of marine copepods and the separation of their ecological niches. In: Symposium on Crustacea. Marine Biological Association of India, 3.

Mullin M.M. and E.R. Brooks (1967). Laboratory culture, growth rate and feeding behavior of a planktonic marine copepod. Limnology and Oceanography, 12, 657-666.

Nival P. and S. Nival. (1976). Particle retention efficiencies of an herbivorous copepod Acartia clausi (adult and copepodit stages); effects on grazing. Limnology and Oceanography, 21, 24-38.

Paffenhofer G.A. and J.D.H. Strickland (1970). A note on the feeding of Calanus helgolandicus on detritus. Marine Biology, 5, 97-99.

Pavlova E.V., T.S. Petipa and Yu. I. Sorokin (1971). The role of bacterioplankton in the diet of marine pelagic organisms. In: Functioning of pelagic communities in the tropical region of the ocean, M. Vinogradov, editor, "Nauka", pp. 142-151.

Perueva E.G. and B.J. Vilenkin. (1970). Nutrition of Calanus glacialis (Jashnov) under different concentration of algae. Doklady Akademii

Nauk SSSR, 194, 943-945.

Petipa T.S., A.V. Monakov, Yu. I. Sorokin, G.V. Voloshina, I.V. Kukina. (1975). Balance of the matter and energy in copepods in trophical upwellings. Trudy Instituta Okeanologii Akademii Nauk SSSR, 102, 335-350.

Piontovsky S.A. and T.S. Petipa (1975). Feeding selectivity in *Acartia clausi* Ziesbr. Biologiya morya No. 33, 3-11.

Poulet S.A. (1976). Feeding of *Pseudocalanus minutus* on living and non-living particles. Marine Biology, 34, 117-125.

Poulet S.A. (1974). Seasonal grazing on *Pseudocalanus minutus* on particles. Marine Biology, 25, 109-123.

Richman S. and J.N. Rogers (1969). The feeding of *Calanus helgolandicus* on synchronously growing population of the marine diatom *Ditylum brightwellii*. Limnology and Oceanography, 14, 701-710.

Semina H.I. (1974). Pacific phytoplankton, "Nauka", 237 pp.

Sorokin Yu. I., I.N. Sukhanova, G.V. Konovalova, E.B. Pavelyeva (1975). Primary production and phytoplankton in the area of equatorial divergence in the eastern part of the Pacific Ocean. Trudy Instituta Okeanologii Akademii Nauk SSSR, 102, 108-122.

Steyaert J. (1974). Distribution of some selected diatom species during the Belgo-Dutch Antarctic expedition of 1964-65 and 1966-1967. Investigacion Pesquera 38, 259-288.

Timonin A.G. (1971). The structure of plankton communities of the Indian Ocean. Marine Biology, 9, 281-289.

Timonin A.G. (1975). Vertical microdistribution of zooplankton in the tropical western Pacific. Trudy Instituta Okeanologii Akademii Nauk SSSR, 102, 245-259.

Timonin A.G. and N.M. Voronina (1975). Distribution of the net zooplankton along the equator. Trudy Instituta Okeanologii Akademii Nauk SSSR, 102, 213-231.

Vinogradov M.E., I.I. Gitelzon and Yu. I. Sorokin (1970). The vertical structure of a pelagic community in the tropical ocean. Marine Biology, 6, 187-194.

Vinogradov M.E., V.V. Menshutkin and E.A. Shushkina (1972). On mathematical simulation of a pelagic ecosystem in tropical waters of the Ocean. Marine Biology, 16, 261-268.

Vinagradov M.E., E.A. Shushkina, I.N. Kukina (1976). Functional characteristics of a planktonic community in the equatorial upwelling. Okeanologiya, 16, 122-138.

Voronina N.M. (1962). On the dependence of the character of the boundary between Antarctic and Sub-Antarctic pelagic zones on the meteorological conditions. Geophysical Monograph. No.7, 160-162.

Voronina N.M. (1964). The distribution of macroplankton in the waters of equatorial currents of the Pacific Ocean. Okeanologiya, 4, 884-895.

Voronina N.M. (1966). Some results of studying the Southern Ocean zooplankton. Okeanologiya, 6, 681-689.

Voronina N.M. (1968). The distribution of zooplankton in the Southern Ocean and its dependence on the circulation of water. Sarsia 34, 227-283.

Voronina N.M. (1969). Plankton of the Southern Ocean. In: Atlas of the Antarctic, E. Tolsticov, editor, Gidrometeoizdat, 2, pp. 496-505.

Voronina N.M. (1970). Seasonal cycles of some common Antarctic copepod species. In: Antarctic ecology, Academic Press, I, pp. 162-172.

Voronina N.M. (1971). The annual cycle of plankton in the Antarctic. In: Fundamentals of the biological productivity of the ocean and its exploration, C.W. Beklemishev, editor, "Nauka", pp.64-71.

Voronina N.M. (1972) a. The spatial structure of interzonal copepod populations in the Southern Ocean. Marine Biology, 15, 336-343.

Voronina N.M. (1972) b. Vertical structure of a pelagic community in Antarctica. Okeanologiya, 12, 492-498.

Voronina N.M. (1974). An attempt at a functional analysis of the distributional range of *Euphausia superba*. Marine Biology, 24, 347-352.

Voronina N.M. (1975). On the ecology and biogeography of plankton in the Southern Ocean. Trudy Instituta Okeanologii Akademii Nauk SSSR, 103, 60-87.

Voronina N.M., I.N. Sukhanova (1976). The composition of food in the mass species of the Antarctic herbivorous copepods. Okeanologiya, 16, 1082-1086.

MAN'S USE OF COASTAL LAGOON RESOURCES

ROBERT R. LANKFORD

CENTRO de CIENCIAS del MARY y LIMNOLOGIA, UNIVERSIDAD NACIONAL AUTONOMA de MEXICO

It has been estimated that of the more than 40,000 km-long shore zone of the world's land -- sea interface, there are some 700,000 km of actual shoreline, much of which is formed by coastal lagoons, estuaries, bays, sounds, fjords and other marginal marine water bodies which indent the margin of the land. This complex system has been, is, and will continue to be one of man's most important natural resources. Considering the coastal lagoon system, which I will define in a moment, as one of the major environmental subdivisions of the sea, it has been the most important marine resource that we have yet utilized for our socio-economic benefit. But coastal lagoons are in danger of becoming a limited, or reduced, or even an exterminated natural resource in certain parts of the world. Today, I will talk briefly about some of man's uses, misuses, and abuses of coastal lagoon resources. I will mention multiple uses and the resulting use conflict of this vital natural resource. Finally I would like to present some ideas and challenges regarding the role of marine scientists and engineers in the rapidly growing demand for enlightened coastal management.

I have repeatedly used the term, coastal lagoon, and I need to define my use of the phrase, considering that all of us have our own definitions firmly in mind. In addition, there seems an inordinate amount of confusion in shore zone nomenclature. My use of the term, coastal lagoon, is all-embracing because it includes estuaries, bays, sounds, coves, drowned valleys, fjords, firths, lagoons, barrier lagoons, etc., as well as equivalent terms in other languages. Quite simply, I am avoiding the multitude of often conflicting definitions for any of the shore zone features that I have just mentioned and I am placing all of these features under the banner of coastal lagoons for communication purposes. My coastal lagoon is broadly defined here as: a tidally inundated depression, located in the shore zone, which is protected from the sea by some type of barrier, but which is in either permanent or ephemeral communication with the sea.

For a number of years, I have been involved in geological studies of coastal lagoons, both modern and ancient, and I have come to think of them in terms of their geological origins and the major processes which shape their environmental behaviour and which produce their similarities and their differences. Homo sapiens is basically a classifying animal and since I am a member of this species, I take little shame in attempting to classify lagoons as I have broadly defined them here. My major criteria for classification are the geologic origins of the tidally inundated depressions and the processes which produce some type of barrier. There are five major classes. See Figure 1.

Type I includes those depressions formed by erosion - or by chemical solution -- and mainly originated during the last Ice Age when sea level had been lowered eustatically. The arrows indicate permanent or intermittent run-off. Type I includes the much-discussed estuary, such as the Thames, where there is run-off; there is no barrier in the physical sense but a hydrodynamic barrier does exist. The presence or absence of a physically solid barrier in Type I coastal lagoons is primarily a function of the coastal oceanographic regime and of available sediment. The present or absence of run-off is climatically and geologically controlled. I have also included in the Type I category, the solution or karst depressions of limestone coasts, the inundated rocky canyons of high relief coasts and the glacially eroded fjords of high latitudes.

Type II coastal lagoons are those produced by differential terrigenous sedimentation and are typically associated with river mouths and deltas. Their age dates from the modern sea level stand. These are not only geologically young features but they are rapidly evolving and highly dynamic lagoons. In addition, they are environmentally more predictable than Type I coastal lagoons due to their close association with river discharge.

The third category, Type III, probably is more typically identified as a coastal lagoon. It has also been referred to as a coastwise lagoon and a barrier lagoon. They come in a variety of shapes and sizes but basically they are all quite similar in that they are the enclosed inner margin of the continental shelf. Consequently they are quite shallow except in tidal channels. The barrier typically is an unconsolidated linear deposit of sand-sized material formed by wave action since sea level reached its

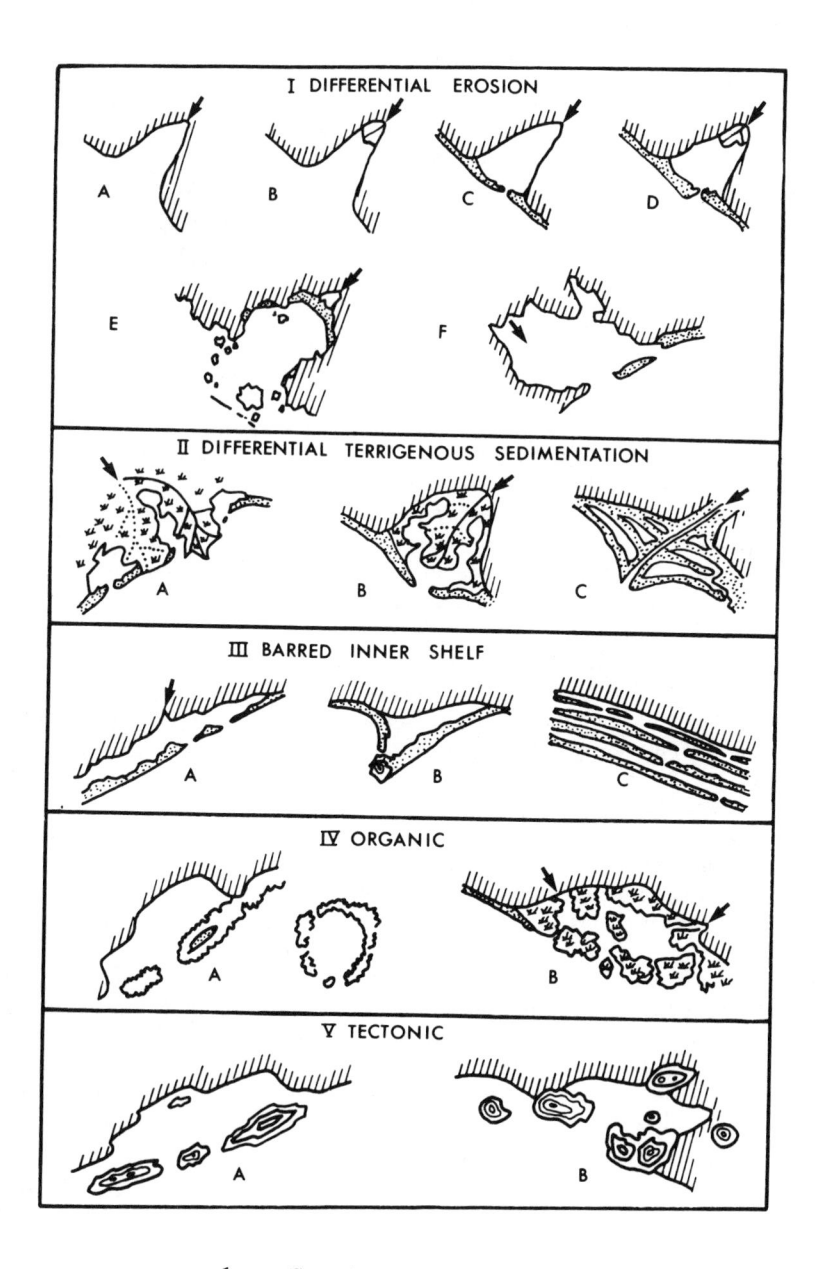

1. Coastal lagoon types.

approximated present level. Runoff usually is not an important attribute of these coastal lagoons. Consequently salinities may be normal or hypersaline. Where runn-off is an important consideration, originally long Type III coastal lagoons may have become segmented by lagoon delta growth into a series of smaller ones which will have altered internal hydrography.

The Type IV lagoons, like Type III, also are inundated and enclosed inner margins of the continental shelf and likewise are geologically young. The protective barrier, however, is organic in origin. The most typical and best known examples of this category are the coral (or coralgal) barred lagoons of the world's tropical oceans. Lagoon depths are variable but salinity is almost always near normal or slightly hypersaline. Organisms other than the coral reef community produce effective barriers also. Mollusk banks or marine grass or algae frequently form effective barriers to marine energy and may secondarily act as sediment traps eventually to form emergent bars. Among the plant barriers, the most wide-spread is the ubiquitous mangrove barrier complex of warm temperate and tropical latitudes. Although mangroves usually are considered as growing within pre-existing coastal lagoons, they not uncommonly form very effective energy barriers on the exposed inner shelf in areas where normal wave energy is low.

The fifth and last category of coastal lagoon depression originates from the earth's crustal activity and is basically independent of marine or terrestrial surface processes. Faulting or folding or volcanic activity produce the depressions and, not uncommonly, all or part of the physical barrier system. Lagoon areas and depths may be highly variable as well as is the degree of protection received from open ocean processes. Although salinity will be a function of the presence of absence of run-off, it has been my experience that tectonic lagoons not uncommonly have relatively great water depths, wide inlets and essentially normal salinity.

I mentioned in the beginning that a relatively large proportion of the 400,000 km of the world's shore zone contains coastal lagoons thus accounting for an estimated 700,000 km of shoreline. A recent survey which included only that portion of the shore zone formed by unconsolidated sediment barriers (that is, parts of my Type I, and all of Types II and III) shows that there are approximately 32,000 km or that 13% of the total coast line of the world is constituted by these types. If the remaining categories are added, then there are some 45,000 to 50,000 km of the world's shore zone composed of coastal lagoon systems.

My reason for giving these figures is to stress the geographic importance of coastal lagoons. Their importance, however, is even more dramatically reflected by past and present world population distribution patterns. To a great extent, mankind has looked to the sea's resources for his benefit and it has been the broadly defined coastal lagoon environment where he has derived the majority of his marine resources. World population maps show a preferred concentration along the coast. Here at this meeting we are gathered in Edinburgh located on the Firth of Forth -- London is on the Thames Estuary, Tokyo is located on Tokyo Bay, New York is on the Hudson Estuary -- and so on. This coastal population distribution trend is not accidental. It is due to the multiple resources that the coastal lagoon complex has provided since pre-historic times and continues to provide for a growing and increasingly demanding world. And how has man used coastal lagoons and what has been his impact on the environment?

Archeological studies indicate that primitive man made frequent visits to coastal lagoon shores to gather fish and mollusks. Small midden heaps of shell with occasionally only the most primitive tools are the only evidence of man's first discovery of the potential resources to be exploited. Barrier sands most probably provided then, as today, vital fresh water, and the gathered seafood supplied man's bodily need for salt.

It is not known how long this simple visitation and gathering era lasted -- possily for hundreds of thousands of years -- but it is probable that primitive man had very little effect or impact on the ecosystem. With the slow rise of mankind on the evolutionary ladder, however, he began to learn to take increasingly from nature for his own sustance. In addition to food resources, the eternal need for salt derived from saline waters drew early man to lagoon shores, first collecting natural precipitates and later discovering the basic principles of salt pan construction and solar evaporation, a technique that continues to be used to the present day. The primitive evaporation pans, filtration mounds and lead canals constructed along the shoreline mark man's first impact on the natural lagoon system. The need for food and for salt, both as a dietary demand and as a method to cure or preserve the gathered food, probably initiated the first permanent settlements along coastal lagoon shores in late pre-historic time thus opening the door for further exploitation and modification of the environment.

In the ensuing millenia, world population increased exponentially and the shore zone became firmly established as a desirable place to congregate, no longer due only to coastal lagoon living resources, but as rapidly expanding urban centers of trade and commerce. Maritime transportation was early raised to an important level by the Phoenecians and the peoples of southern and southeastern Asia. Ports and harbors were built. Tidal energy was harnessed in Europe by the 11th Century as a precursor to the present-day power plant at La Rance, France. Today, coastal lagoons are in varying stages of use and modification. In the highly developed areas of the world, lagoon exploitation is intense; elsewhere the use has been less. In some remote areas, one can still find coastal lagoons in a primitive state of exploitation and, rarely, they are essentially in a pristine state.

What are the uses which we of the modern world impose upon the coastal lagoon system? See Table 1. We can immediately identify the living resources, the food -- shellfish and finfish, even algae -- which are taken from lagoon waters and other organisms which are indigenous to the lagoon borders. Only recently have we recognized that many food species, although captured elsewhere, use coastal lagoons for their larval development -- or that migratory species, ranging from giant whales to water fowl, depend on quiet lagoon waters for reproduction or wintering grounds. Consequently we are belatedly beginning to set aside certain portions of coastal lagoons as biological preserves to ensure a continuation of lagoon-dependent species.

Lagoon borders not only harbor migratory water fowl, but man as well. Urbanization is rapidly claiming shore zone space as is industrialization; tourism and recreation facilities are demanded; land transportation unites with shipping; energy production in all of it forms is either conveniently or mandatorily located on lagoon shores; non-living resources such as salt, or fossil fuels, or construction material (shell, sand or gravel) are extracted from coastal lagoon environs. All of these uses produce waste in one form or another - chemical contaminants, sewage, dredge spoil and thermal pollution, to name a few. And all too frequently, the coastal lagoon becomes an economically convenient waste receptacle for industry, commerce and urbanization.

I think it is immediately apparent that such diverse uses run a head-on collision course and use conflict develops. I have prepared a schematic illustration of a coastal lagoon suffering such multiple use conflict. See Figure 2. We first note that great and small urban centers have grown up; the rivers which once discharged into the lagoon are now dammed; ports, complete with cargo docks, fuel docks and rail heads, meet a complex system of dredged shipping channels. The channels are bordered by spoil banks which are ideally designed to inhibit lagoon circulation. I have included a fuel-burning power plant with water heat exchangers; subsurface hydrocarbon production and refining are here also; highways, causeways and bridges cross parts of the lagoon. Real estate developers provide bulk-headed canal lots so that the recreation-minded buyer may dock his small craft at his front door -- a delightful setting were it not for the fact that the fringing marsh and submerged grass flats are destroyed as nursery grounds for aquatic species. Adjacent agricultural areas with their

Table I. Coastal lagoon use.

LIVING RESOURCES	NON-LIVING RESOURCES
WILDLIFE PRESERVES	ENERGY PRODUCTION
URBANIZATION	INDUSTRIALIZATION
TOURISM-RECREATION	TRANSPORTATION
	WASTE DISPOSAL

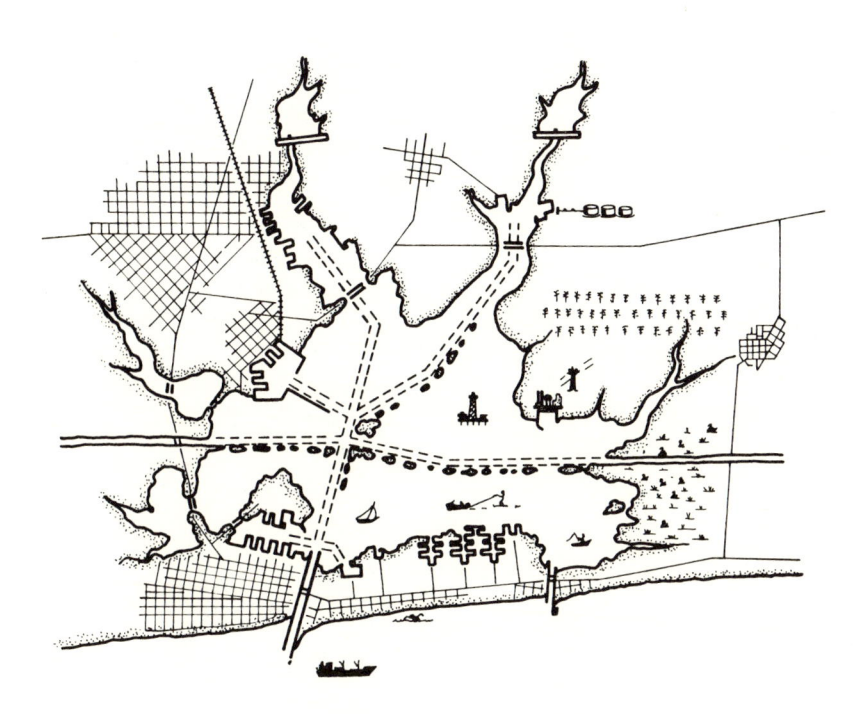

2. Schematic map of a coastal lagoon suffering multiple use conflict.

fertilizer and pesticide run-off have their impact. In the midst
of all of this, there is still the harvesting of some commercial
sea food, a biological preserve has been set aside, and tourists
and sportsmen flock in great hordes.

I said that this illustration was schematic but I am sure that
it could be identified as any of hundreds of modern coastal lagoons
undergoing similar multiple-use conflict. A glance at history
and the socio-economic demands of modern population growth does not
indicate a pleasant future for coastal lagoons. Altruistically
we all want to do something about it. But we want our cake and
we want to eat it too. "Having and eating" implies the antithetical
concepts of <u>conservation</u> and <u>consumption</u>. To some, conservation
means to keep forever, to keep intact, to keep safe. But what
good to us is a tree full of fruit that eventually will fall to the
ground and rot? As a species, we have evolved to our present
status because we learned to use the benefits of our natural resources.
We cannot conserve our coastal lagoons in their pristine
state. We need to use them but we must learn to use them wisely
-- to practice an enlightened conservation much as the housewife
conserves the summer fruit for later use in the winter.

Proper or wise use and conservation of coastal lagoon resources
bring us to the doorstep of resource management which involves
itself directly with the conservation and wise use conflict -- to
have our cake and to be able to eat it also. Coastal lagoon
management -- and for that matter, the management of any resource
-- simply is a matter of making wise, informed decisions for
appropriate action. There are three sectors or areas of influence
in coastal lagoon management and the decision-making process. See
Table II.

First is the information base, the information provided by
trained analysts, the marine scientists and engineers who know the
problems and recognize the need to have them resolved. The information
base forms the input necessary for intelligent planning
and, once plans are established, they must be acted upon by responsible
and responsive government. This appears to be very
logical and a workable system of resource management. In all
fairness, I must admit that there are isolated cases where the
three areas of management have fortunately co-existed and there

Table II.   Coastal lagoon management.

RESPONSIBLE GOVERNMENT
|
PLANNING
|
INFORMATION BASE

are some coastal lagoons which are convalescing slowly. But these cases are all too few. Too many of our valuable coastal lagoons are all but destroyed and others are on the endangered species list. Why? I believe it is, mainly due to unworkable legal tangles, to faulty planning, to poor communication, and to the ever-present conflicts of interests.

Since coastal lagoon management is invariably the responsibility of the public rather than the private sector, that is to say, governmental responsibility, there frequently are problems here such as unresponsive local, regional or national government. It is admittedly difficult for the concerned marine scientist to have immediate impact in this problem area. In such situations, coastal lagoon deterioration most probably will persist. And even when responsible and responsive government does exist, many still will lack a proper infra-structure or legal system for the implementation of wise-use programs.

There are also problems in the planning sector of resource management. Too often, planning boards are poorly staffed, poorly directed or poorly motivated. This does not appear to be a particularly new character of planning. One of the first well-known planners in marine affairs is said to have been Christopher Columbus. When he left, he did not know where he was going; when he got there, he did not know where he was; and the whole affair was paid for by the government.

And finally -- and this applies directly to the marine science community -- there are problems in the information base sector. All too often, scientific information either is not the sort required or, if it is, it has not been presented in the form which is readily assimilated by planning and government. I quite frankly feel that we, as both scientists and citizens, can afford to abandon, at least occasionally, the heights of our ivory towers, and to contribute our talents in the most effective manner possible to the resolution of problems inherent in coastal lagoon resource management. I will go further and suggest that we all have the responsibility to do more than make our research meaningful and intelligible. We also must take the responsibility to make our voices heard, to serve on planning commissions, to approach local, or regional, or national governmental bodies with our ideas and knowledge and to work for that which we believe in.

If it sounds as though I am chastizing the marine science community, it is because I myself feel a sense of guilt for having too long remained content to study the taxonomy of an obscure group of lagoon Foraminifera. If I and if you wish to continue to enjoy the varied benefits of coastal lagoon resources, we all must accept our responsibilities inherent in each phase of resource use and management. Without our effective participation, we will surely be the losers, and "our thing", the Cosa Nostra will no longer be viable.

NON-RENEWABLE RESOURCES OF THE SEA

F. NEUWEILER

PREUSSAG AG, HANOVER

ABSTRACT

This paper summarizes the natural concentration processes acting on seabed minerals in the course of generating deposits of economic interest. Apart from the hydrocarbon and coal deposits of the offshore not considered here, the non-renewable resources of the sea include the huge potential of the manganese nodules, the important construction materials sand and gravel, placer deposits supplying a large part of the tin and titanium requirements, metalliferous sediments such as the Red Sea occurrences and the phosphorites of the shelf.

As to the role of the marine resources for the future raw material supply it is felt that the common comparisons of proved reserves with projected demands are not an entirely satisfactory basis for a country's commodity policy. If the interruption of supplies or other political or economic-political restrictions have to be feared, countries may be forced to develop alternative supplies of raw materials. Technologies for production and metal extraction have not yet matured for allowing deep ocean mining on a commercial scale. Steps for a final assessment of the economic exploitation have been initiated, however, for the manganese nodule and the Red Sea metalliferous mud potentials.

Whereas the legal regime necessary for mining placers and construction material has already been in existence for some time, and a special arrangement was recently concluded by the Sudan and Saudi Arabia to develop the metalliferous mud deposit in the Red Sea, no such basis exists for the deep seabed and the nodule potential. A regime for its orderly development is urgently required, however.

1. INTRODUCTION

The material contents, the very substance of our Earth have been an almost precisely determined and unchanged value since the days of its origin with little extraterrestrial influence or addition - meteorites omitted as comparatively negligible. Since then transition processes take place with varying velocities.

It is only the living substance which can grow or shrink on the inorganic substrate, depending on the environmental situation. Favorable conditions of growth for organisms and their incomplete decomposition form the important prerequisites for the origin of petroleum, natural gas and coal which substances shall constitute until the year 2000 and beyond the energy raw materials of paramount importance. Since the formation of hydrocarbon and coal deposits takes place over geological times, almost infinite compared to the relatively short period of utilization, such deposits may be classified here as non-renewable resources, in spite of the ever continuing slow processes of origin.

The scope of my following paper should therefore encompass the petroleum and gas deposits, as well as the coal fields, covered by oceans.

In the course of a development of the offshore techniques for about 25 years ever improved prerequisites evolved for a systematic prospecting for new mineral deposits in the shelf areas of the continents. At present almost one fourth of the total production of petroleum and natural gas is derived from offshore wells. For the year 2000 one half is expected from the seabed area. I should like to close with these general remarks to the energy raw materials and now refer to the types of mineral deposits that constitute the subject of this paper. I shall have to begin this by shortly describing concentration processes leading to the formation of economic deposits.

2. CONCENTRATION PROCESSES

In general normal marine sediments do not represent mineral resources. Only by passing of mostly several enrichment stages can certain minerals be concentrated in restricted environments to such a degree as to render them suited as potential ore deposits. For the manganese nodules the cumulative enrichment factor for their value metals from the original terrestrial or volcanic source rock to the ore type lies in the order of magnitude of $10^2 - 10^3$.

The mechanical concentration is restricted in general to the shore and the shallow shelf sea. Waves and currents can continue to work or to rework a mixed sediment until all argillaceous substance has been removed. Sand and gravel, of importance to the construction industries, thereby remain as residues. Differences in hardness and specific weight of quartz compared to the disseminated metals and their oxides such as gold, cassiterite, zirconium, rutile, magnetite or diamonds may lead through continued selective action in the shore area to an improved concentration and eventually to economic mineral sands or placers.

Biological-chemical processes contribute in a number of cases to an important enrichment of certain elements. A selective extraction of specific compounds from the sea-water and their incorporation in the bodies and the skeletons of organisms, respectively, are commonly observed. Following their death the mineral substances become embedded in the sediments.

The largest concentration of this type is the forming of calcareous skeletons of foraminifera and other marine animals. Silica is extracted by diatoms, radiolaria and sponges in rock-forming quantities. Several elements present in seawater only in traces become enriched in organisms by a factory of 10^5 and more. These elements include V, Fe, Pb, Sn, Zn, and Cu. Other elements, concentrated in marine organisms, comprise J, Ni, Co, U and Mn.

Another process of concentration leads to selective sedimentation in parts of the deep ocean. It is mostly based on differences of temperature, pressure and dissolved carbon dioxide. In general the deep sea sediment of basins far from terrestrial influence consists, in the tropical and moderate latitudes, overwhelmingly of calcareous remains of organisms. As the solubility of $CaCO_3$ increases with decreasing temperature and increased CO_2 content and as the deepest ocean basins are characterized by a high CO_2 rate and low temperature (between 0 and 4° C) the calcareous remains become dissolved again in depth of 4000 - 5000 meters (carbonate compensation depth). This process leads to a residual deposit with an extremely low sedimentation rate, the Red Clay, apart from silicious shells, consisting of very fine argillaceous substance and exhibiting a distinct enrichment in certain elements (Al, Mn, Ti, V, Fe, Co, Ni, Cu, Pb).

The forming of concretions is a widely observed process of enrichment in the marine environment. This implies the spheroidal growth around any nucleus precipitated from substances dissolved in the sea or the sediment pore water. Concretions of sand particle size from within the sediment or turbulent water; larger nodules mostly at the water/sea-floor interface or within the sedimentary deposits. The mode of origin has not been fully understood. Chemical and physico-chemical, in some cases also microbiological effects have been taken as explanation. The 0.1 - 1 millimeter ⌀ ooliths of calcium carbonate originating in the near-shore marine environment are the most common concretionary form. There is ample evidence of deposits of this mode of origin in the geologic column. The iron colite of the European Jurassic, known as "minette", has been identified in recent deposits only in a few African lakes.

Another common concretionary process leads to the formation of calcium phospate. As ooliths or nodules such concretions originate preferably below a water depth between about 40 and 400 m near the shelf margin or on submarine ridges with upwelling water.

The concretions of presently interesting economic value are represented by the so-called manganese nodules. The concretionary part, apart from the core (mostly volcanic material, rarely organic remains such as sharks' teeth), consists predominantly of manganese and iron oxides. The nodules of certain deep-sea basins growing at an extremely slow rate (1 centimeter per 1 million years) exhibit concentrations of a number of elements commonly present in sea water only in traces, such as Ni, Cu, Co, Zn, V and Pb. The distribution of the elements depends, on one hand, on the primary sources of enriching agents, respectively. Apart from terrigenous and volcanogenic deposits the residual deposits from dissolved organic shells constitute the suppliers. On the other hand, water depth, sedimentation rate and currents also play an important role for the mineral content.

Hydrothermal-volcanogenic enrichment is another process of forming additional marine mineral resources. Vulcanism is generally active along the mid-ocean ridges where continental margins are under thrust by oceanic crust. Solid and volatile constituents of the earth's mantle are thereby ejected into the sea. In addition the ocean floor exhibits high heat flow values in such zones. Water entering the sea-floor, especially as a consequence of eruptions and sea-quakes becomes heated and reacts with the wall rocks, such as basalt flows, sediments and their pore waters. Such percolating water becomes charged with metals which, in turn, when re-entering into the sea, precipitate as metalliferous mud. Such enrichment metals include Fe, Mn, Cu, Zn, Pb, Cd and Ag. Since silica, iron and manganese predominate in the solutions, deposits from such sources within the open sea consist mainly of precipitates of iron-rich silicates and various iron and manganese oxides and oxidic hydrates, respectively.
A deposit of economic interest or value can only originate where a restrictive environment is formed through a morphological trap and the supply of hydrogen sulphide. Such a deposit was investigated in great detail in the case of the Atlantis-II-Deep of the Red Sea. A partial differentiation of the commercially interesting metals copper and zinc from iron and manganese was observed. Following the deposition of the metalliferous mud, a post-sedimentary concentration and improvement in value can take place. In the case

of the manganese nodules the dentritic area becomes subsequently mineralized. Metalliferous muds commonly are subjected to a post-sedimentary supply of metal solutions from the substratum. An increase in ore quality in phosphorites is also observed, leading through epigenetic upgrading of original grades (though only a few percent of P_2O_5) from the potential resources to a mine-grade ore.

3. THE NON-RENEWABLE RESOURCES

3.1. Manganese Nodules. This is the most important mineral resource on the sea-floor so far known. Nodules occur in most ocean basins, and preferably in the deeper parts far away from land, and generally form one layer. Their size is mostly between 1 and 10 centimeters.

Manganese nodules contain 10 - 35% manganese; 1.5 - 27% iron; 0.1 - 2.0% nickel; 0.1 - 1.3% cobalt and 0.03 - 1.6% copper. They are considered potential ores of nickel, copper and cobalt, and, to a certain extent, manganese.

The estimated reserves, distributed over the three oceans, amount to more than 10^{11} tons. These enormous reserves will certainly represent a major factor in the metal supply of the future. At present, however, only a limited portion is of economic interest.

Taking the partly unsolved technical problems of collecting, hoisting and processing of nodules into account, only those nodule fields can be considered, which show a high nodule distribution density and metal grades substantially better than comparable oxidic ores on land. High-grade nodules have so far been found mainly in the North-Equatorial belt of the Pacific and the South-Equatorial belt of the Indian Ocean.

Concerning the nodule density in connection with high metal grades satisfactory results have been obtained only in the North Pacific. Exploration work with the German research vessel "Valdivia" has shown, however, that even the most promising areas have complex characteristics which must be explored thoroughly before exploitation can start.

This concerns mainly the nodule distribution and the surface and subsurface characteristics, but also metal distribution or size and form of nodules.

Different from reserves on land, the dimension of future mining operations cannot be derived from the size of the deposit, but the size of the deposit has to be chosen according to the requirements of the processing plant.

A plant treating about 4.3 million tons of wet nodules per year

(= 3 million tons dry) is considered an economic size. Taking the presently known deposit characteristics into account, a surface at least of the size of Scotland 80 000 km^2) has to be explored in detail to secure a 20 years supply for the plant. This is an important aspect when discussing the legal problems connected with the mining. Four international consortia have been formed to explore, mine and process manganese nodules. Others are likely to be formed. Additionally there are several scientific and technical institutions involved in development of nodule utilization.

Exploration is going on since 1965 and has been intensified considerably since specialized ships and equipment were built and put into operation a few years ago. Such equipment includes freefall samplers and television cameras. The principle of processing techniques could be derived from existing technologies and adapted to the new type of ore. Inexpensive physical methods of concentration have to be excluded because of the fine grain and intergrowth of the minerals. Several hydro and pyrometallurgical processes have been tested and found suitable for nodule treatment. Three companies made already pre-pilot size tests. For full scale pilot operations larger amounts of nodules from the mining site must become available.

Mining itself is still the main problem. There are at least two approaches. One is a continuous line bucket system already tested three times. Further development is supported by a large number of companies outside the existing consortia. The other approach is continuous collecting and transport through a pipestring to the surface whereby the vertical transport of the nodules is effected either by pumping or by the injection of air into the pipestring (airlift principle). Such a system has been tested at intermediate water depths, but the results cannot be easily extrapolated to the Pacific deepsea conditions. New equipment will be tested within the near future.

3.2. Construction material. In densely populated areas it becomes increasingly difficult to provide construction material like sand, gravel or carbonates from onshore deposits. Such material is needed for concrete structures, road building, or the establisment of new ports and artificial beaches. Sand and gravel are successfully dredged since a few years on the shelves of the United Kingdom, the U.S. and some Mediterranean places. Carbonates for cement production may be produced offshore if such deposit are missing on land.
There are additional potential resources not yet examined in detail, e.g. diatomites.
Others are exploited locally without previous systematic exploration, like big erratic boulders in the Baltic Sea.

3.3 Placers. Placers are heavy mineral concentrates of sand. Typical placer minerals are rutile and ilmenite (for pigment and titanium production), zirconium (refractory sand), magnetite (iron production), cassiterite (tin), monazite (rare earth and thorium) and some gem stones and gold. These placers concentrate in coastal plains, on the beaches, in river beds and on the continental shelf. Most occurrences presently mined are beach or near-beach sediments.

Australia furnishes most ilmenite and nearly all rutile. At Richards Bay on the South African Coast, a huge deposit is under development (expected volume of world titanium supply: 15%). Another large deposit has been surveyed by the German research vessel Valdivia offshore Mozambique. Magnetite sand production is known from New Zealand, Java and Japan, but there are many other prospects. Zirconium is produced as a by-product with the other mineral sands, mainly in Australia. Tin sands are already mined offshore Indonesia and Thailand in drowned river beds.

Placer exploitation is an important industry, since some elements and compounds are nearly entirely produced from these sands. Since the land bound deposits can easily be detected, most of them are already known, and future exploration will focus on the offshore deposits. Such deposits can be expected worldwide, because Pleistocene coast lines and beach sediments are frequently covered by water at present.

The quality requirements for economically interesting deposits depend on their size and the value of minerals contained. Large rutile deposits may be workable with only 0.5% TiO_2, cassiterite deposits with only 0.05%.

3.4. Metalliferous sediments. Fine-grained metalliferous sediments have been found during the last few years at several places of the ocean. They are recent equivalents of the fossil iron-chert formations and massive sulfide deposits, and they are potential resources of iron, manganese, copper, sinc, silver and lead.

There is an apparent connection with rifting and volcanism. Tectonically unstable areas with volcanism are, therefore, the most prospective areas for metalliferous sediments. Most known occurrences are containing only iron or manganese.

They will probably not be competitive with land deposits, unless they are of very high grade and found near-shore. The most promising deposits, so far known, occur in the Red Sea. They are covering the sea bottom to a maximum of about 25 meters in certain submarine brine pools. The interesting metals are zinc, copper, and silver. One deposit, the Atlantis-II-Deep, contains more than 2 million tons of zinc, 800,000 tons of copper and 9,000 tons of silver.

Exploration methods have been developed by PREUSSAG during recent years. No systematic survey for metalliferous sediments has yet been undertaken so far outside the Red Sea.

Processing and mining methods are now under study. There are still many problems to be solved. Such problems are connected with the low grade, extremely fine grain (80% < 0.002 millimeters) and high salt water content of the sediments and their position within a rough bottom morphology under 2000 meters of water. PREUSSAG has recently concluded a contract with the Saudi Sudanese Joint Commission for an economic assessment of these metalliferous deposits.

3.5. Phosphorites. Phosphorites are important raw materials for fertilizer production. Apart from the phosphate they contain interesting amounts of fluorine and uranium. Most phosphorites currently mined on land are of marine origin. Efforts have therefore been made to recognize them also in the present oceans. A few occurrences had already been known for several years, but many new ones were detected recently.

Phosphorite nodules or sands are now known in the offshore areas of California, the South American West coast, North Carolina, the African West and South coast, some Australian places, the Chatham Rise East of New Zealand and some other localities.

Most of these phosphorites appear to be reworked older sediments, but some of them are still forming now. They are connected with nutrient rich upwelling water and are deposited in 30 to 500 meters of water depth. There are a few estimates of total tonnages available. The Californian deposits seem to contain about 1 billion tons of nodules. Quantity, however, is not the main factor in the question, if marine nodules should be mined.

Commercial land phosphates generally contain 30 - 40% P_2O_5. Material with less than 12% is not considered an ore. Marine phosphates from South Africa are reported to contain 15 - 21% P_2O_5, those from the Chatham Rise 16 -26%. Most deposits seem to have average contents less than 30% and therefore less than land deposits actually mined.

The question whether the marine phosphate resources will be used depends largely on the price policy of the two main producers, Morocco and the US(Florida). Since 1964, phosphate prices rose temporarily to nearly 5 times the original prices and are still kept high artificially. If this policy continues, many developing countries will come into difficult conditions in procuring their fertilizers necessary for their agriculture.

Countries with offshore phosphate potentials should therefore seriously consider its use. The reduced transport costs could probably balance higher production costs.

No special mining system has been presented for this kind of deposit. Continuous line bucket dredging would probably be feasible.

4. The role of the marine resources in the future raw material supply

The question whether or not our raw material reserves suffice for meeting the increasing demand has resulted in controversial answers at all times. This controversy has become especially acute in recent years. So far the optimists proved right. However since about 1950 the demand rises at an increased rate and the projections for several metals, such as zinc and tin, suggest that their reserves will be exhausted before the turn of the century. The value of such projections is but limited since potentials of these metals beyond the reserves in question were not considered and since the major factor for the demand, the development of the non-industrialized countries, can hardly be assessed quantitatively. The range of possibilities is well illustrated by the ratio of per capital energy consumption between the USA and India of 50.1.

Such uncertainties about the future consumption are accompanied by those of political or economic-political nature. Proven reserves can thus be withheld from the consumers' supply through political restrictions and a price policy of single or groups of states. The oil crisis and the price regulation for phosphates serve well as examples from the recent past.

The question whether or not marine mineral resources should be utilized must therefore be decided for some countries on the basis of their own domestic situation, independently from a theoretical global supply assessment.

Further cuts in the availability of terrestrial raw materials result from environment protection especially within heavily populated areas and high land prices. Disregarding the political aspects and considering the issue at shorter term, it is noted by comparison with potential land resources that the major difficulties in utilizing seabed minerals rest in the immense technological problems of their production techniques and the extraction of metals, and the resulting costs. In the case of the hydrocarbon potential the transition from onshore to offshore exploration and production was more gradual by nature and therefore fast, although at the expense of highly increased costs.

Two important technological trends facilitate plans for utilizing seabed resources. One is the rapid development of offshore installations, devices etc. and the electronics. The second one stems from the mechanizing of terrestrial mining and the improvement of processing plants. For land mines there is a distinct trend from high-grade and small sites to low-grade and large ones. The cut-off grade of copper mines has thus fallen from 5.2% at about 1890 to below .5% today. From their mode of origin most seabed mineral potentials are of low grade and of large extension. The metal contents of several marine resources, such as the manganese nodules, are well above the cut-off grade of many terrestrial mines of the respective metals. Whereas the technical progress benefits both land and seabed mining, the terrestrial activities will undoubtedly gain an advance. In the case of land development an important progress can be expected from deeper prospecting; for ocean mining the major problem rests in the exploitation technique. For some substances there is a competitive situation as for manganese nodules and lateritic nickel ore. Bigger exploitation problems of the nodules compare to a more complicated processing of the laterites.

Disregarding resources of the seabed's subsurface, such as petroleum and natural gas and coal, as well as the near-shore mineral sands, the present marine mining activities are restricted to the construction materials sand and gravel and the tin placers. The installations necessary for their recovery had already been developed for the requirements of the dredging industry.

Utilizing the resources of the deeper ocean necessitates new technologies whose development is still in its initial stage requiring very large funds.

The metals to be recovered from deep-sea mineral occurrences comprise nickel, copper, zinc, cobalt, silver, manganese and molybdenum.

As in the case of metalliferous mud only one site has become known of possible economic order of magnitude world-wide, any influence of its utilization on world markets need not be discussed. In the case of manganese nodules with their huge potential, an influence on the global economy might, however, well be assumed.

The impact of metals or compounds of the nodules on the world market depends, apart from demand, largely on the development of landbound lateritic ores and on the questions whether the manganese in the nodules be extracted or not. At present only one consortium considers the winning of the manganese using a hydrochlorination process. Output of manganese from more than one operation will probably influence the manganese price.

The additional nickel output of three operations would considerably improve the supply of the countries involved, but would be less than the additional world demand, assuming a 5-8% annual growth rate.

No impact at all is expected from the copper production from nodules.

Cobalt production from one operation will be about 1/10 of the current world production and might, therefore, have an influence on the price. Anyway cobalt will be produced as a by-product of nodules as well as in most onshore productions, and a decreasing price and larger output will probably lead to new applications.

5. The Political and Legal Situation

Depending on the distance of the mineral resources from the coast different legal regimes exist or may be expected to arise in order to meet with the requirements for orderly development. All placer deposits occur in shallow water close to the coast. (They could not yet be exploited at a water depth exceeding 100 m). Therefore they fall under the sovereign rights of the coastal state.

Offshore petroleum and natural gas may also occur within the seabed of the territorial sea or the area beyond as already settled by the 1958 Geneva Convention. The further seaward extension of hydrocarbon deposits will probably become regulated within the rights of coastal states in 2 200 nm economic zone or the outer edge of the shelf.

The unique occurrence of the metalliferous mud of the Red Sea caused the two governments of the adjacent coastal states of Saudi Arabia and the Sudan to conclude in 1974 an agreement establishing a common central zone and providing for the joint exploration and exploitation of the mineral resources of its seabeds (by the way Preussag recently concluded a contracts with the Red Sea Joint Commission for a final assessment of these mineral deposits).

With regard to the enormous potential of the manganese nodules of the deep sea the conclusion of an internationally accepted agreement has not been achieved. The establishment of a regime including the utilization rights of nodules has been part of the endeavours by a number of individuals, private and government institutions ever since the sixties. The Maltese Ambassador Pardo forwarded in 1967 the idea to have the bed of the deep sea declared a "common heritage of mankind", to be administered by a newly formed international body. This principal was accepted by the United Nations in 1970. Following the preliminary work of a Seabed Committee the subsequently convened U.N. Conference on the Law of the Sea with sessions in Caracas (1974), Geneva (1975) and New York (Spring and Fall 1976) has failed to bring about a solution. The developing countries, about 110 strong, insist that the international authority ruled by their majority of votes should have the exclusive right to exploit the seabed minerals. The industrialized countries, on the other hand, being the only ones capable through their financial and technological capabilities to exploit the seabed deposits want to have free access to the sites and be given the same chances for recovering nodules and producing the metals as the international authority. This proposal of a so-called parallel system, already a compromise compared to the original position of the industrial states, has failed to convince the developing countries. The prospects for an agreement within the United Nations are therefore not bright. The USA have been considering for several years the introduction of an interim legislation to allow US-companies to take up ocean mining. It is expected that 1977 will be the decisive year for the type of regime adopted for ocean mining.

SALINITY POWER, POTENTIAL AND PROCESSES, ESPECIALLY MEMBRANE

PROCESSES

[*]SIDNEY LOEB, M. RUDOLF BLOCH and JOHN D. ISAACS [∅]

[*]RESEARCH AND DEVELOPMENT AUTHORITY, BEN-GURION
UNIVERSITY OF THE NEGEV, BEER-SHEVA, ISRAEL

[∅]INSTITUTE OF MARINE RESOURCES, UNIVERSITY OF CALIFORNIA,
LA JOLLA, CALIFORNIA, U.S.A.

Salinity power refers to the extraction of energy from the salinity difference of two saline solutions, each of which is naturally available in large quantities. The total planetary supply of energy in this perpetually renewable form is immense, and in this respect it ranks with other great natural energy sources. Salinity power is unusual among ocean power sources in possessing very high energy densities represented by 250 to 5000 meters of water head for seawater versus fresh water and seawater versus brine, respectively.

Two membrane processes are being examined for salinity power service. In Pressure Retarded Osmosis (PRO) a brine acquires potential energy as hydraulic pressure by virtue of water permeation from the dilute to the concentrated solution, against the hydraulic pressure gradient. The volume-enhanced brine would be subsequently depressurized through a hydroturbine generator. With the Dialytic Battery (DB), also called reverse electrodialysis, voltage and current are produced directly by diffusion of ions across permselective membranes separating the concentrated and diluted solutions.

Membrane replacement cost, which accounts for half the total energy cost, has been estimated as a function of solution-pair type and the membrane process employed. The data are based on the use of presently available membranes tested for possible salinity power service, but developed for other applications. The estimated minimum replacement costs range from $0.10/kwh for osmotic membranes to $1.5/kwh for dialytic membranes. Clearly the costs are too high.

FIG. 1

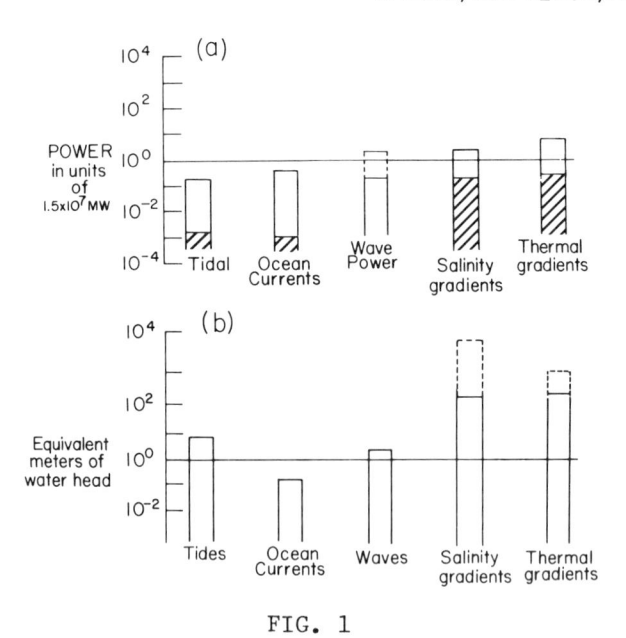

OCEAN POWER SOURCES

The results indicate the two most fruitful lines of development for energy cost reduction, namely: the membranes must be cheaper and must have less resistance to the permeable species.

The operation of a turbine by means of the vapor pressure difference between concentrated and dilute solutions was also examined. This is a salinity power variation of the original Claude process in which warm (surface) and cold (deep) ocean water were employed similarly. However, both of these approaches may invoke fundamental thermal and thermodynamic limitations which do not apply to the previous approaches.

I. SALINITY POWER AND ITS POTENTIAL

Salinity power refers to the extraction of power from the free energy made available when a dilute and a concentrated saline solution are mixed. In Fig. 1a is shown a comparison of the potential of salinity power with other possible sources of ocean power, in units of 1.57×10^7 megawatts, the projected world power need in the year 2000.* It is apparent that salinity power ranks very high (Wick and Isaacs 1976).

---

*The hatched areas of the bars represent the more readily available fraction of the power from each source.

The lower bar graph (Fig. 1b) depicts the energy density of these renewable sources, in terms of equivalent meters of water head. Energy density is, of course, an important parameter, for it is an inverse function of the dimensions and capital cost of installations for converting such energy sources into useful form. As can be seen the energy density of salinity power is some orders of magnitude greater than that of the three forms (or rather frequencies) of water motion - tides, waves and currents, and at its most dense even greatly exceeds maximum ocean thermal power.

Why should such a large and dense potential power source have received so little attention? Perhaps it is because other forms of energy are so much more conspicuous. A great fleet lifted by the tide some meters in a few hours is impressive. The same fleet lifted every few seconds by the waves of the open sea is even more so. Similarly boiling, freezing, and temperature (sensible heat) changes are other obvious manifestations of the presence of available energy. With regard to the expression of salinity power nature has played a trick on us. The principal solute of natural brines is sodium chloride. When dilute NaCl brine or water is mixed with concentrated NaCl brine very little sensible heat change occurs. Instead the free energy of mixing is consumed primarily in increasing disorder, i.e. entropy increase.* Had the Great Salt Lake, the Dead Sea, or any of the innumerable salt pans along desert coasts been dominated by almost any other electrolyte (say magnesium sulphate, or calcium chloride) the tributary streams would have steamed as they entered, or for others almost frozen! The free energy of mixing would be common lore.

The dilute and concentrated components of the binary system under consideration will be referred to herein as a solution-pair. We have considered the solution-pair river water-sea water for which the power dissipation is 2.6×10^6 megawatts (Wick and Isaacs, 1975). Aside from the potential power available, river water-sea water is attractive because its energy density is about 250 meters of water.

A second solution-pair considered was sea water - Dead Sea brine, an example of a perpetually renewable concentrated brine. Finally sea water-sodium chlorine brine was examined. Such brine is made by solution from surface or subterranean NaCl deposits. Brine from surface deposits would be renewable if obtained from salt pans along desert coasts. For the solution-pair sea water - Dead Sea brine, the free energy of mixing is equivalent to a hydrostatic head difference in the order of 5000 meters of water. (The osmotic pressure of Dead Sea brine is in the order of 500 atmospheres).

---

*Recall that, by definition, Gibbs free energy is equal to enthalpy minus the product of temperature and entropy.

II. PRESSURE-RETARDED OSMOSIS

A) Principle of Pressure-Retarded Osmosis (PRO)

In Fig. 2A is shown brine separated from water by a semipermeable membrane, i.e. a membrane permeable to water only. With such a membrane, the water naturally permeates from the water side to the brine side in the process of <u>osmosis</u>. The process can be stopped if a sufficiently high hydraulic pressure is applied to the brine side (Fig. 2C). The hydraulic pressure required to maintain this osmotic equilibrium is called the osmotic pressure, π

If the volumetric permeation rate of the water through the membrane, ΔV, is divided by membrane area, we have the water permeation flux, J_1. The relation between water permeation flux and applied forces, in the ideal case, is given by:

$$J_1 = A (\pi - P) \tag{1}$$

where P is the hydraulic pressure applied and A is the water flux constant, primarily a characteristic of the membrane used.

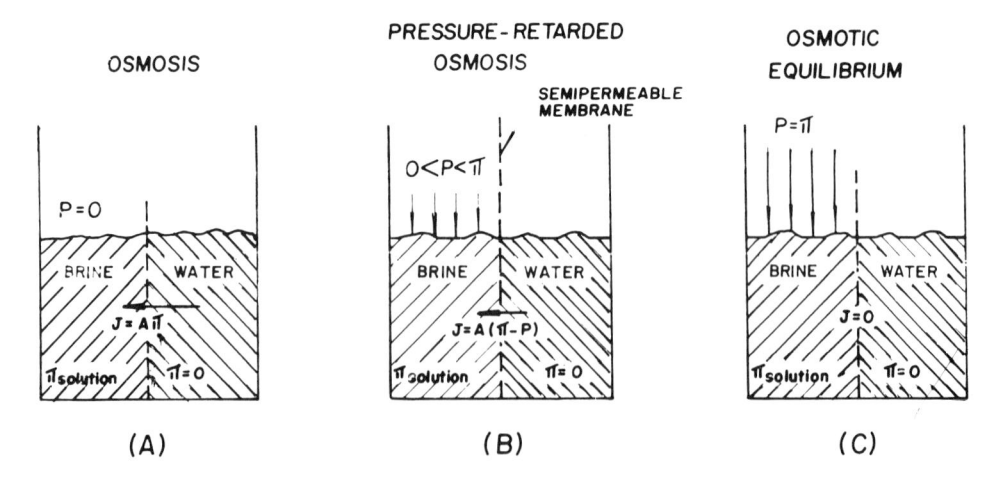

Fig. 2. The place of pressure-retarded osmosis (pro) in osmotic processes.

It is clear from Eq. 1 that if P=0 we have osmosis, for which $J_1 = A\pi$, and that if $P=\pi$ we have osmotic equilibrium, i.e. $J_1 = 0$. However, if P is between zero and π (Fig. 2B) we have Pressure-Retarded Osmosis (PRO), so defined because the direction of permeation flux is still the same as in osmosis but the flux is decreased as the hydraulic pressure increases.* It is important to understand that in PRO water permeates against the hydraulic pressure gradient, i.e. the flux is "uphill". The subsequent depressurization of the permeate through a hydroturbine-generator set would produce power by what may be described as an "osmotic waterfall", and the magnitude of the power produced at any instant would be the product of the hydraulic pressure head and the volumetric permeation rate, i.e. would be:

$$\text{Power (Ideal)} = P \Delta V \quad (2)$$

The PRO principle would still apply for the case that the liquid in the right hand compartment of Fig. 2B is a pressurized solution, i.e. also possesses both hydraulic and osmotic pressure, as long as:

$$J_1 = A(\Delta\pi - \Delta P) \quad (3)$$

We will always be considering this more general situation in which water permeates from a Permeate-Donor** solution into a Permeate-Receiver** brine (the solution-pair).

B) Application of Pressure-Retarded Osmosis (PRO) to Continuous Power Production

As shown in Fig. 2B, Pressure-Retarded Osmosis (PRO) suffers from the fact that it is an intermittent or batch-type operation. Large scale osmotic power production could be carried out in a continuous and steady-state fashion, utilizing one of the solution-pairs described below.

1. River Water - Sea Water as the Solution-Pair

Referring to Fig. 3, we see that sea water at an osmotic pressure, π_{PR} (PR = Permeate-Receiver), of 25 atmospheres is pumped to a hydraulic pressure, P_{PR} of 10 atmospheres and at a rate $\dot{V}$ cubic meters per day (m³/day). The sea water enters the PRO

---

*This is in contrast to reverse osmosis, a well-developed technology for which P is greater than π and in which an increase in P increases the flux.

**For the case of the Dialytic Battery, to be discussed, the permeate is not water but electrolyte. To avoid confusion we use the terms "Permeate-Donor" and "Permeate-Receiver" to refer only to water permeation in PRO, and not to salt permeation with the Dialytic Battery.

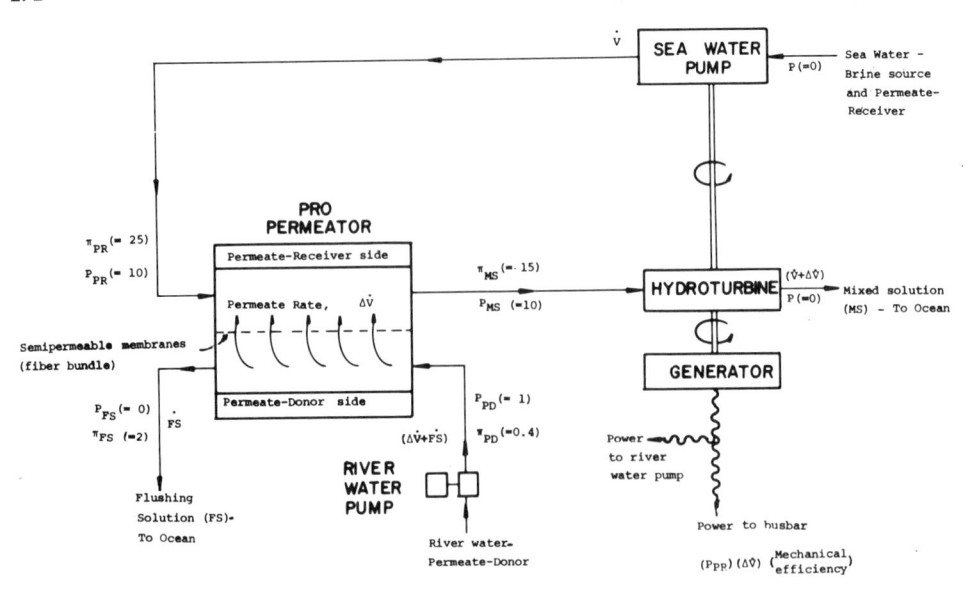

Fig. 3. Power production by pressure-retarded osmosis with river water-sea water as the solution-pair.

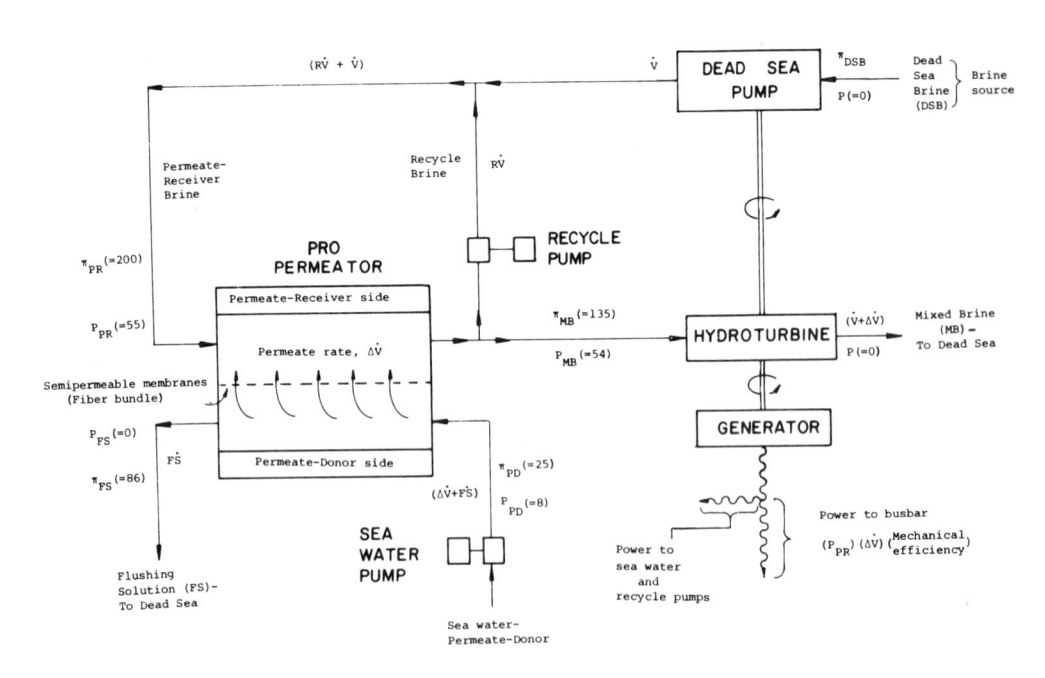

Fig. 4. Power production by pressure-retarded osmosis with sea water-dea sea brine as the solution-pair.

SALINITY POWER

permeator on the Permeate-Receiver side, in which the volumetric rate is enhanced by $\Delta\dot{V}$, the water permeating through the membrane from the Permeate-Donor side, without loss of pressure. This Mixed Solution is sent to the hydroturbine where it is depressurized, producing power for the sea water pump and the generator. The generator supplies power for the river water pump, and the remaining power is sent to the busbar as permitted by the fractional mechanical efficiency, ME, defined by:

$$ME = \frac{\text{Busbar power}}{(P_{PR})(\Delta\dot{V})} \tag{4}$$

It will be noted that the river water pump supplies both the water permeate, at a rate $\Delta\dot{V}$, and the Flushing Solution, at a rate $\dot{FS}$. The Flushing Solution is necessary to prevent accumulation and possible precipitation of salts on the Permeate-Donor side.

2. Sea Water-Dead Sea Brine as the Solution-Pair*

The flow diagram for sea water-Dead Sea brine** is shown in Fig. 4. A new feature is the Recycle brine, circulating at a rate, $R\dot{V}$, where R is any number greater than zero. Recycling may be desirable because saturated Dead Sea brine (or sodium chloride) might damage the membranes.

It should be noted in Fig. 4 that the hydraulic pressure, P_{PR}, can be much higher than for the case of river water-sea water.

---

*The case of river water as the dilute solution has been considered by Loeb (1976a) and by Loeb et al. (1976b). However in the more general application sea water would be used together with such concentrated brines. In addition (Jordan) river water is an ever-dwindling supply to the Dead Sea. Consideration is being given to bringing Mediterranean water to the Dead Sea for a hydroelectric project. The head difference is in the same order of magnitude as P_{PR} in Fig. 3. Hence PRO at the Dead Sea terminus could in principle double the power output.

All solution-pair cases discussed herein in which sea water is the Permeate-Donor are considered in more detail by Loeb et al. (1976c).

**Dead Sea brine has a density of 1.23 and a concentration, in grams per liter solution, as follows: $MgCl_2$ - 165, $CaCl_2$ - 48, NaCl - 105, KCl - 14.5, $CaSO_4$ - 1.

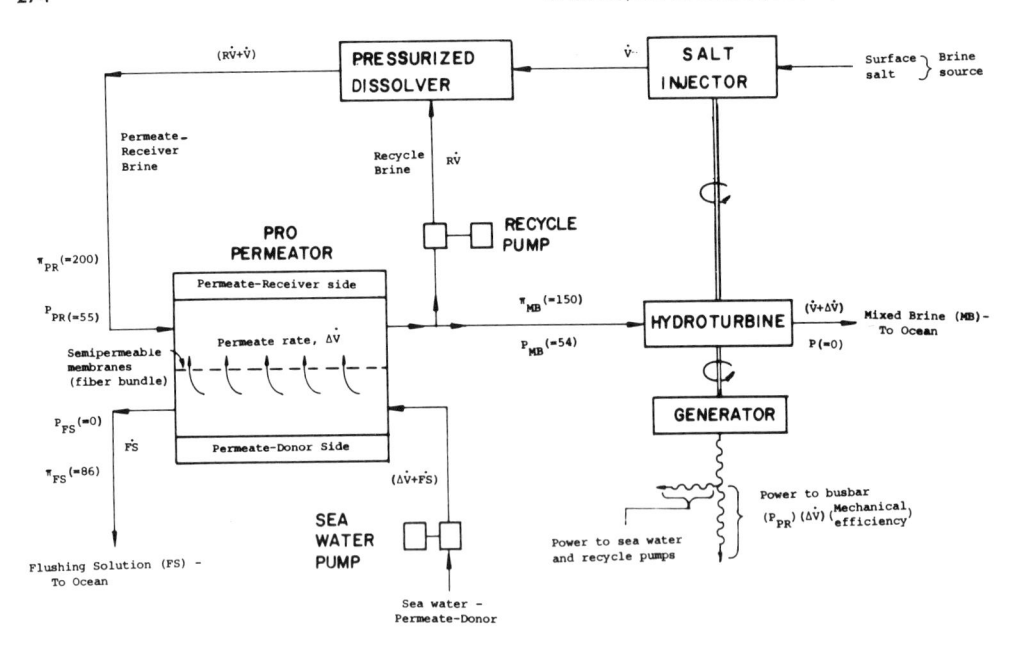

Fig. 5. Power production by pressure-retarded osmosis with sea water-sodium chloride brine as the solution-pair, and surface salt as the brine source.

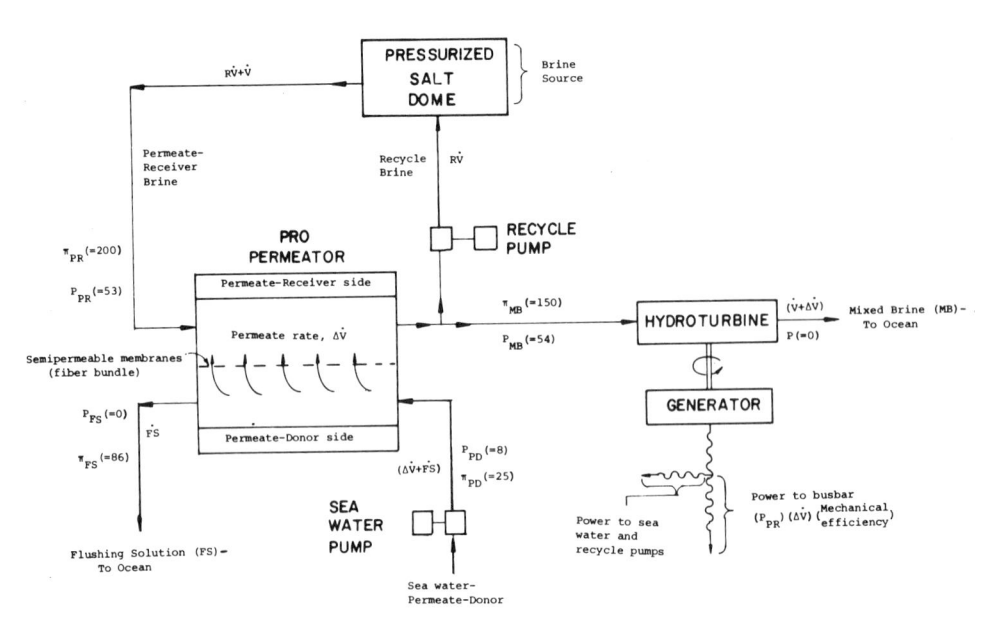

Fig. 6. Power production by pressure-retarded osmosis with sea water-sodium chloride brine as the solution-pair and underground salt as the brine source.

3. Sea Water-Sodium Chloride Brine as the Solution-Pair, Where the Brine Source is Solid NaCl

The flow diagram would be as in Fig. 5. A salt injector must pump the solid salt into the pressurized dissolver. The Recycle Brine now has the added function of dissolving solid salt, and the osmotic pressure π_{PR}, will be a function of the magnitude of R.

We may also think of using underground salt deposits such as occur in salt domes. In this case, as shown in Fig. 6, the Recycle brine would be pumped to a pressurized salt dome to dissolve more salt. The salt injector would not be necessary. The production of brine by pumping water into an underground NaCl deposit is a well-developed technology, and an appreciable fraction of sodium chloride is provided in this manner. Also in many places in the world, salt domes are adjacent to or underneath the ocean (Kaufman 1960).

C) Test Equipment and Procedure

In PRO, liquid streams must flow freely on both sides of and in close interfacial contact with a membrane across which a large hydraulic pressure drop exists. These dual requirements can be met by very fine hollow fibers fabricated to be essentially thick-walled tubes in terms of the ratio of outside to inside diameter.

The hollow fiber used by Loeb et al. (1976 a,b,c) was the du Pont B-10 Permasep hollow fiber. This fiber, made for sea water desalination by reverse osmosis, is fabricated from an aromatic polyamide fiber to have outer and inner diameters of 98 and 42 microns respectively. Furthermore the fiber possesses a thin but dense skin on the outside surface, the remainder of the fiber wall having a relatively porous substructure, i.e. the membrane is asymmetric. Such asymmetry can be of value in minimizing resistance to permeation flux.

A hollow fiber "Minipermeator" is shown in Fig. 7. Testing consisted of passing Permeate-Receiver brine (partially diluted Dead Sea Brine or a sodium chloride brine) at a hydraulic pressure of 53 atmospheres on the outside of the bundle of hollow fibers, while the Permeate-Donor, filtered sea water, was pumped into the inside of the fibers at an inlet pressure sufficient to supply the permeate rate ΔV and the desired Flushing Solution rate FS. Because of the small inner diameter of the fiber bores, the required inlet pressure was appreciable, in the order of 5-15 atmospheres. The relation measured was water permeation flux as a function of driving force, $(\Delta\pi - \Delta P)$, and subsequent calculation of the water flux constant, A (Eq. 3).

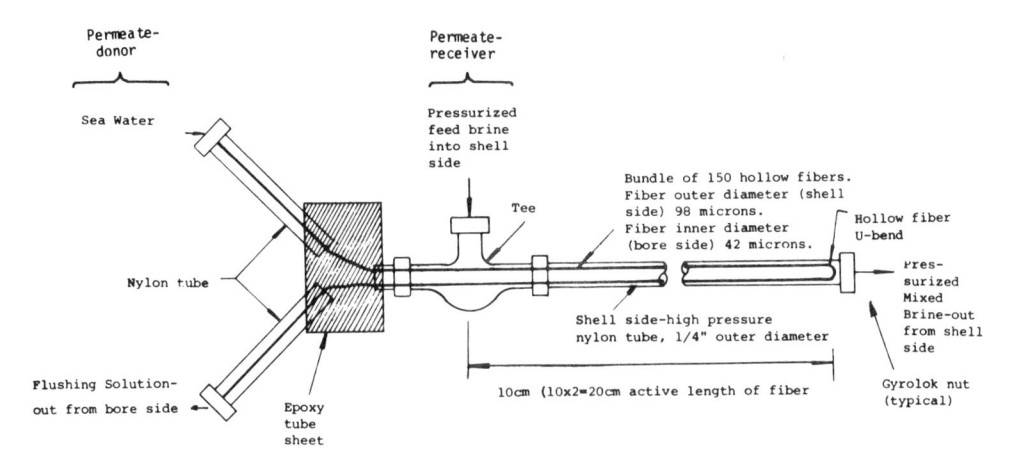

Fig. 7. Schematic drawing of minipermeator.

D) Energy Costs with PRO

1. Present Costs

In Table I are shown membrane replacement costs $/kwh, which represent about half the total cost of the energy. These data were obtained from the tests described above and by the analytical methods detailed by Loeb (1976a). These methods are given in condensed form in APPENDIX A.

As can be seen in Table I, the membrane replacement cost is 0.22, 0.13, 0.10 and 0.13 for the solution-pairs river water-sea water, sea water-Dean Sea brine, sea water-NaCl brine, from surface salt and sea water-NaCl brine from a salt dome.

2. Prospects of Lower Membrane Replacement Costs with PRO

From consideration of the factors, largely discussed in APPENDIX A, contributing to membrane replacement cost, it is possible to say that:

$$\frac{\$}{kwh} \Big/ membrane \sim \frac{\$\ day/m^3}{(A)(\Delta\pi - \Delta P)_{mean}(P_{PR})(ME)(membrane\ operating\ life)} \quad (5)$$

where $\$\ day/m^3$ is the membrane original cost per daily cubic meter of water permeate. The values of the various terms in Figs. 3, 4, 5, 6 and in Table I-A of APPENDIX A are such that the product $(\Delta\pi - \Delta P)_{mean}(P_{PR})(ME)$ is fairly close to maximum as given therein. However it does seem possible ultimately to cut membrane original cost in half and to double the permeability, i.e. the A value.

TABLE I

MEMBRANE REPLACEMENT COSTS WITH PRESENTLY AVAILABLE MEMBRANES

| Solution-pair || Membrane replacement cost, $/kwh ||
Dilute Solution (Permeate Donor)	Concentrated Solution	Pressure-Retarded Osmosis	Dialytic Battery
River water	Sea water	0.22	1.8
Sea water	Dead Sea brine	0.13	1.5
Sea water	NaCl brine (available from salt deposits)	0.10 (surface salt) 0.13 (salt dome)	1.5

TABLE II

CONSUMPTION RATES OF FEED STREAMS IN A 100 MEGAWATT PRO POWER PLANT

| Solution-pair || Approximate consumption rate of feed stream for 100 megawatt power plant, metric tons/day ||
Dilute Solution (Permeate Donor)	Concentrated Solution	Permeate-Donor	Solid Sodium Chloride
River water	Sea water	20,000,000*	–
Sea water	Dead Sea brine	5,800,000	
Sea water	NaCl brine	5,900,000	440,000

*This may be conceptualized as one-eighth kwh produced per m^3 of permeate.

The total effect would be to reduce membrane replacement costs by a factor of four, giving about 0.05 $/kwh for river water-sea water and about 0.03 $/kwh for the other two solution-pairs. Assuming that membrane replacement costs are half the total energy costs, the total energy costs would be 0.10 $/kwh for river water-sea water and 0.06 for sea water-Dead Sea brine and sea water-NaCl brine.*

E) Consumption Rates of Feed Streams with 100 Megawatt Plant

It is important to realize that, because of the relatively low value of the free energy of mixing, the consumption rates of feed streams entering a PRO complex would be quite large. In Table II we give consumption rates of pertinent quantities for a 100 megawatt power plant. The rates given are obtained by use of Figs. 3,4,5,6 by APPENDIX A, and by the formula:

$$\Delta V \ , \ m^3/day = \frac{(100000)\ (24)}{(P_{PR}\ (0.028)(ME)} \qquad (6)$$

where 100000 is kilowatts, 24 is hour/day, and 0.028 is kwh/m^3 atm.

These rates are, as can be seen in TABLE II, very large. For a 100 megawatt plant the river water rate (Row 1) is about 5 times the water consumption rate of New York City. Also the sodium chloride consumption rate (Row 3) would be in the same order of magnitude as the world production rate of sodium chloride.

III. THE DIALYTIC BATTERY (REVERSE ELECTRODIALYSIS

Instead of generating a pressurized liquid, one may also think of generating voltage and electric current directly by means of another membrane process referred to as the Dialytic Battery, or Reverse Electrodialysis. This method, described below, was first considered some years ago, and recently by Weinstein and Leitz (1976), whose work is described below. Forgacs (1975) has a very active program to produce energy economically by means of the Dialytic Battery, and appears to be approaching the membrane improvements cited below as necessary for economical energy production.

A) Principle and Description of the Dialytic Battery

Assume that a membrane permeable to ions of one sign, say to cations, (cation exchange membrane) separates river water and sea water, and that Ag/AgCl electrodes, i.e. reversible, are at each end of the container. If these electrodes are connected through

---
*At present, energy costs in the United States are in the order of 0.03 $/kwh.

SALINITY POWER

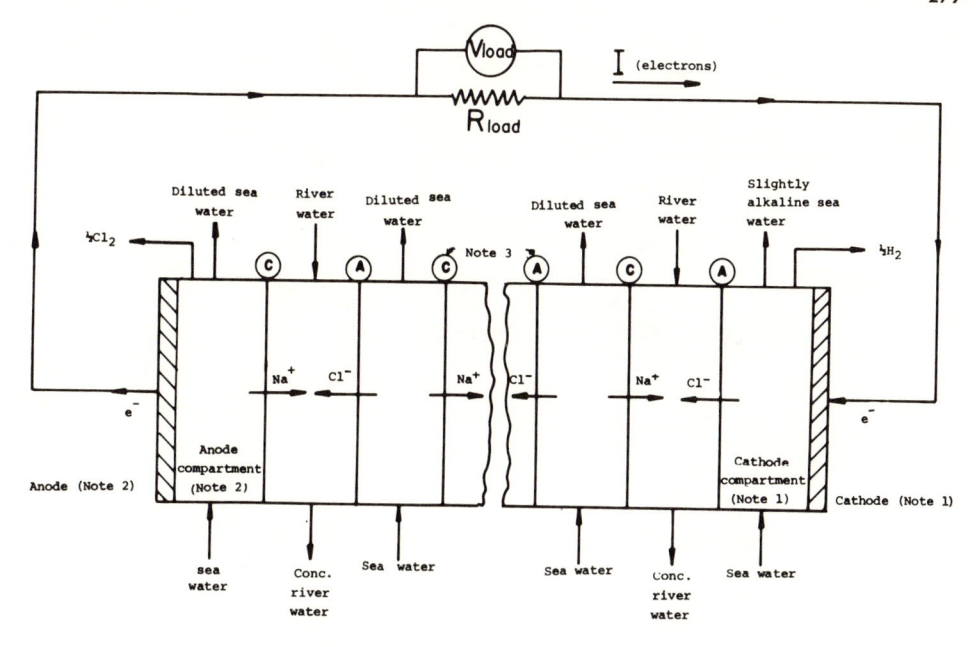

Fig. 8. Dialytic battery.

an external resistance, current will flow and useful energy will be obtained. However, the voltage due to the concentration difference will be very low even at the maximum value, i.e. at zero current. The voltage can be multiplied merely by increasing the number of membranes, using alternate cation and anion exchange membranes to separate the river water and the sea water in the membrane stack, shown in Fig. 8.

In large scale application it would not be appropriate to use Ag/AgCl electrodes. Irreversible permanent electrodes would be used, thus causing electrochemical reactions at each electrode. These reactions consume energy but this loss would be minimized by using a large number of membranes. A more serious source of energy loss is the electrical resistance of the stack given as the sum of the resistance of the membranes and the solution-pair compartments.

B) Test Equipment and Procedure

Weinstein and Leitz (1976) conducted Dialytic Battery tests on a membrane stack containing 31 cation and 30 anion-exchange membranes each 0.6 mm thick and with a total membrane area of 1.4 m^2. Other equipment details are given in APPENDIX B. Weinstein and Leitz used 0.57 molar NaCl solution to simulate sea water and measured power delivery as a function of the salt concentration of the "river" water, consisting of dilute NaCl solutions. The tests confirmed their theoretical analysis, and enabled calculation of minimum energy costs, as given below.

C) Energy Costs with the Dialytic Battery

1) Present Costs

As given in Table I and calculated in APPENDIX B, the membrane replacement cost with presently available membranes, used primarily for electrodialysis, is 1.8 $/kwh for river water-sea water and, surprisingly enough, 1.5 $/kwh for sea water with either Dead Sea brine or concentrated NaCl brine made from solid salt. The high value of 1.5 $/kwh with concentrated brines can be explained by reference to Eq. 1-B in APPENDIX B. There it is seen that the maximum power output is a direct function of the _ratio_ of the salt activities and not of the difference in the activities. This ratio is actually less for the solution-pair sea water - concentrated brine than for river water-sea water (although compensated for by the lower resistance of the stack when the former solution-pair is used).

2) Prospects of Lower Membrane Replacement Costs with the Dialytic Battery

Following Weinstein and Leitz we used an original membrane cost of 10 $/m^2, an areal resistance of 11 ohm-cm^2 for the membranes, and 40 ohm-cm^2 for the river water compartment in determining membrane replacement costs with present membranes. Furthermore we assumed a 4 year membrane life. Weinstein and Leitz consider it within the realm of possibility to reduce original membrane cost and overall resistance* each by a factor of ten by using very thin membranes and river water compartments. They also believe that membrane life can be doubled (compared to electrodialysis, a more stringent application. In electrodialysis, both polarization and membrane failure are more serious than in reverse electrodialysis). If all these improvements are attained the membrane replacement cost will be in the order of 1.80/200 $\simeq$ 0.01 $/kwh, a very respectable figure.

---

*The overall resistance to the passage of ions is the reciprocal of the conductance. The latter would be analogous to the water permeability constant in PRO.

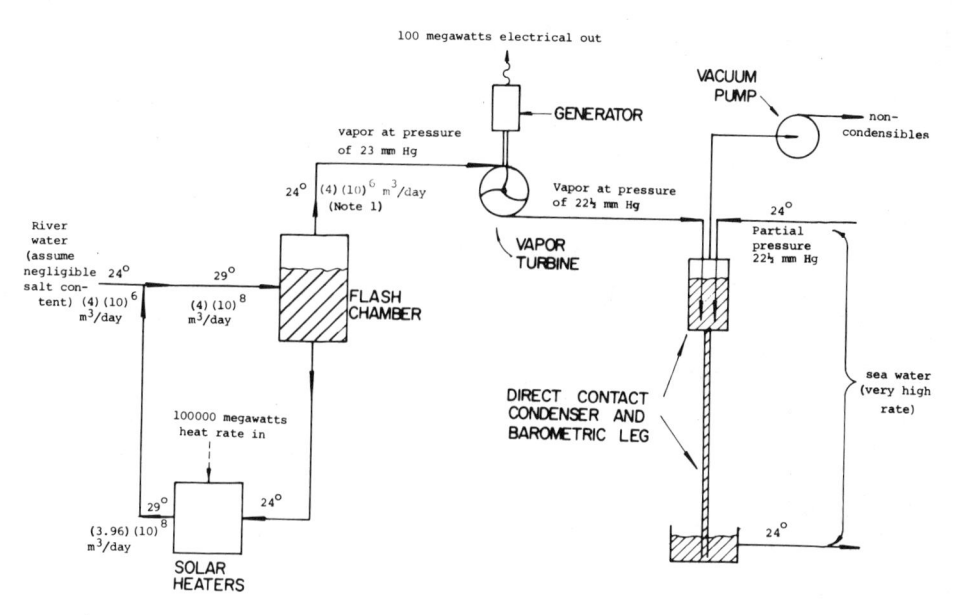

SALINITY POWER

Fig. 9. 100 megawatt vapor pressure difference power plant driven by solution-pair: river water-sea water.

D) Consumption Rates

It can be expected that the consumption rates with the Dialytic Battery would be similar to those with PRO (Table II).

IV VAPOR PRESSURE DIFFERENCE ENGINES DRIVEN BY SALINITY POWER

In the early part of this century Claude (1930) tested a proposal, first made by D'Arsonval, to drive a vapor turbine by means of the vapor pressure difference (vpd) between two sea water solutions at different temperatures. This process is now being re-examined in the light of the vast energy resources potentially available in the temperature differences between surface and deep sea water.

The process is a form of heat engine and hence is limited by the magnitude of the available temperature differences (and ultimately by the Carnot cycle restrictions). In principle, at least, it would appear that the thermal limitations could be removed by utilizing a dilute and a concentrated solution at the same temperature to achieve the desired vpd. The principle is shown in Fig. 9

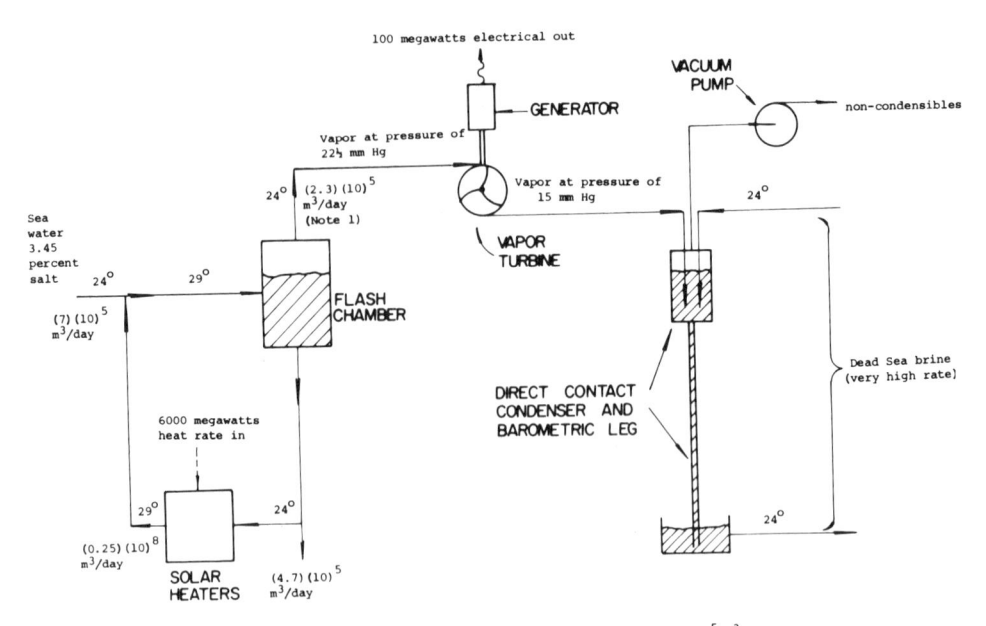

Fig. 10. 100 megawatt vapor pressure difference power plant driven by solution-pair: sea water-dead sea brine.

for the solution pair: river water-sea water. As can be seen, river water enters a flash chamber where saturated vapor is produced at a temperature of 24°C and a pressure of 23 millimeters of mercury (mm Hg). The latent heat for the evaporation is provided by circulation of the river water through solar heaters. The vapor passes through the turbine in which the vapor pressure is reduced by 1/2 mm Hg to a vapor pressure equal to the partial pressure of water in a sea water solution also at 24°C. The vapor is condensed in a direct contact condenser by contact with this sea water flowing at a very high rate. Non-condensibles are removed by a vacuum pump and a barometric leg enables raising of the emerging sea water from near vacuum conditions to atmospheric pressure.

Consideration of Fig. 9 reveals three major engineering difficulties:

1) The vpd is only 1/2 mm Hg (compared with perhaps 12 mm Hg in the Claude process). The low vpd would enormously increase the size of the turbines required* for a given power plant.

---

*The low absolute values of the vapor pressures involved would also contribute greatly to the size of turbines and attendant piping in any vpd power plant.

2) For each kilowatt hour of useful energy produced, about 1000 kilowatt hours of heat must be added from solar (or other) heaters and transferred from evaporating to condensing streams. Thus a very large capital expense would be incurred in solar heater and heat exchanger capacity.

3) For an assumed temperature rise of 5°C, a 100 megawatt plant would require a river water circulation rate of 4×10^8 m^3/day, about 1/4 of the Mississippi river flow rate.

The above difficulties can be diminished by using a solution-pair having a greater driving force. In Fig. 10 is shown a plant driven by the solution-pair sea water - Dead Sea Brine. The vapor pressure difference is about 8 mm. This is an improvement by a factor of 16 over the previous example. This same factor occurs in the lowering of the latent heat/useful work ratio to $1000/16 \simeq 60/1$ and a decrease in the circulating stream rate to $4 \times 10^8/16 \simeq 0.25 \times 10^8$ m^3/day. These improvements would make the process more attractive but the basic problems remain of large turbine size, large solar heater and vapor condenser capacity, and large circulation rates.

Since essentially the same amount of heat must be added at the flash chamber and then removed in the direct contact condenser, one might think of circulating Dead Sea Brine between these two units, thus eliminating an external source of heat. However in order to transfer heat to the flash chamber a lower temperature level would be required in it then in the condenser. Unfortunately this direction of temperature drop would decrease and possibly even eliminate the all-important vapor pressure difference across the turbine.

It appears then that a vpd power plant driven by salinity differential would still be subject to serious thermal limitations.

V ACKNOWLEDGEMENT

This research was sponsored in part by a grant from the United States-Israel Binational Science Foundation (BSF), Jerusalem, Israel.

APPENDIX A

CALCULATION OF MEMBRANE REPLACEMENT COSTS IN PRO WITH PRESENTLY AVAILABLE MEMBRANES

I. With River Water-Sea Water as the Solution-Pair

By means of the calculation methods of Loeb (1976a) it is possible to give the membrane replacement cost with this solution-pair as

$$\left.\frac{\$}{kwh}\right|_{\text{membrane replacement}} = \frac{0.051}{(J_1)(P_{PR})(ME)} \quad (1\text{-}A)$$

where

$$J_1 = (0.0062)(\Delta\pi - \Delta P)_{mean} \quad (2\text{-}A)$$

$$ME = 0.75 - 0.20 \, V/\Delta\dot{V} \quad (3\text{-}A)$$

The term 0.051 is arrived at on the assumption that original hollow fiber cost is 35 dollars per daily cubic meter of permeate ($/day/m^3) at a water permeation flux of 0.04 m^3/m^2 day, a fiber life of 3 years and 90% on-stream time. These values are in accordance with test results in PRO and membrane properties.

The term $(\Delta\pi - \Delta P)_{mean}$ is the log mean driving force and has the value 7 as determined from Fig. 3. The coefficients in Eq.3-A are arrived at by allowing for various irreversible energy losses. Pump, generator, and hydroturbine efficiencies are assumed to be 0.90, 0.98 and 0.93 respectively. The river water side is also a consumer of energy because of the hydraulic pressure needed and because of the flow rate of Flushing Solution. These are given as ratios: $P_{PD}/P_{PR} = 0.1$ and $FS/\Delta\dot{V} = 0.2$.

The components required in Eqs. 1,2,3-A are given in Table I-A and we find that the membrane replacement cost is 0.22 $/kwh.

II. With Sea Water-Dead Sea Brine as the Solution-Pair

For this solution-pair the pertinent equations are

$$\left.\frac{\$}{kwh}\right|_{\text{membrane}} = \frac{0.098}{(J_1)(P_{PR})(ME)} \quad (4\text{-}A)$$

$$J_1 = 0.0047(\Delta\pi - \Delta P)_{mean} \quad (5\text{-}A)$$

$$ME = 0.70 - (0.22)(\dot{V}/\Delta\dot{V})(1+0.1R) \quad (6\text{-}A)$$

SALINITY POWER

For the term 0.098, the original hollow fiber cost is 68 $/day/m³. The coefficient 0.0047 is the experimentally determined value of A multiplied by a factor of 0.85 to allow for long-term membrane deterioration. The coefficients of the mechanical efficiency equation were obtained in a manner similar to those of Eq.3-A, but with allowance for the recycle factor, R, and for a 2% friction loss on the Permeate-Receiver side.

When the terms are inserted, as given in Table 1-A, we find that the membrane replacement cost is 0.13 $/kwh.

III. With Sea Water -NaCl Brine as the Solution-Pair

$\frac{\$}{Kwh}\big|_{membrane}$ (surface salt or salt dome), given by Eq.4-A

$$J_1 \text{ (surface or salt dome)} = (0.00040) \, (\Delta\pi - \Delta P)_{mean} \quad (7\text{-A})$$

ME(surface salt), given by Eq. 6-A

$$ME(\text{salt dome}) = 0.70 - 0.067 \, \dot{RV}/\dot{\Delta V} \quad (9\text{-A})$$

Membrane replacement cost is found to be 0.10 $/kwh for brine made from surface salt and 0.13 kwh for brine made from a salt dome.

APPENDIX B

CALCULATION OF MEMBRANE REPLACEMENT COSTS WITH THE DIALYTIC BATTERY WHEN PRESENTLY AVAILABLE MEMBRANES ARE USED

I. With River Water-Sea Water as the Solution-Pair

According to the development of Weinstein and Leitz (1976) at optimum operating conditions, the maximum power per square meter of membrane is given by

$$\frac{Watt}{m^2}\bigg|_{max} = \frac{(\{\alpha RT/F\} \ln a_s/a_r)^2 \cdot 10^4}{2(R_a + R_c + h_r/\Lambda_r \, C_r)} \quad (1\text{-B})$$

where:

α is the average membrane permselectivity, defined as the ratio of desired ion permeation rate to total ion permeation rate. α is taken as 0.64.

R is the gas constant, 8.314 $\frac{watt\ sec}{mol°K}$

T is 298° Kelvin

F is the Faraday number, 96500 $\frac{coulomb^*}{equivalent}$

---

*Implicit in Eq. 1-B is the assumption that we are dealing with uni-univalent electrolytes.

TABLE 1-A

Components Determining Membrane Replacement Costs in PRO

Dilute	Concentrated solution or brine	Osmotic and hydraulic pressures as in Fig.	$\dot{V}/\Delta\dot{V}$ ratio of brine source rate to water permeate rate	R Recycle factor (Note 1)	ME Fractional Mechanical efficiency (Eqs.3,6, 8-A)	$(\Delta\pi-\Delta P)_{mean}$ Mean driving force in module, atm.	A_{ave} Average water flux constant, m³/m²day atm (Eqs.2,5, 7-A)	J_1 Water flux m³/m² day	Membrane replacement costs $/kwh (Eqs.1, 4-A)
River water	Sea water	3	1/1	0	0.55	7.0	0.0062	0.043	0.22
River water	Dead Sea brine	4	1/1	2	0.44	67	0.00047	0.031	0.13
Sea water	NaCl brine From surface deposit	5	0.17/1	20.5	0.59	69	0.00040	0.028	0.10
Sea water	NaCl brine From salt dome	6	0.17/1	20.5	0.47	69	0.00040	0.028	0.13

1) R is determined by π_{PR} and π_{MB} (Figs. 4,5 and 6). Calculation method not given herein.

SALINITY POWER

10^4 is cm^2/m^2

R_a and R_c are the areal resistances of the cation- and anion-exchange membranes, assumed to be 11 ohm-cm² each, for a membrane thickness of 0.6 millimeters.

h_r is thickness of river water compartment, assumed to be 0.1 cm.

Λ_r is equivalent conductance of river water, assumed to be 0.096 liter/(ohm)(mol)(cm).

C_r is river water concentration. This is subject to optimization. W and L found an optimum value at 0.0259 mol/liter.

$h_r/\Lambda_r C_r$ is the areal resistance, R_r, of the river water compartment (that of the sea water compartment being considered negligible) and is 40 ohm-cm² with the values given above.

a_s/a_r is the ratio of NaCl activities. W and L used $C_s = 0.57$ mol/liter to give $a_s/a_r = (0.57/0.0259)^{0.92} = 17$.

When the above values are placed in Eq. 1-B we find that

$$\left.\frac{watt}{m^2}\right|_{max} = 0.174$$

The original cost of the membranes per kilowatt is

$$\left.\frac{\$}{Kw}\right|_{membranes} = (10)(1000)/(0.174) = 57500$$

Where:

10 is the assumed membrane original cost, $/m^2$

1000 is watt/kilowatt

We assume four year life and 90% on-stream time to give

$$\left.\frac{\$}{kwh}\right|_{\substack{membrane \\ replacement}} = (0.25)(57500/(365)(24)(0.90) = 1.8$$

where (365)(24) is hour/year

II. With Sea Water-Concentrated Brine (Either Dead Sea Brine or NaCl Brine) as the Solution-Pair

We still use Eq. 1-B with the following changes:

$a_{brine}/a_s = (26/3.5)^{0.92} = 6.3$

where 26 is the salt concentration in the Dean Sea brine wt.%

The resistance of both liquid compartments is considered negligible (in Eq 1-B the resistance of the river water compartment is not negligible).

When these values are placed in Eq. 1-B we find that membrane replacement cost is 1.5 $/kwh.

REFERENCES

Claude, G. (1930). Power from the Tropical Seas. Mechanical Engineering 52, 1039-44.

Forgacs, C. (1975). Generation of Electricity by Reverse Electrodialysis (RED). BGUN-RDA-69-75. Research and Development Authority Ben-Gurion University of the Negev, Beer-Sheva, Israel.

Kaufmann, D. (1960). Sodium Chloride. Reinhold, New York.

Loeb, S. (1976a). Production of Energy from Concentrated Brines by Reverse Osmosis. I. Preliminary Technical and Economic correlations. Journal of Membrane Science 1, 49-63.

Loeb, S., van Hessen, F., and D. Shahaf (1976b). Production of Energy from Low Concentrated Brines. II. Experimental Results and Projected Energy Costs. Journal of Membrane Science 1, 249-63.

Loeb, S., van Hessen, F., Levi, J. and M. Ventura (1976c). The Osmotic Power Plant. In proceedings of "Energy 11" Eleventh Intersociety Energy Conversion Conference, State Line, Nevada, Institute of Electrical and Electronic Engineers, Vol. 1, 51-57.

Weinstein, J., and F. Leitz (1976). Electric Power from Differences in Salinity. Science, 191, 557-9.

Wick, G., and J. Isaacs (1975). Salinity Power. IMR Reference No. 75-9 Institute of Marine Resources, University of California, La Jolla, California.

Wick, G., and J. Isaacs (1976). Utilization of the Energy from Salinity Gradients. Wave and Salinity Gradient Energy Conversion Workshop, University of Delaware, 24-26 May.

PHYSICAL OCEANOGRAPHY

J.D. WOODS

OCEANOGRAPHY DEPARTMENT, SOUTHAMPTON UNIVERSITY

It is no easy task to summarize the wide variety of material presented by physical oceanographers during this Joint Oceanographic Assembly, where the main emphasis lies on the multi-disciplinary unity of our subject. How can one provide an overview in which the contribution of the physicist is seen in relation to that of his colleagues in sibling disciplines? One way is to classify each contribution in terms of a single parameter. Choosing "time", I ascribed to each paper a range of time scales that appeared to be of primary interest to the author. Thus, those concerned with the evolution of the ocean basins were given the range 100 to 1000 million years; those concerned with changing patterns of climate and fisheries populations, years and decades; those concerned with turbulence in the ocean, minutes and seconds; and so on. Then I plotted a frequency distribution showing the number of times each occurred, to the nearest half decade on a logarithmic time scale. The result is shown in fig. 1. I should mention that this histogram is based on the abstracts, and some of the speakers appeared to have changed their interests between abstract and final presentation, but I hope the resulting shifts in time scale do not too seriously alter the shape of fig. 1.

Looking at the distribution we notice right away that it can be divided into three main bands, with spectral gaps at 10,000 years and one season. The centre of interest lies at a time scale of a few decades - the human (or should I say fish) time scale - and the range extends eight decades on either side, up to a few milliard years and down to a fraction of a second. We see that the palaeo-oceanographers and the micro-oceanographers, both relative newcomers to our community, are (to use a colloquial expression) equally "way out".

1. Distribution of numbers of topics discussed at JOA vs. time scales in the range 10^{-7} – 10^9 years.

My task in this summary lecture is to identify some of the highlights of the physical oceanography contributions, to identify trends and to make predictions about where we might be going between now and the next Assembly. Let me begin by going back six years to 1970, to the closing session of the Tokyo Joint Oceanographic Assembly.

Micro-oceanography

In summing up the physical oceanography contributions during the Tokyo JOA, Walter Munk highlighted the rapid progress made during the previous quinquennium in exploring the micro distributions of temperature, salinity and velocity in the Ocean. In 1970 our knowledge of micro-structure depended almost exclusively upon measurements with instruments (STD, XBT) designed to explore larger scale structures. Since then there has been developed a new generation of purpose-built microstructure instruments capable of resolving structure so small that they are directly attenuated by molecular transport. Armed with these new instruments, the micro-structure enthusiasts are now able to study distributions on scales of a few decimetres and less, and to deduce from them what are the processes responsible for smearing out spatial irregularities in the temperature salinity and velocity fields. These processes provide the ultimate sink of variances created initially at larger scales by atmospheric exchange. Having learnt how to measure the rate at which variance is being destroyed by these sinks, the next step will be to study their distribution with depth and geographical location, and with respect to the larger motions in which they are imbedded and to which they respond. We heard a preliminary report on a major Soviet effort to explore these global distributions - the first step towards a World climatology of ocean microstructure.

Synoptic-scale eddies

The past ten years saw the final penetration to the molecular limit and the resulting establishment of micro-oceanography as a recognized sub-discipline. As this "decade of the variance sink" comes to an end, the emphasis has switched to much larger scales, and in particular to the spectral peak of turbulent kinetic energy, now identified as occurring at 100 km and a month, plus or minus an octave in each dimension. The first major attempts to explore the motions responsible for this spectral peak, which embraces 99% of the kinetic energy of the Ocean, took place during the six years since the Tokyo JOA, and we have heard a number of reports concerning the Soviet, British and American expeditions; POLYGON, MODE and POLYMODE, the last of which is currently in the field phase.

The key development exploited by these expeditions has been the ability to measure horizontal velocity directly, by moored current meters and by drifting Swallow floats, rather than to estimate it indirectly by the dynamic method, or even less directly by the patterns made by temperature, salinity, oxygen, etc. The velocity fluctuations in the interior of the ocean are now known to be so much more rigorous than the mean currents and so intermittent and of such short scale (∼100 km), that it is inconceivable that a sufficient number of direct current measurements could be made to produce accurate maps of the mean circulation.

Mean circulation

It is interesting to speculate what state physical oceanography would be today if, instead of Nansen bottles and reversing thermometers, we had depended upon current meters and Swallow floats for exploration of the deep ocean during the century since HMS "Challenger". We would surely have had a highly stochastic image of the ocean, and I doubt whether we would have known very much about the mean circulation. Reality was, of course, different, and descriptive physical oceanographers have greatly contributed to our understanding of the World Ocean by their Atlases of scalar distributions. I suggest that such integrated mean pictures have even greater value today, now we know of the chaos on the 100 km scale. The synoptic scale eddies may contain 99% of the kinetic energy, but the slow residual flows moving at one-tenth of the speed of the eddies dominate the transport of scalars on scales of a thousand kilometres and more, as we would predict from simple dimensional arguments.

	L	U	$K \propto LU$
Synoptic scale eddy	100	10	10^8
General circulation	5000	1	5×10^8
	km	cm/s	$cm^2 s^{-1}$

Sir George Deacon pointed out over breakfast yesterday that the Bible has something to tell us in this connexion. If one turns to Matthew 18, 10 and, as is appropriate in Edinburgh, reads from the King James translation:

> "If a man have a hundred sheep, and one of them be gone astray, doth he not leave the ninety and nine, and goeth into the mountains and seeketh that which is gone astray? And if so be that he find it, verily I say unto you, he rejoices more of that sheep than of the ninety and nine which went not astray".

Deep circulation

The pictures of mean circulation are by no means finally established. We heard, for example, of a recent analysis of the deep circulation in the North Atlantic, where (it is now suggested) there is a major anticyclonic gyre transporting 60 million tonnes per second in a concentrated flow under the surface Gulf Stream. What drives this deep circulation? There are suggestions that the 100 km eddies must be included in any dynamical solution to that question. But computer models of the Oceanic mean circulation appear to produce quite realistic results for heat transports even though the effect of the mesoscale eddies is treated in terms of horizontal (not even isentropic) transport coefficients, equal for heat, salt and momentum, and invariant with position and time. Perhaps before the next JOA it will be possible to perform numerical experiments to discover how sensitive the mean circulation and associated transport are to spatial inhomogeneties in the 100 km eddies.

Geosecs

Meanwhile we can look to a greatly improved understanding of the mean circulation thanks to the development of methods to analyse the concentration of radioactive elements in seawater. "The Ocean World", proceedings of the Tokyo JOA, contains only one passing reference to this revolutionary development, which adds the fourth dimension, time, to the integrated picture of classical - or DC - oceanography. The techniques must be extremely difficult and the interpretation of results is still in its infancy, but I suspect that the results gained by radio-chemistry in general and its application, together with chemical analysis of other trace elements, during the recent Geosecs expedition will be featured prominently in the general symposia of the next JOA. These new distributions should prove a happy hunting ground for physical oceanographers during the next five years, who will no doubt concern themselves with the question of precisely what is water mass formation.

Fluctuations in the mean circulation

Turning back to fig. 1, it will be seen that I have drawn a tentative extension of the physical oceanography realm occupied by motions whose characteristic dimensions in space and time obey simple dynamical rules, into the realm of climate, where they do not. The study of fluctuations in the large-scale distributions of temperature, salinity, etc. and the corresponding fluctuations in the general circulation on horizontal scales an order of magnitude larger than those of the eddies in the 100 km spectral peak, has scarcely begun. Yet the effect of such fluctuations on fish distributions and on the rate of heat supply to the atmosphere, and hence on climatic fluctuations, is being increasingly appreciated, not least because we are becoming increasingly aware of significant changes of climate and fish distributions in our own lifetime.

The special symposium on the effect of the ocean on climate and weather showed how difficult it is convincingly to establish cause and effect on the basis of statistical analysis alone. What are urgently needed are theories of climatic change involving the ocean surface temperature, heat transport, salinity and ice cover. We heard of progress in this direction based on three main approaches:

(1) General circulation models involving long term computer integrations of the primitive equations.

(2) Much simpler computer models in which the whole effect of the atmosphere is treated as random noise acting on the ocean.

(3) Ingenious laboratory models designed to reveal instabilities in an environment containing air, sea and sea ice.

It will be interesting to see in five years' time which approach proves most fruitful.

Equally important is the collection of long time series of observations of the ocean surface conditions. These are crucial in testing theoretical models of the coupling of atmosphere and ocean to produce changes in climate, and it is hoped that the establishment of the Climate Dynamics Programme of GARP (The Global Atmospherical Research Programme) will stimulate further extension of such ocean monitoring. But, as we heard in a number of papers, such observational sequences are vulnerable to interruption and even termination by War and decisions of governments under economic pressure.

In many ways, the climate problem poses the biggest challenge of all to physical oceanographers, but it would be unduly optimistic to expect much progress by the next JOA.

Waves

The study of waves in the ocean has been the mainstay of dynamical oceanography for well over a century, but the collection of adequate observations at sea for the detailed testing of wave theory is a relatively recent development, with important advances since the last JOA, in 1970. We have heard papers describing expeditions mounted during the past six years to measure the development of the wind wave spectrum (JONSWAP), to sample the mean internal wave spectrum in mid ocean for a month (IWEX), to sample baroclinic Rossby waves in the Pacific over a period of six years, and so on. The availability of improved observations from these and other similar investigations has stimulated theoretical oceanographers to develop models of the wave fields - and especially of their spectral form - more closely matched to the local environmental conditions. Remembering Samuel Johnson's well-known epigram: "The sure knowledge that in fourteen months there will be data on which to hang his ideas clarifies the mind of a theoretician wonderfully". Relating field observations to the predictions of this new generation of wave theories by means of elaborate statistical analysis has developed, during the past six years, into a major industry employing students and post-doctorate fellows, and such "model fitting" will be an important feature of the physical oceanography scene from now on. Furthermore, the closer relationship so engendered between experimenter and theoretician represents a major advance for our subject.

Conclusion

My time has been too short to refer to many of the fascinating topics in physical oceanography presented verbally or as posters during this Assembly. Some of the Special and Association symposia for example those devoted to the skin of the ocean and to the mixed layer and seasonal thermocline, contained fascinating glimpses of complex phenomena, but left one with the feeling that no clear theoretical framework has emerged around which one can construct a unified picture, incorporating the wide variety of observational material. Perhaps these regional studies, like those of climatic variations, will require a longer gestation time before they can be highlighted in a future JOA.

BIOLOGICAL OCEANOGRAPHY

MARTIN V. ANGEL

INSTITUTE OF OCEANOGRAPHIC SCIENCES
WORMLEY, GODALMING, SURREY

During this Assembly the multi-disciplinary approach illustrated the diversity of biological oceanography, furthermore almost every advance in other branches of oceanography offers the biologist a chance of new insights into the processes at work in the ocean biotope. The type of insight will vary with the scale of the phenomenon being observed, whether it is microscale, mesoscale or megascale in space or in time. It will also vary with the level of biological organisation studied ranging from the effect of temperature on a specific biochemical reaction to the broad-brush-work of zoogeographic distributions and their relationships to the oceanic gyres and previous tectonic movements.

Reviewing so complex a set of papers and picking out the highlights of the Assembly presents exactly the same problems of scale and levels of organisation. In addition what may now appear to be a trivial contribution on a specific organism may eventually prove to be the catalyst for a whole new field of interest adding significantly to our general understanding. Conversely an *in vogue* study which may now be considered most important, may instead lead into a blind alley squandering our limited resources of manpower, time and materials.

Biological oceanography tends to be the poor relation of physical, geophysical and chemical ocean sciences in its level of funding and instrumentation so we must continually ask ourselves if it is properly organised. In such a diverse discipline advances tend to be on a broad front. Concerted attacks on specific problems are comparatively rare and rapidly become diffuse again as each investigator wanders off along avenues of personal interest which are not necessarily relevant to the main theme. So the programme tends to degenerate into sand-pit science each person working in the same area, but building their own castles. Maybe there are selective processes which favour research to be carried out by individuals or small groups who temporarily amalgamate into larger loosely structural associations for particular experiments or expeditions. These selective processes may contain elements which are totally unrelated to scientific merits, such as the ability to raise funding and other political skills. Results even within experiments may prove to be incompatible as a consequence of logistic differences in the tasks of studying different groups of organisms or different processes. A simple example being the numbers of phytoplankton in a single water bottle sample may exceed the numbers of zooplankton in a net haul which may in turn exceed the numbers of fish collected on the whole of an oceanographic cruise. This difficulty in creating and maintaining coherence within biological programmes may be one reason why, in contrast to the papers in the physical and chemical sciences, few biological papers have dealt with the results of large cooperative programmes. Even some of the obvious exceptions such as the papers resulting from CUEA involve biological parameters at a high level of organisation which can be treated in an analogous way to physical parameters. This lack of coherence between biological programmes tends to generate a massive accumulation of heterogeneous data. Attempts to use data from a wide variety of sources are seldom successful except at the coarsest of levels.

Much of the work tends to be infilling - repeating proven or sometimes unproven observational and experimental techniques on other species, or in other oceans, or during other seasons, or at other depths, or under new sets of controlled or semi-controlled conditions. Advances are of two main types, firstly through the development of new techniques of sampling, experimentation or analysis, secondly through the development of generalisations and theories through the integration of infilled results derived from both old and new techniques. We should be developing theories which are falsifiable, rather than concepts which since they deal with tautologies are not falsifiable (PETERS, 1976). Biological oceanography, in common with ecology in the broadest sense, desperately needs a unifying field theory analagous to the theory of relativity that will give us some ability to predict events. Throughout the Assembly we heard how some measures of predictability are emerging within limited areas of interest. Perhaps the most depressing aspect of the Assembly is that we appear to lack any individual or group of individuals with the intellect or rather genius who have the ability to slice through our Gordian Knot of complexity to produce a unifying theory.

However, I must return to my remit which is to review the biological oceanography contributions to this Assembly.

Firstly, what have been the most significant technical advances? The simplest observational technique is just to look and see. Before this Assembly most participants would have considered observations in the deep oceans required expensive logistic back-up of cameras, television, holography, underwater habitats or submersibles. However since the Tokyo Assembly a group of divers have been making direct observations on near-surface organisms (e.g. HAMNER, MADIN, ALLREDGE, GILMER and HAMNER 1975), and we have heard several more such contributions here. Out of such a simple direct technique has emerged results on animal associations and the importance of gelatinous structures and mucus, which are substantially modifying our perspectives on the trophic structure of pelagic communities. SCUBA diving itself is by no means a new technique, but we are only just beginning to use it properly as a research tool. Similarly for many years the use of submersibles has centred around the technology and has largely ignored their scientific application. Now we hear less about the undoubtedly high technology of the submersibles and more of the science. CHILDRESS (1975) suggested that every blue-water biologist should have at least one dive in a submersible to help correct their obvious misconceptions about the environments they are studying.

Out of an accidental submersible observation arose another field of investigation which has led to the development of a set of novel techniques. The "ALVIN lunch" experiment indicated that microbial degradation, even at quite shallow sea-bed depths relative to the average depth of the world's ocean, is one or two orders of magnitude slower than at shelf depths. The recently developed techniques of culturing and sampling these microbial communities and maintaining them at high pressure have already yielded results necessitating a radical re-think about the re-cycling of organic material in the deep sea, and about the structure of abyssal food webs. However, there always must be the slight feeling of apprenhension that on the sea-bed as in midwater, the size of the culture vessel will have a significant effect on the results.

Small culture vessels though have provided invaluable results on the physiology and autecology of individual species. However, many problems cannot be tackled without being able to manipulate whole systems. Canadian limnologists have the benefit of hundreds of virtually identical lakes that they can experiment with, but the marine biologist has to create his own enclosed bodies of water. As we have heard the engineering and other technical problems associated with such projects are formidable, and there is the likelihood that the population changes inside may never duplicate events outside, but at least results replicate between enclosures, so that a new research tool has been added to our array of techniques.

Analytical techniques illustrate a recurrent problem in

biological oceanography, that is as the techniques become more sophisticated so they are liable to become an end in themselves. Mathematical analysis and modelling use language and concepts that are outside the ken of the average biologist. With a few notable exceptions few biologists are able both to generate biological data and to process it, but as soon as there is a division of labour, there is a danger of a break down in communication. However, misinterpretation of data is far more likely to be made by a non-mathematically trained biologist blindly following the formulae, who is unaware of the basic underlying assumptions of the methods he is using or rather misusing. The analytical methods and the use of computers are invaluable in extracting information from large and complex data sets. Similarly models and simulations can give important insights into the possible functioning of complexly interacting physical and biological systems. Even bearing in mind CONOVER'S perfectly valid warning that mathematical models often oversimplify or fail to recognise the importance of feedback, modelling provides a method of testing whether or not intuitively derived interpretations of sampling results are sensible. Modelling forces the biologist to be rigorous in his thinking and helps to generate theories that can form the basis of fresh experiments and observations. The modellers know what they want from the biologist, but I feel that few biologists know what they want from the modellers. However, efforts are now being made to get verification data such as the paper by TYLER on quantum efficiency in phytoplankton.

The stimulus of technological advance in accelerating progress has probably been best illustrated by the recent advances in paleo-oceanography as summarised by IMBRIE and HAYS. The advances have been in the ways in which cores are examined so that the data extracted can be cross-correlated between cores and also in the analytical techniques that transform the data into an historical record of surface water temperatures. Here is the meeting point between geology and biology giving the interaction of climatic history and biological processes on what approaches an evolutionary time scale. All present day communities carry the marks both of their contemporary physical environment and their recent and geological histories, as shown in several papers. Is the richness of the Antarctic fauna compared with the Arctic merely a function of the greater productivity of the area or is its longer geological history as important? Are the communities in the great oceanic gyres saturated with species? Or have the fluctuations in the current patterns resulting both from tectonic movements, as for example during the gradual opening up of the Atlantic, and from the effects of the glaciations as demonstrated by RUDDIMAN and McINTYRE, left some communities under-saturated making further speciation possible? On the shorter time scale we have heard a plethora of papers on the changes occurring in fisheries and plankton in the last century or so; changes often correlating with climatic fluctuations in a sufficiently repetitive pattern to make one believe a full comprehension of the casual linkage is within our reach.

At the zoogeographic or megascale considerable advances have occurred in the last five years resulting from the painstaking accumulation of compatible data. REID, McGOWAN, BRINTON, FLEMINGER and VENRICK beautifully illustrated the interlinking of the physical and biological patterns on the geographic scale. Whereas in the past indicator species were sought, but rarely found, it is now clear that indicator assemblages of species are a reality. The variations in numbers of species within a 'province' might well prove a useful topic of investigation to see if any of MacARTHUR'S theories of island biogeography can be extended to oceanic environments (MacARTHUR, 1972) in other words do the gyral provinces in the Pacific which are larger in extent carry more species than the corresponding but smaller provinces in the Atlantic? An essential pre-requisite of answering such a problem is for more studies in the Atlantic with the comprehensive breadth of geographical coverage involved in the study by BACKUS, CRADDOCK, HAEDRICH and ROBISON on Atlantic midwater fishes. These gyral studies illustrate the value of the close liaison between the physical oceanographers and the biologists, and this was further stressed in JANSSON'S paper on the Baltic.

Similarly at the mesoscale level without the support of chemical and physical oceanography, the studies on biological processes associated with upwelling would have been impossible. Our knowledge of the mechanisms of coastal upwelling, such as what factors determine whether or not dense phytoplankton blooms develop and just how patches are generated, has advanced considerably with programmes like CUE. SMITH described the physical aspects and BARBER the biological aspects of coastal upwelling. WROBLEWSKI and also WALSH showed how it is possible to build on these sorts of results and to model many features of the production cycle and generate realistic estimates. Equatorial upwelling, however, has received far less attention and so VINOGRADOV'S descriptions of the oscillations in the balance between high plant production followed by the build up of high grazing pressure, and then high nektonic predation were particularly interesting. However, the investigation of the biological significance of mesoscale features with dimensions of about 100 km and time scales of months, as described by MUNK will force biologists into organising multi-ship sampling programmes.

At the microscale we reach the limits in precision of so much of our instrumentation. Biological sampling runs into the limit problems of inadequate numbers of larger organisms being caught as progressively smaller samples are collected whereas the smaller organisms which still occur in adequate numbers for meaningful analysis, present considerable processing and taxonomic difficulties. Resort has to be made to automatic instrumentation such as fluorometers and particle counters. Biological progress is also hampered by the inadequate knowledge of the physical processes dominant at such small scales. These are scales equivalent to an individual grazer's ambit at which diffusion, turbulence and internal waves are all important. Fronts, Langmuir circulation cells and cyclic countercurrents generated by internal waves all appear to be patchiness generating mechanisms. Patches may be generated by the interaction of vertical migration behaviour and internal waves as shown by KAMYKOWSKI and ZENTARA, or merely by the growth of phytoplankton if a patch of enriched water is extensive enough. DUBOIS has shown mathematically how the combination of diffusion, the growth by phyto- and zooplankton, and their predator/prey relationships will be expected to generate plankton waves. The task of demonstrating the actual existence of such waves offers another challenge to anyone using the range of sampling techniques presently available.

The existence of patches of high concentration of food particles was shown by LASKER to be essential for the survival of the early larvae of the anchovy Engraulis mordax. An ill-timed upwelling event in the Los Angeles Bight can disperse a dense patch of phytoplankton, so change the feeding conditions from being good to poor in a matter of hours. Walsh showed how recruitment of the anchovetta may be lowered during the El Nino off Peru because of the dispersal of high food concentration by storms. Recruitment of many commercial species may be determined by similar chancy occurrences. HARDEN JONES suggested that the migrations by many commercial fishes is an adptation towards abundance by maximising larval survival and subsequent growth. An interesting example of a breeding migration was given by LENZ for the myctophid fish Luciosudis normani. LEE pointed out that increase in phosphate levels in the North Sea is correlated with recruitment in various species of roundfish, but such improvements possibly resulting from sewage disposal will be negated if the main nursery areas of many demersal fish, are reclaimed from the sea or destroyed by gravel extraction. It is most important we wrestled with these problems of conflicting useage and conserve marine ecosystems so that they survive and continue to be of value to man.

Rational exploitation of marine resources is urgently needed. Even if biologists could provide all the information needed, demands of industrial investment and political expediency usually override scientific logic. Over-exploitation and pollution are all too frequently blamed for decline in stocks and subsequently the decline is shown to be the result of the fundamental variability of oceanic ecosystems. The tragedy will be if this undermines the scientifically valid cases where over-exploitation and pollution are the real causal agents. At the Tokyo Assembly a heavy emphasis was placed on the acute threat of pollutants to marine environments. Yet at Edinburgh little mention was heard of organo-chloride residues, PCB's, heavy metals or chronic oil pollution. Has some Hercules cleaned the Augean Stables of the World's oceans, if so why has he not been honoured with a Nobel Prize? Which is the healthier approach the alarmism of Tokyo or the relative disinterest or was it a balanced view at Edinburgh? If pollution studies are out of fashion, is biological oceanography too subject to fads and fashions created by the fancies of the funding agencies. If so then biological oceanographers also constitute a resource that needs rational exploitation. However, I may be rash in assuming that the papers offered at this JOA reflect the balance of research throughout the World's oceanographic laboratories.

The second part of my remit, in summing up the biological content of the Assembly, is to forecast where significant advances are to be made. It is tempting to wander into avenues of science fiction by postulating new technical innovations, but I am limiting myself largely to present day technology. It is always useful to be able to stand back and see one's problems in a wider context. To some the satellite is carrying this approach just a little too far, but we heard from APEL what types of features can be observed from space. The precise location of such features could make their study more practical as ship's time would not be wasted in searching. Ground truth observations are often needed but improved communications between the satellite centre and research vessels at sea, could result in immediate investigation of exceptional phenomena seen from space, phenomena which may turn out to be important natural biological experiments.

Man's technological advances are beginning to further threaten deep-living oceanic communities. The new threat coming from deep ocean mining and the dumping of toxic and radioactive wastes. Benthic communities were describe as being fragile and so the revival in interest in their study is timely. Only with the study of fish communities does progress appear to be rapid enough. This is the only group for which the systematics of the benthic organisms. I remain unapologetically convinced that a full taxonomic foundation is a basic requirement of the full understanding of an ecosystem. However, taxonomy is not the only bottleneck in benthic research, as with all biological sampling the collection of the organisms is expensive and time consuming, but no one seems to have come to grips with improving the rate of converting the jumbled heap of material in the net's cod-end as it arrives on deck into usable data. Sorting and enumeration is just as much a problem in pelagic sampling. Computer pattern analysis is now at a stage of technical development that would allow mechanical initial rough sorting of plankton and nekton samples. However, the cost of developing such a system is likely to keep sorting a labour intensive occupation.

Systematics and identification are also major problems in processing microzooplankton and phytoplankton, only here accentuated by manipulative difficulties. Even the advent of the first sea-going scanning-electron microscope is unlikely to give the marine ecologist the tool he requires for processing large quantities of material. So we have to rely on integrative techniques such as fluorometry and particle counting, which give results that are a one step abstraction away from the individual organisms and species. BANSE'S paper illustrated some of the problems associated with such methods. However, such techniques when used in conjunction with undulating samplers like the batfish, are working at the discriminating limited of spatial pattern for any physical oceanographic instrumentation. It is here at the interface between physical processes and biological processes that big advances are being made and will continue to be made. Further areas which will be productive are the biological importance of frontal zones and in the effects of the 3-dimensional chemical microstructuring of the ocean. In this context a technique enabling us to make productivity profiles would not only make production estimates for any position more accurate but would also allow a better understanding of biological-physico-chemical interactions. Such interactions will only be interpretable through the aid of mathematical analysis and modelling. Already you can see that such programmes would involve multidisciplinary participation.

At present problems of physical oceanography are defined to answer purely physical problems, but as the CUEA programme has indicated by redefining physical problems within a biological context new insights into ecological processes emerge. How the questions are asked determines the type of answers that emerge. Biologists are sometimes considered weak-kneed because they do not mount large experiments separately from the physical oceanographers. The trouble is we need the physicists but they do not need biologists. It is for example a tragedy that no biological programme was planned to run in parallel with POLYMODE, since this could have given a first order assessment of the importance of the sea's physical mesoscale structure to its biology. We have the techniques available for such an experiment but not the vision and determination to seize such an opportunity. Interdisciplinary interaction will provide us with our greatest advances.

Having said that, I am equally certain that ecological theoreticians who so far have been unable to see further out to sea than the inter-tidal or possibly the nearest mangrove swamp, will have to focus more attention on the deep-sea, where it is possible to sample without irreparably damaging the environment, where boundaries and micro-habitats are few, where sample size can be as large or as small as needed. The potential quality and quantity of the data obtainable from the ocean far exceeds in its comprehensiveness of environmental coverage any of the data available from terrestrial environments. Perhaps it is in the realm of trophic relationships that marine biological processes are relatively the least well understood. Yet ISAACS theoretical studies on food web structures and reproductive strategies are beginning to throw light on fundamental differences between the functioning of marine and terrestrial ecosystems. However, this is not to argue that we all ought to turn into ecological theoreticians overnight. If we make the rough assumption that the theoreticians factual assimilation efficiency is approximately 10% then a ratio of one theoretician to ten field biologists is possibly about right. There is an analogous biological oceanographic pyramid to Elton's ecological pyramid. On the backs of a broad phalanx of field biologists ride a smaller number of theoreticians, modelers, administrators and planners. Without this broad inflow of real field and experimental data the pinnacle of the pyramid is liable to become top-heavy and to spin off into a simulated world of make-believe. The relevance of our future advances to the real world is totally dependent on maintaining both the quality and the quantity of the field data.

One further caveat, the collection of data for data's sake is as scientific as philately. Sampling programmes and experiments must be properly designed to answer precisely defined questions, and executed with the maximum precision possible. The more controlled the sampling regime, the greater are the chances of cross-compatability with other programmes. Out of a broadly spread data base will spring a more comprehensive understanding of the ecology of the World's Oceans, the information derived by physiologists can then be placed in a proper context, functional morphology will become more meaningful, distribution patterns will be more readily interpretable, seemingly-dubious data will be shown up as valid or invalid, extrapolation from one area to another will be more likely to be accurate, and the ability to make sensible predictions, the fundamental aim of true Science, will move nearer to becoming a reality.

REFERENCES

CHILDRESS, J.J. 1976. Physiological approaches to the biology of midwater organisms. In 'Oceanic sound-scattering prediction' Edited ANDERSEN, N.W. and ZAHURANEC, E.J. Marine Science Volume 3 Plenum Press.

HAMNER, W.M., MADIN, L.P., ALLDREDGE, A.L., GILMER, R.W. and HAMNER, P.P. 1975. Underwater observations of gelatinous zooplankton. Limnology and Oceanography, 20, 907-917.

MacARTHUR, R.H. 1972. Geographical Ecology, Harper and Row, New York 269 pp.

PETERS, R.H. 1976. Tautology in evolution and ecology. The American Naturalist, 110, 1-12.

THE MARINE GEOSCIENCES

JOHN G. SCLATER

DEPARTMENT OF EARTH AND PLANETARY SCIENCES

MASSACHUSETTS INSTITUTE OF TECHNOLOGY CAMBRIDGE MA 02139

INTRODUCTION

Marine geosciences can be split into three major topics:

1. Tectonics, the study of how the earth moves,
2. Sedimentology, a study of where and how sediments are deposited on the sea floor and
3. Petrology, a study principally of the rocks created by ocean floor processes.

These topics are very broad and require a great deal of time to be covered adequately. Sensibly, the Geosciences Symposia at this Congress did not attempt to cover all of them. This is fortunate for me as I certainly do not have the background to give a review of the whole field. So with apologies to those of you who are petrologists for the short shift I will give to this field what I will present is my personal view of the major advances in Marine Geosciences in the past four or five years. Most of what I will concentrate upon was presented in the symposia organized by Eric Simpson on the History of the Oceans, Gerry van Andel on Paleoceanography, Bill Hays on Paleostratigraphy and by Tom Gaskell on Marine Geology. In order to give some order to my presentation I will discuss tectonics first, follow with recent sedimentation, tie this to older sedimentary processes and finally finish with paleoceanography and paleosedimentation as it is influenced by our now more refined view of the tectonics.

TECTONICS

Without question the major advance in marine geosciences is the theory of plate tectonics. In this theory the earth is envisioned as a set of rigid plates in constant interaction. The interaction of these plates which are 50 to 100 km thick gives rise to earthquakes which in turn define the plate boundaries (Figure 1). This theory stresses three concepts, the rigidity of plates, the idea of relative motion and the continuity of the plate boundaries (Wilson, 1965; McKenzie and Parker, 1967; Morgan, 1968; LePichon, 1968). The plates are created at mid-ocean ridges or spreading centers. Plates move past each other along transform faults and are destroyed in trenches. As a consequence of plates being formed by the intrusion of molten magma which cools through the Curie temperature they acquire a surface magnetization in the direction of the earth's field. As this field changes and we know how it changes with time (Cox et al., 1964) we can deduce from the magnetic anomalies on the sea floor the age of the oceans (Vine, 1966). Identifiable magnetic anomalies have been recognized in all the oceans and Pitman et al. (1974) have compiled these on a chart of the world (Figure 2). By assuming rigidity and superimposing isochrons on either side of the ridge axis one can reconstruct the past position of the continents and end by reconstructing the original positions of quite complex plates. As an example of this I have shown (Figure 3) a reconstruction of the North Atlantic (Sclater et al., in press). This was based on airborne magnetic lines in the Arctic and Norwegian seas presented by Phillips et al. (1976) and upon data gathered further south by Kristoffersson et al. (1976) and Pitman and Talwani (1972).

Another simple prediction of the theory of plate tectonics is that hot molten magma which is injected at a spreading center cools by conduction as it moves away from this center. As the upper surface of the plate is in contact with seawater, which can convect, the plate loses heat vertically much more easily than it dissipates horizontally. As a consequence, there is a simple relation between age and heat loss and between age and contraction of the plate or elevation of the sea floor (Figure 4).

In the simple theory it has been shown by Parker and Oldenburg (1973) and Davis and Lister (1974) that for young crust, the depth should be proportional to the square root of time. Parsons and Sclater (in press) have plotted the data presented in Figure 4 against the $\sqrt{t}$ (Figure 5). Note that for regions out to 70-80 Ma the depth follows the predicted relationship well. However for older ocean crust the match is not good. The reason why the depth should flatten in the older regions is not completely understood. At present it is thought that the flattening is related to heating from below in older regions (Parsons and Sclater, in

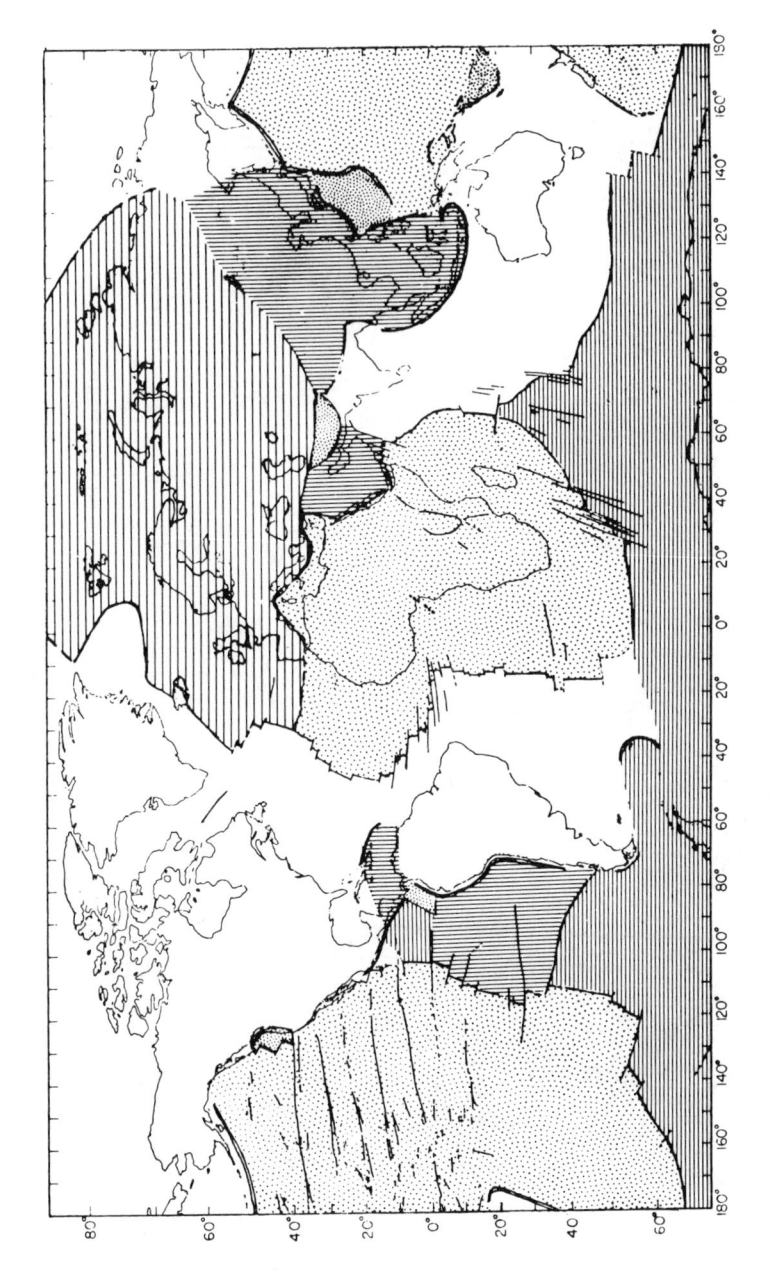

Figure 1. The crust of the earth is divided into units that move as rigid blocks. The boundaries between them are rises, trenches (or young-fold mountains) and transform faults (Morgan, 1968).

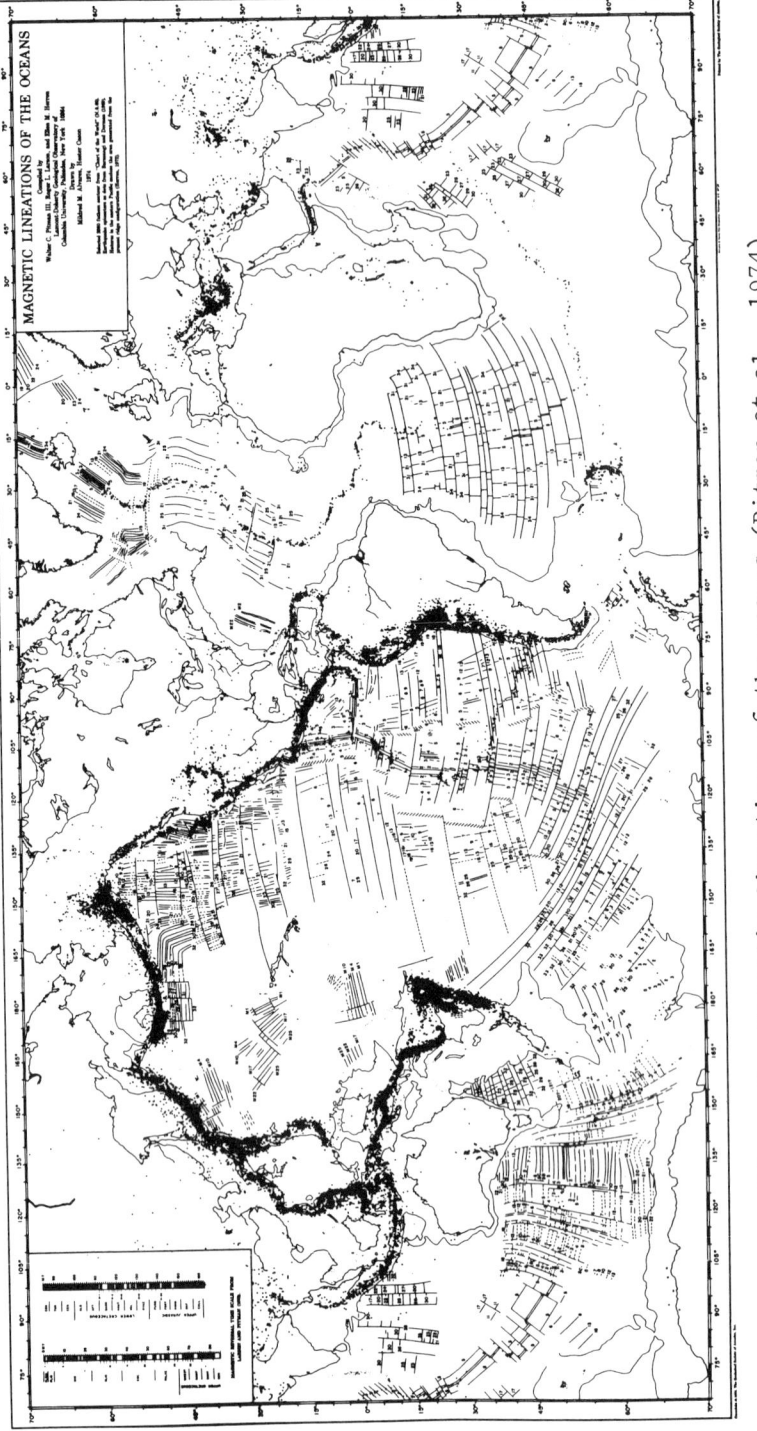

Figure 2. Magnetic lineations of the oceans (Pitman et al., 1974).

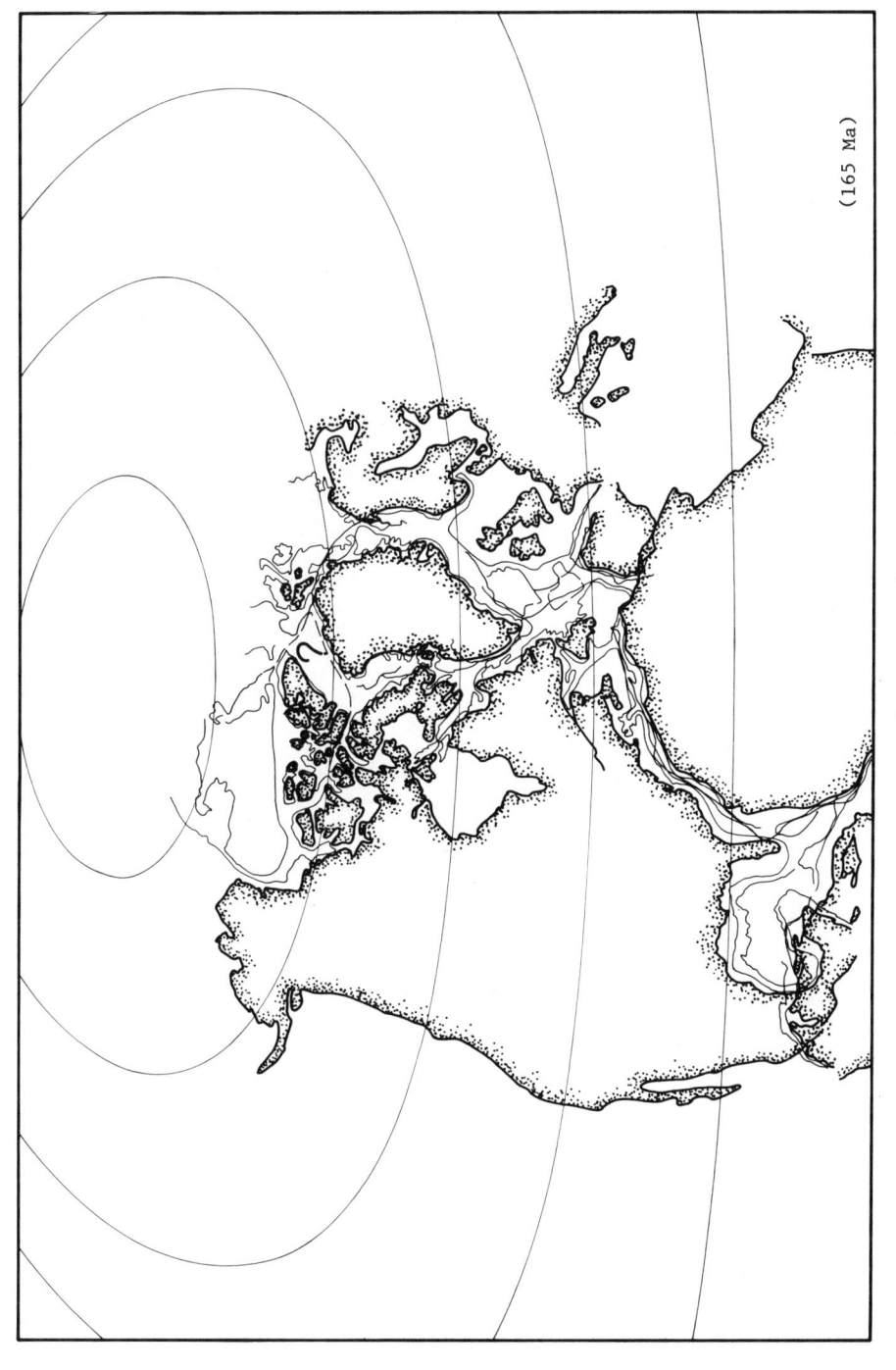

Figure 3. A reconstruction of the continents bordering the North Atlantic (Sclater et al., in press).

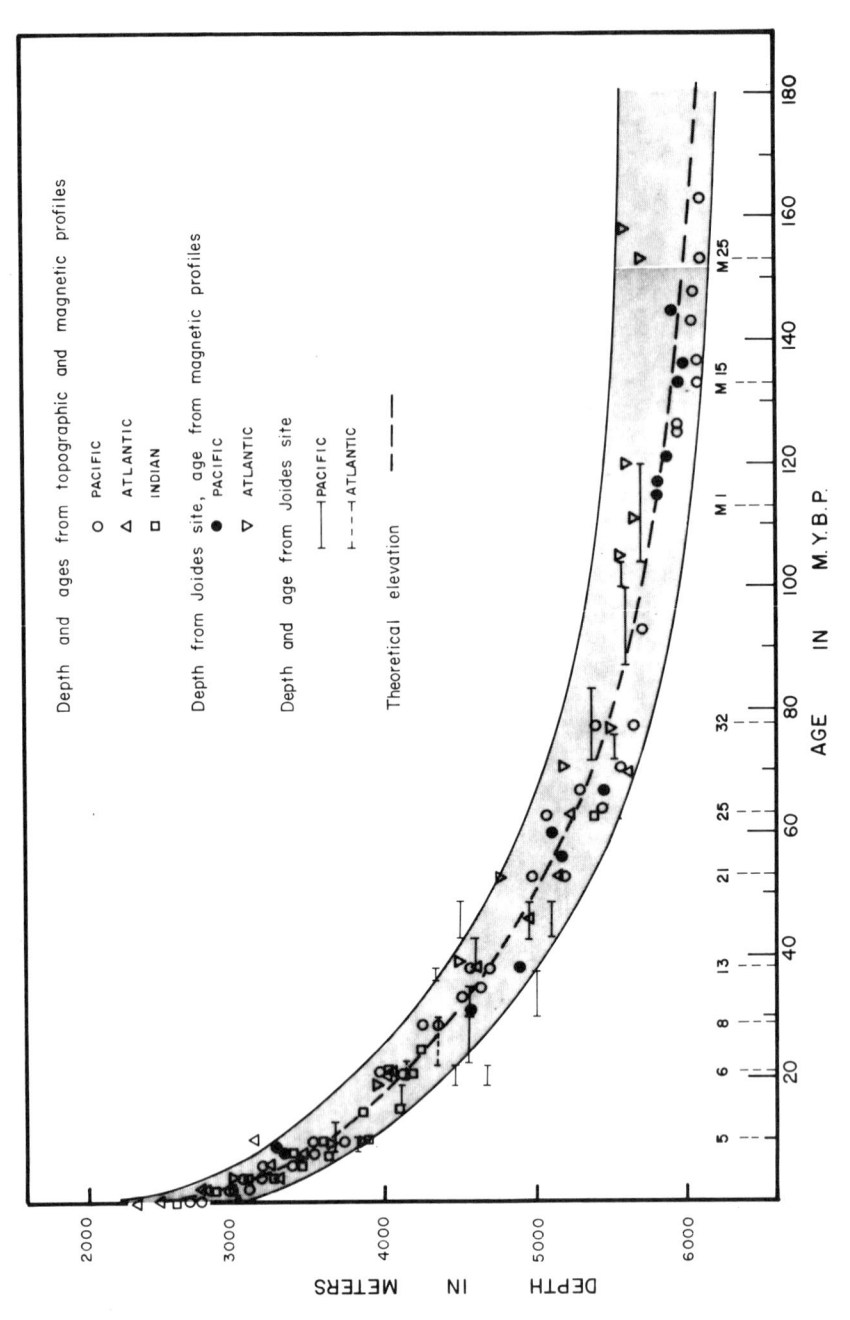

Figure 4. All the various mean depths in normal ocean floor plotted as a function of age. The shaded band marks out a ±300 m error band which includes nearly all the measurements. The theoretical curve is calculated for the plate model using the lithospheric parameters of Parsons and Sclater (in press).

THE MARINE GEOSCIENCES   313

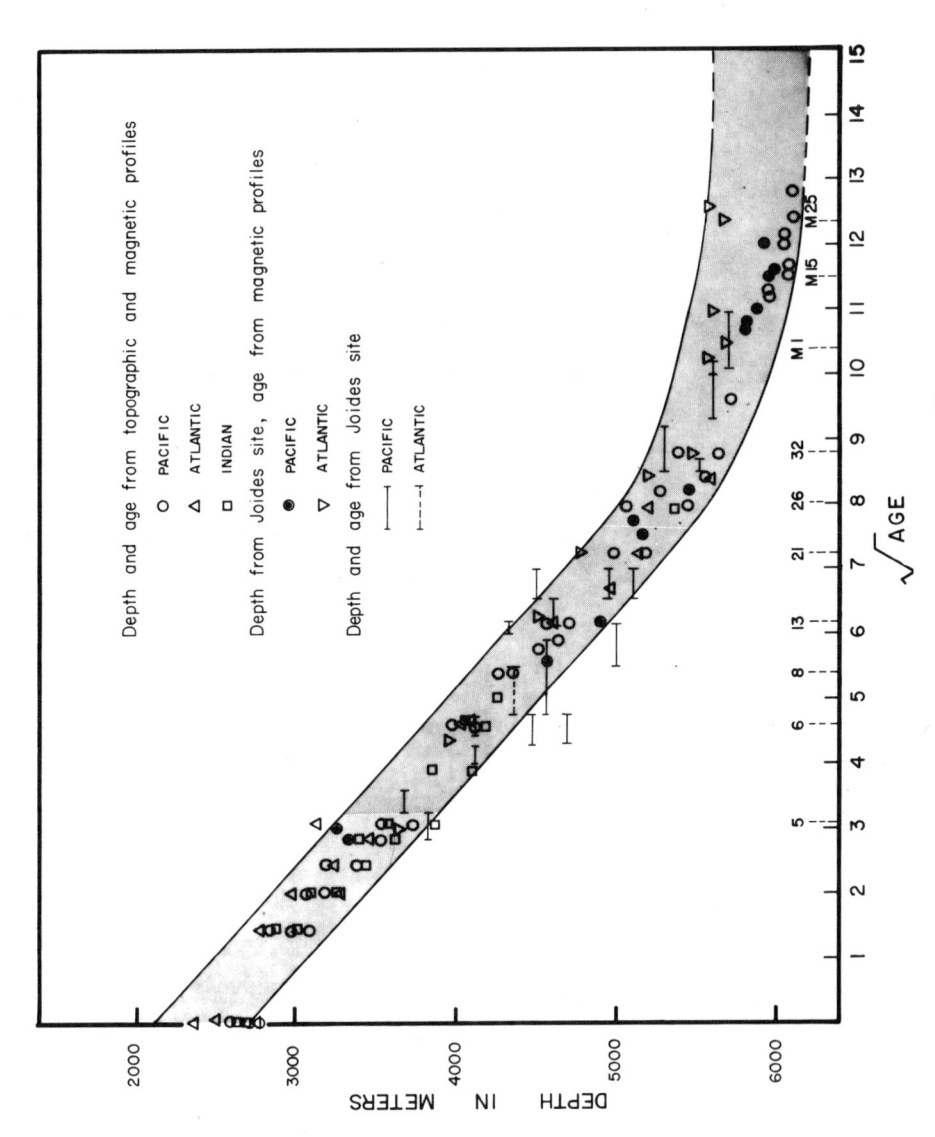

Figure 5. The same data set as in Figure 4 plotted against $\sqrt{t}$. Note that the age range 80-160 Ma is compressed relative to that for 0-80 Ma (Parsons and Sclater, in press).

press). However, the relation for the youthful regions is exactly as predicted by the simple theory and is a direct result of the creation of the plates.

PALEOBATHYMETRY

As a consequence of the depths/age relation, if one knows the age of the ocean floor, one can predict the depth. Further, if one also knows the tectonic history of the plates, one can predict the depth as a function of time and position. From the depth/age relation the 3000 m contour lies on the ± 2 Ma isochron, the 4000 m contour on 20 Ma isochron and the 5000 m contour on 50 Ma isochron (Figure 5). Thus by determining the position of these isochrons on the reconstructed history of the plates one can determine the past gross bathymetry. Unfortunately, there are anomalies in the depth age relation, the most important being the aseismic ridges, which have to be modelled before the charts can be considered quantitative. However the aseismic ridges proved relatively easy to model. From sediment recovered in the Deep Sea Drilling holes it appears that they were all formed at the ridge axis close to sea level and have the same subsidence rate as the sea floor to which they are attached (Detrick et al., in press). Thus if one knows the tectonic history of the plate to which they are attached the subsidence history of the aseismic ridges is fairly easy to work out. As an example of paleobathymetric reconstructions I have chosen to show four charts of the South Atlantic from Sclater et al. (in press). The charts were constructed assuming rigid plates and the paleolatitude of the plates was determined from the North Atlantic paleomagnetic pole averages of Phillips and Forsyth (1972). I have started from the reconstructed position of the continents bordering the Atlantic (Figure 6, 165 Ma) and worked forwards in time.

The second chart shows the position of the continents 125 Ma (Figure 6, 125 Ma). Also shown is the paleobathymetry in the North Atlantic. The South Atlantic has not yet opened. Note the black triangles which are Joides sites and the lines on the ocean floor which are bathymetric contours. At this time the North Atlantic is the only sea open and trends Northeast-Southwest.

The following reconstruction presents the position of the continents and the proposed bathymetry 80 Ma (Figure 6, 80 Ma). Note that both the North and South Atlantic are now open. However, the North Atlantic is closed to the north and the Argentine Basin is closed to deep water to the south by the Falkland Plateau. The gap between Iberia and Africa is quite large. Also, the Walvis and Rio Grande Rises separate the North and South Atlantic into two quite separate basins.

THE MARINE GEOSCIENCES 315

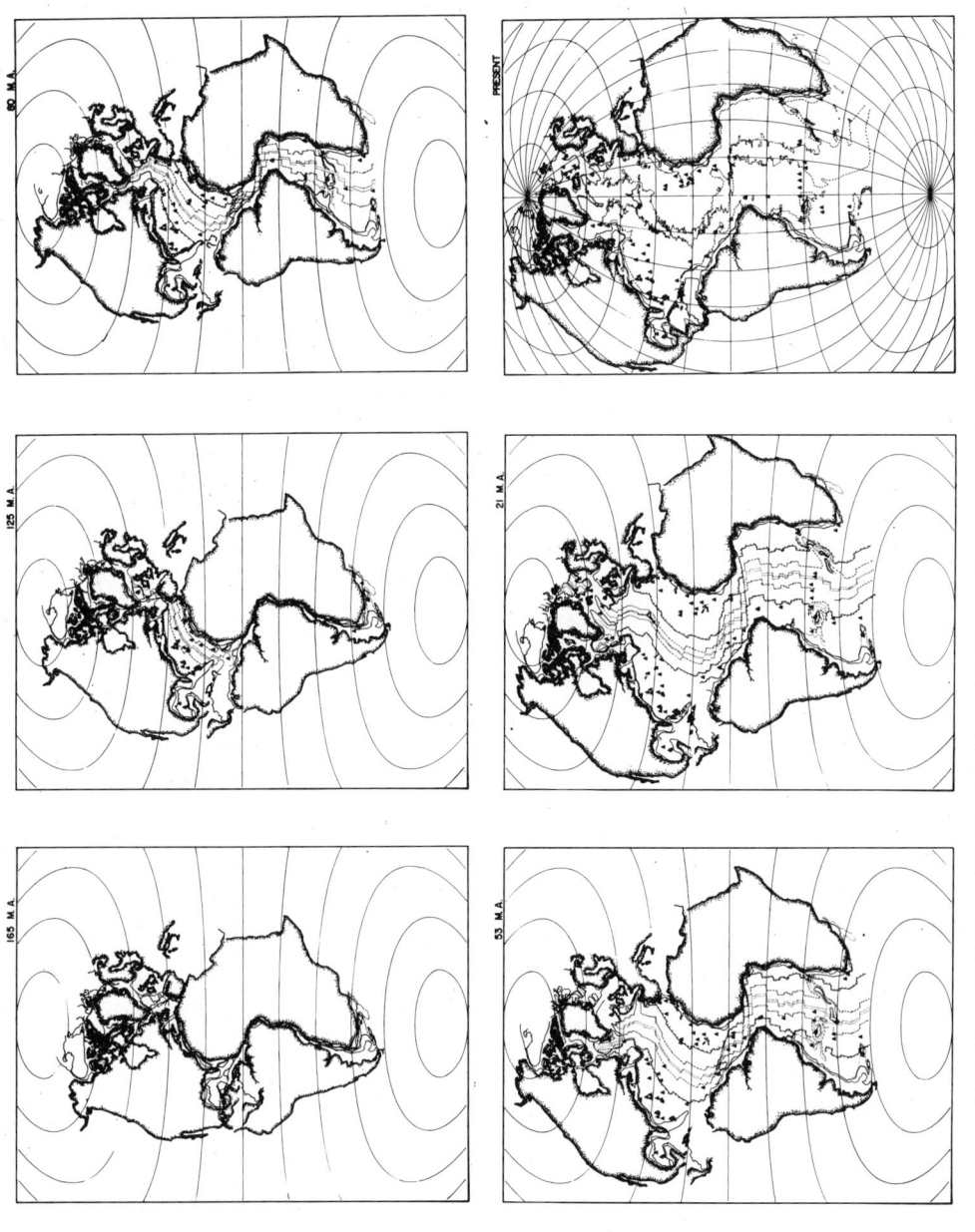

Figure 6. 165 Ma, 125 Ma, 80 Ma 53 Ma, 21 Ma and present bathymetric charts of the Atlantic (Sclater et al., in press). Lambert equal area projection.

By 53 Ma (Figure 6, 53 Ma) the Atlantic is open to deep water both from the north and from the south. The Walvis and Rio Grande Rises are sufficiently deep so that the basins in the South Atlantic can be considered connected. To the North, Africa and South America are well separated and the North Atlantic is a fully developed basin.

The last two reconstructions shown on Figure 6 are the paleobathymetry of the Atlantic 21 Ma and the present bathymetry. They are shown for two reasons. The first is to examine how the various basins continue to develop and second to point out how similar the 21 Ma paleobathymetry appears to what we see today. This not only supports the suggestion that there has been little change in the past 20 million years but also the close match gives one confidence that the earlier reconstructions are reasonably accurate.

RECENT SEDIMENTATION

I have emphasized paleobathymetric charts in this review for two reasons: (a) because these charts were the major tectonic topic presented at this Congress and (b) because they tie so well into the discussion of paleosedimentation with which I will end this review. However, they were only one of many new topics with important implications for sedimentotogy that were presented. For example one important aspect of the study of sediments is to attempt to interpret the record recovered in sediment cores as a function of simple processes which have affected the past oceans. At present we have two major data sets which we can use. One is the large number of surface cores taken in the recent past by sediment coring techniques from oceanographic research vessels. The second is the 400 odd cores taken by the Deep Sea Drilling Project which extend this areal near surface record much further back in the past (Figure 7).

The first major advance in examining the sediments as a function of depth is to be able to set up both a gross and also relatively accurate short-term time scale. Figures 8a and 8b show the correlation obtained by Ryan et al. (1974) and Hardenbol and Berggren (in press), between the geomagnetic reversals, the basic geologic time zones, and the nannofossil and foram zones for the neogene and paleogene. It is these zones and their relation to radioactive ages and to the geomagnetic time scale that gives us our basic time horizons in the cores.

Another major interrelated advance in sedimentology is the advance in our understanding of the Pleistocene that has resulted from the CLIMAP program (CLINE and HAYS, 1976). We now have a fairly extensive core coverage in the North Atlantic. This core

THE MARINE GEOSCIENCES   317

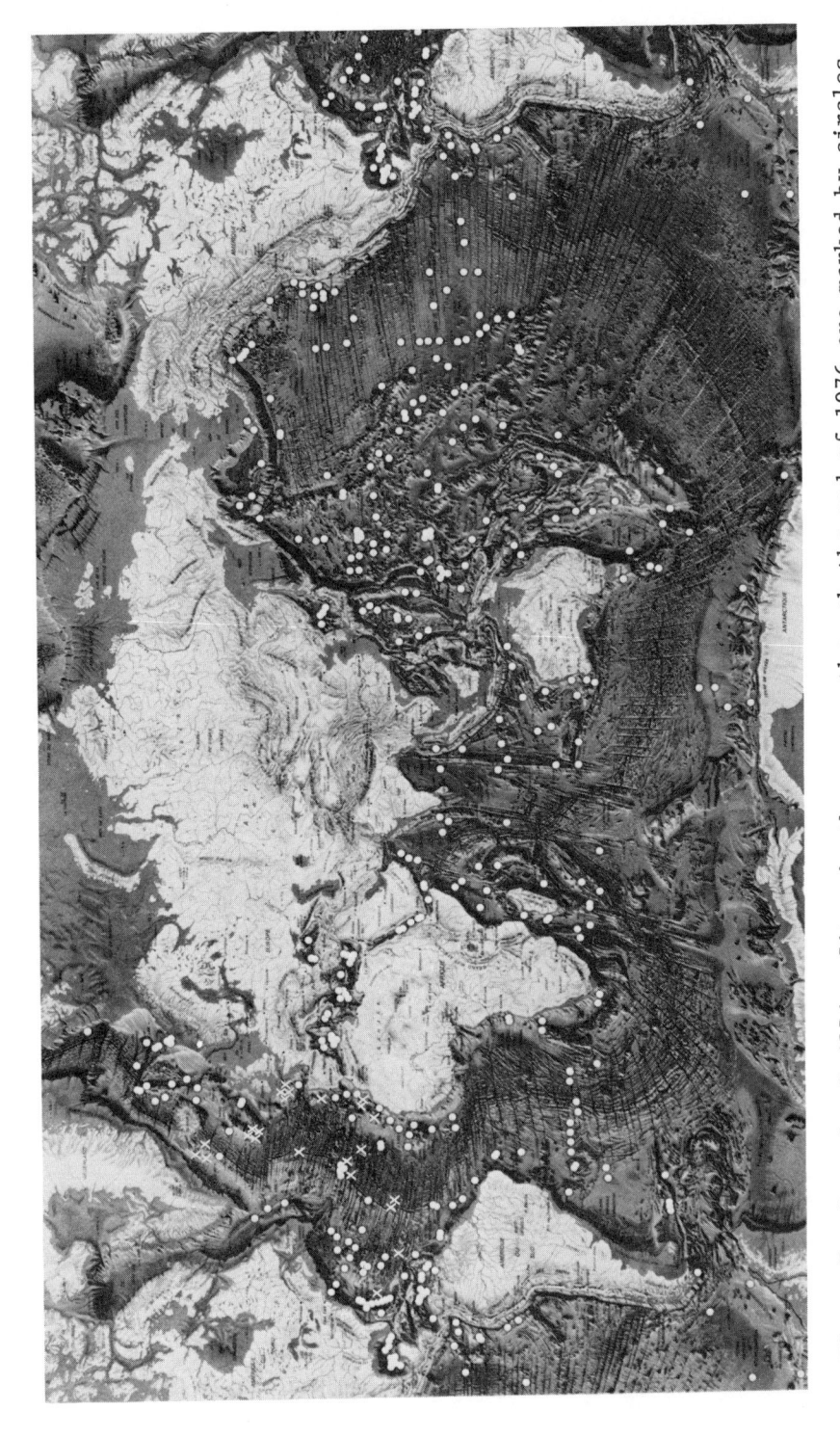

Figure 7. Deep Sea Drilling Sites in the oceans through the end of 1976 are marked by circles. IPOD sites are marked by crosses (Moore, personal communication).

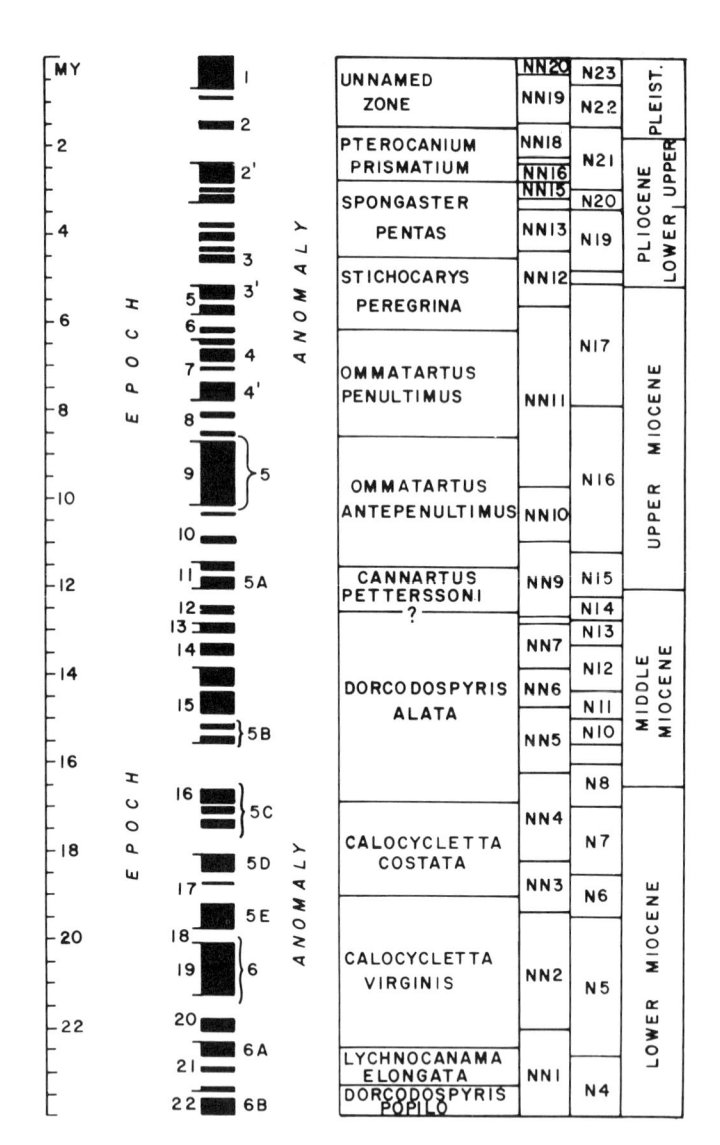

Figure 8a. Paleomagnetic assignment of Neogene stage boundaries (Ryan et al., 1974).

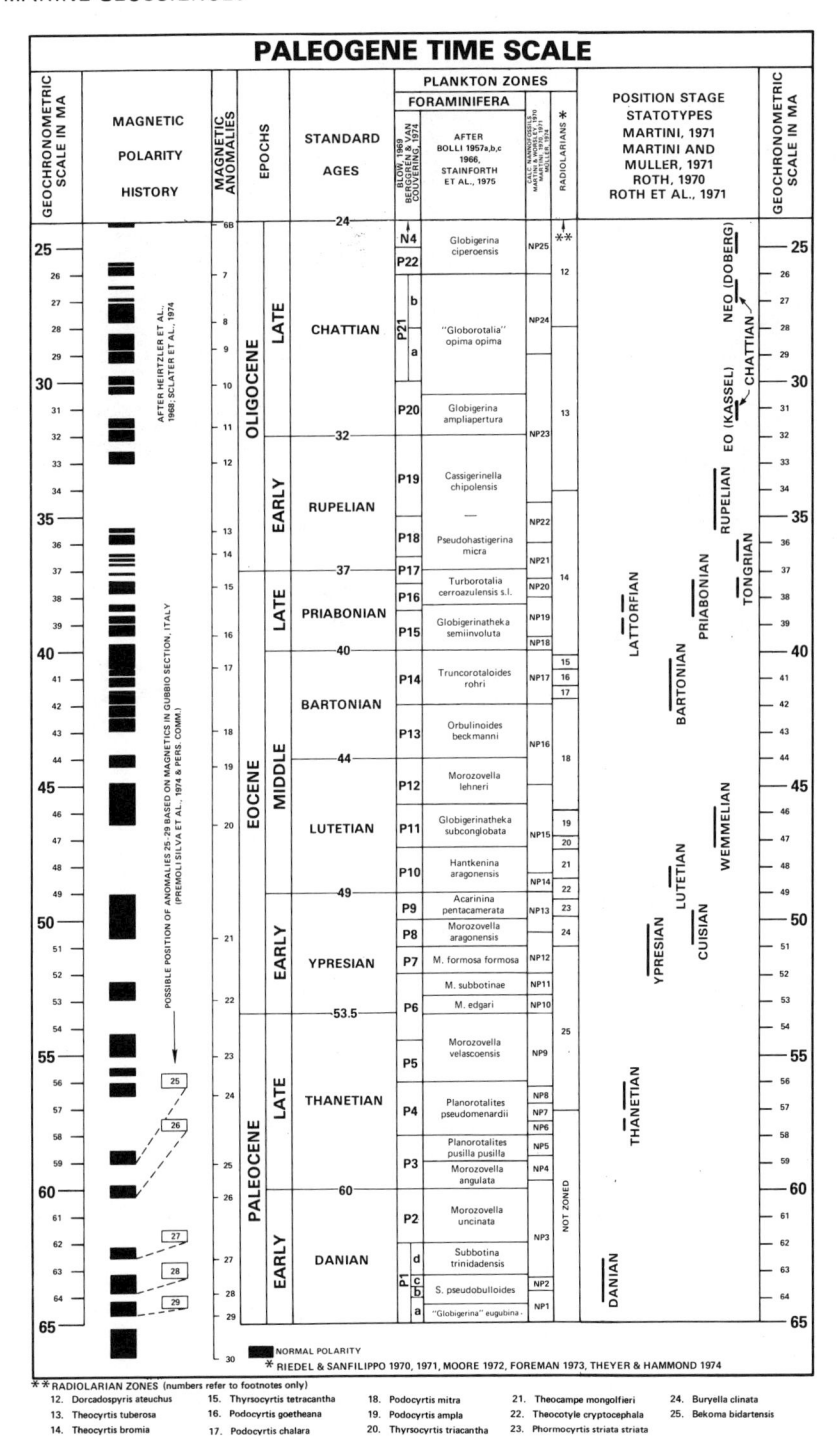

Figure 8b. Paleomagnetic assignment of Paleogene stage boundaries (Hardenbol and Berggren, in press).

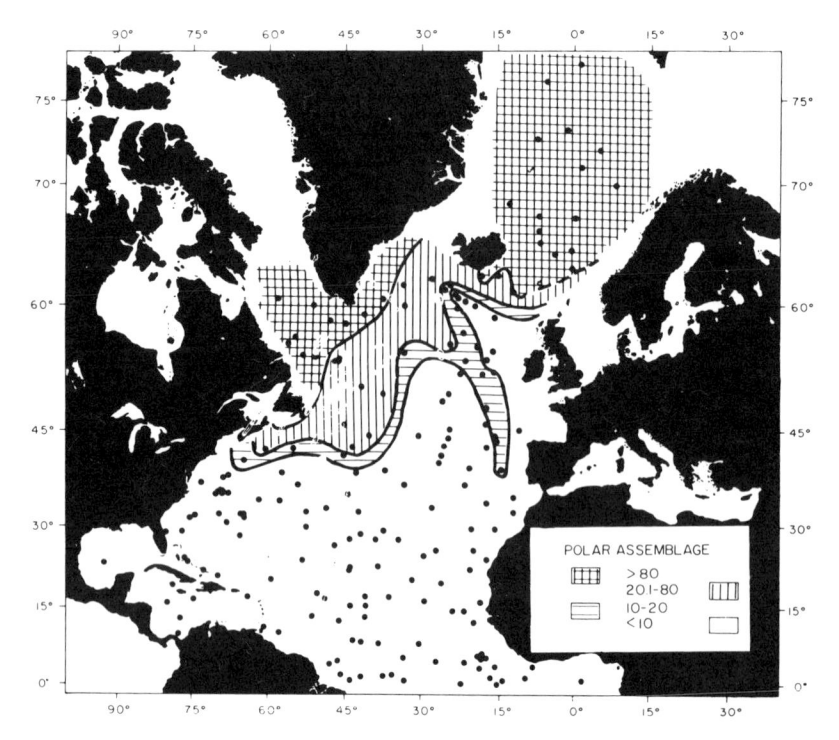

Figure 9. Distribution of polar assemblages on North Atlantic sea bed (Kipp, 1976).

coverage and the interpretation of the planktonic foraminiferal species in terms of assemblages has shown dramatic changes in the distribution of these assemblages in the recent past. For example at present the polar assemblages predominate in a line running from Newfoundland south of Greenland and north of Iceland (Figure 9). However 18,000 years ago these assemblages predominate as far south as a line between Cape Hatteras and Spain (Figure 10).

Imbrie and Kipp (1971) have shown that the percentage of given assemblages in a core is related to a series of forcing functions; such as temperature, salinity, oxygen, etc., in the surface waters. They have also demonstrated that if the data base is sufficiently large it is possible to use transfer function analysis to separate the strongest of these forcing functions. It appears that the strongest and most interesting is temperature. From an analysis of the variation of the assemblages in the North Atlantic, McIntyre et al. (1976) have constructed a sea surface isotherm map for 18,000 Ma (Figure 11). Note how the 2° isotherm mirrors the polar assemblages in the previous figure and shows that during the past ice age the polar front was much further south than

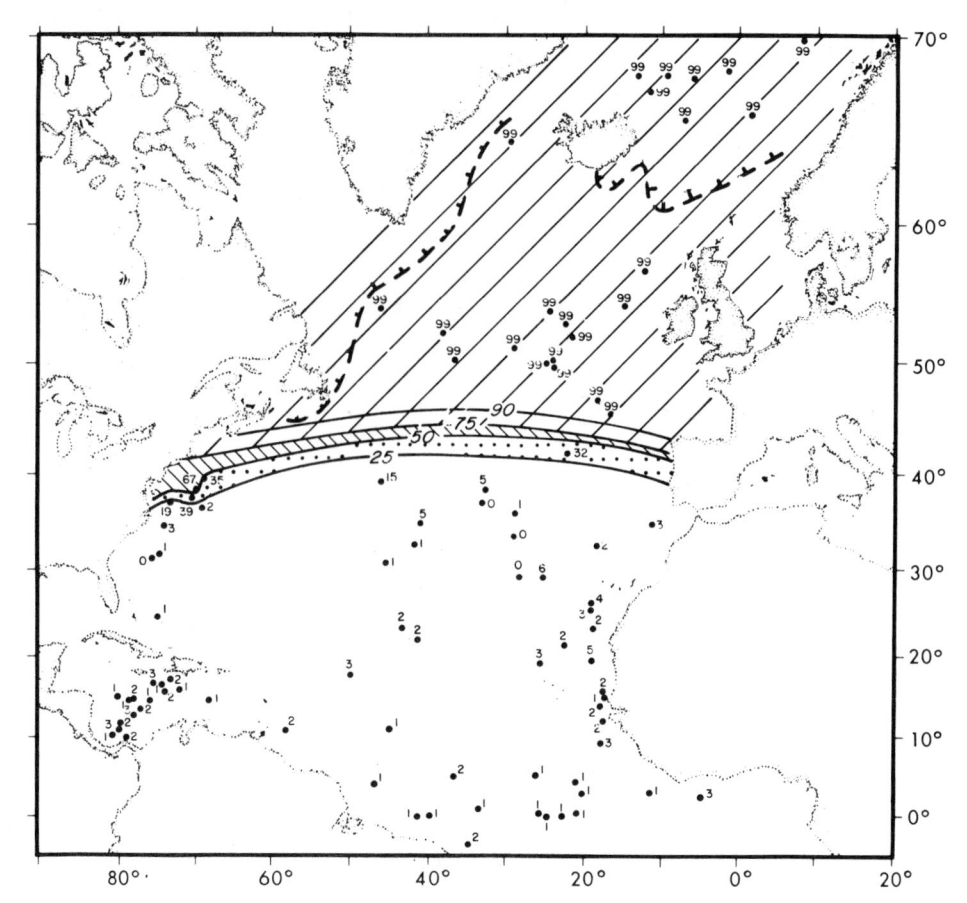

Figure 10. Polar foraminiferal assemblage distribution at 18,000 Ma. The center of concentration of the assemblage in modern sediments has been marked by the line of the truncated "T". The shank of the "T" points towards the maximum concentration (McIntyre et al., 1976).

it is today. These results are exciting as they demonstrate how in the immediate past the sedimentary record gives a very clear indication of near surface oceanographic conditions some 18,000 years ago.

Another recent advance in sedimentology has been a refined understanding of the oxygen-isotope composition of deep sea sediments. At first it was thought that this composition was predominantly related to temperature (Emiliani, 1955). However more recently Shackleton and Opdyke (1973, 1976) have shown that the primary mechanism giving rise to changes in the δO^{18} composition

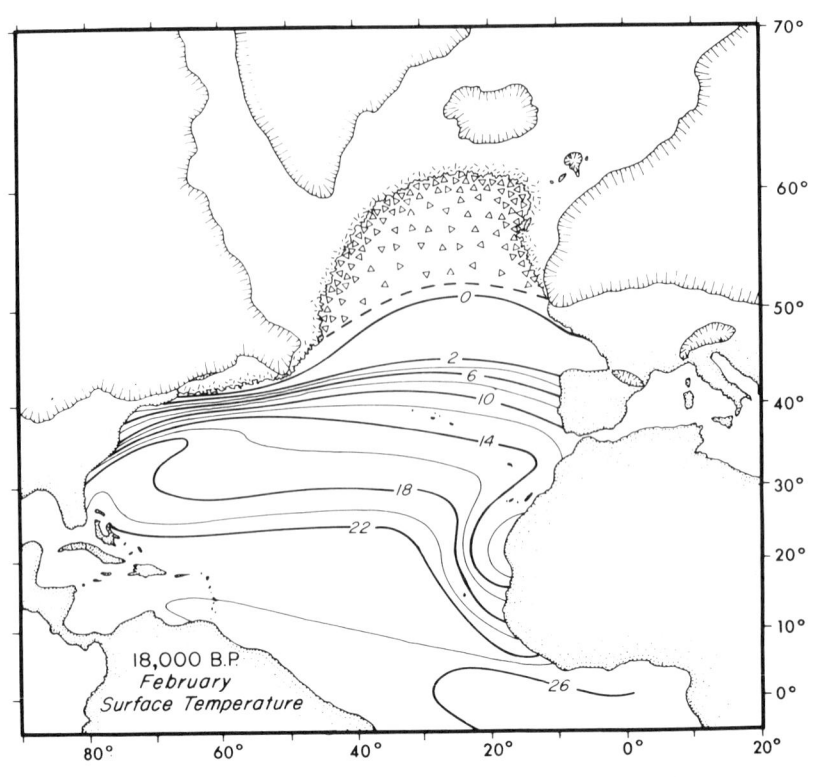

Figure 11. Surface-water isotherm map for February 18,000 Ma in degrees Celsius. Contour interval is 2°C. Dashed isotherms are interpretative. Major continental ice masses are delineated by hatched borders, permanent pack by granulate borders, loose pack by triangles. Glacial shoreline is drawn using present bathymetry to a lowered sea level of 100 m (McIntyre et al., 1976).

in pleistocene cores is the growth and retreat of the continental ice sheet in the northern hemisphere. The δO^{18} changes correlate with the ice ages and by tying the oxygen-isotope record to the paleomagnetic record in two Deep Sea Pacific cores, Shackleton and Opdyke (1976) have determined an accurate relative time scale for the Pleistocene (Figure 12). This has led to some quite surprising results.

Hays et al. (1976) examined two cores from the southern Indian Ocean. They looked at variations in δO^{18}, sea surface

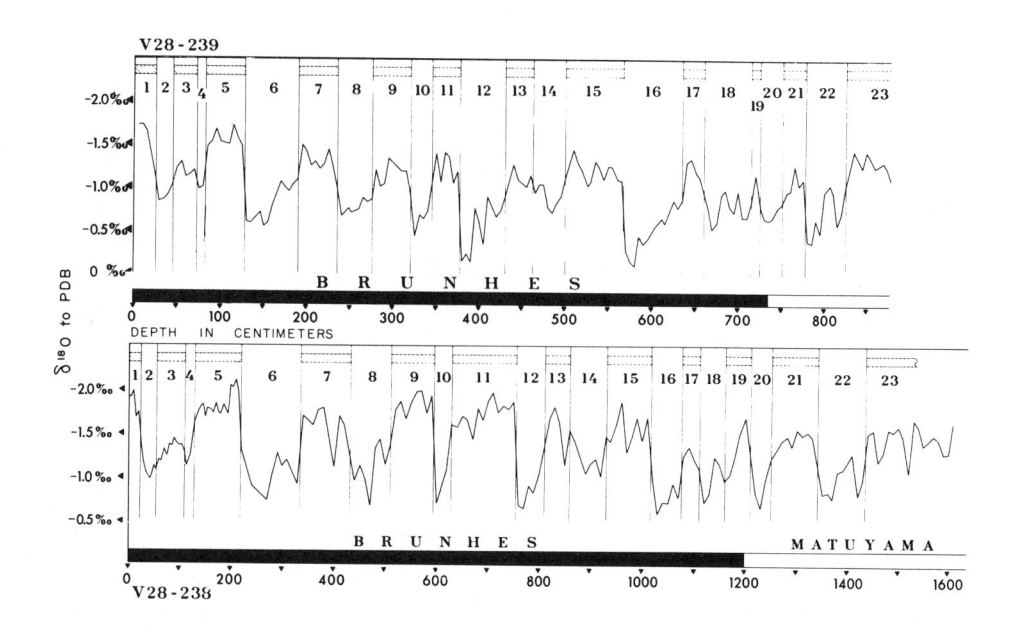

Figure 12. Comparison of oxygen-isotope and palaeomagentic record in upper 880 cm of core V28-239 (above) and in upper 1600 cm of core V28-238 (below) (Shackleton and Opdyke, 1976).

temperature as predicted by the transfer function approach and the concentration of a certain assemblage as a function of depth (Figure 13). They used the principle peaks of the δO^{18} plot to give them time, normalized everything from depth down core to time and digitized the resulting data. They took the Fourier transform of the data first without change and then pre-whitened. In both cases, they found significant peaks corresponding to a 100,000 year period, a 43,000 year period, a 24,000 year period and in some cases a 20,000 year period (Figure 14). The 43,000 and 24,000

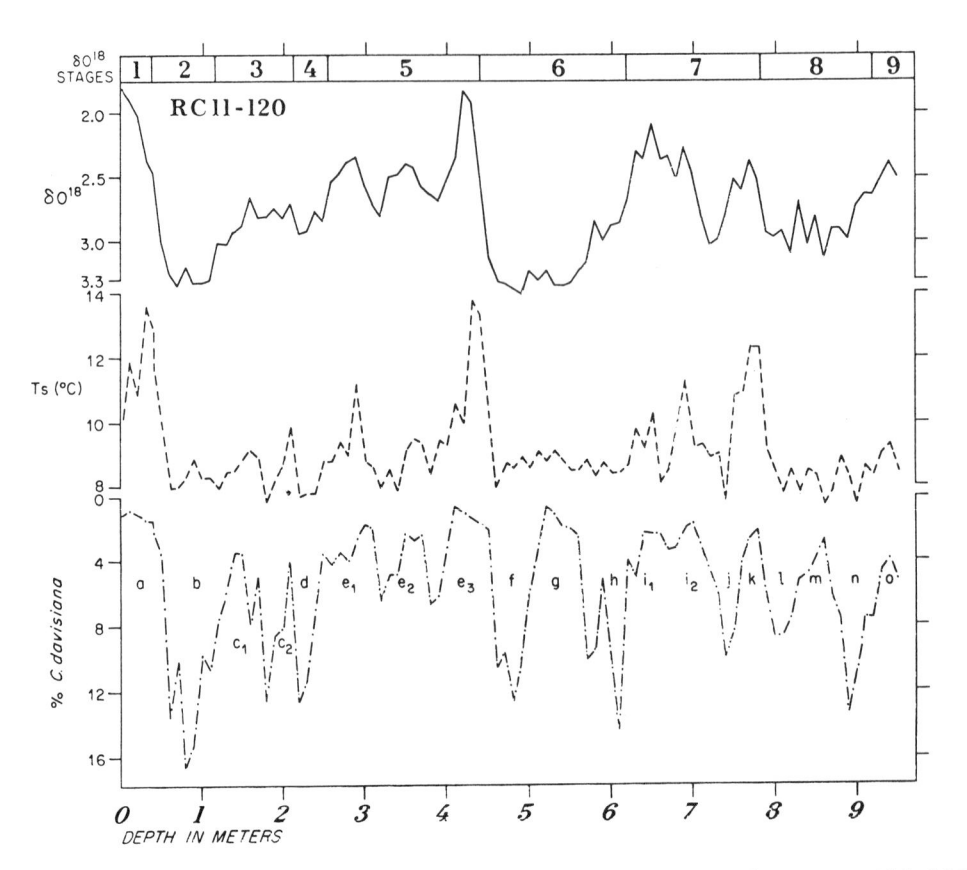

Figure 13. Depth plot of three parameters measured in core RC11-120: δO^{18}, solid line (top); Ts, dashed line (middle); and % C. davisiana, dash-dotted line (bottom) (Hays et al., 1976).

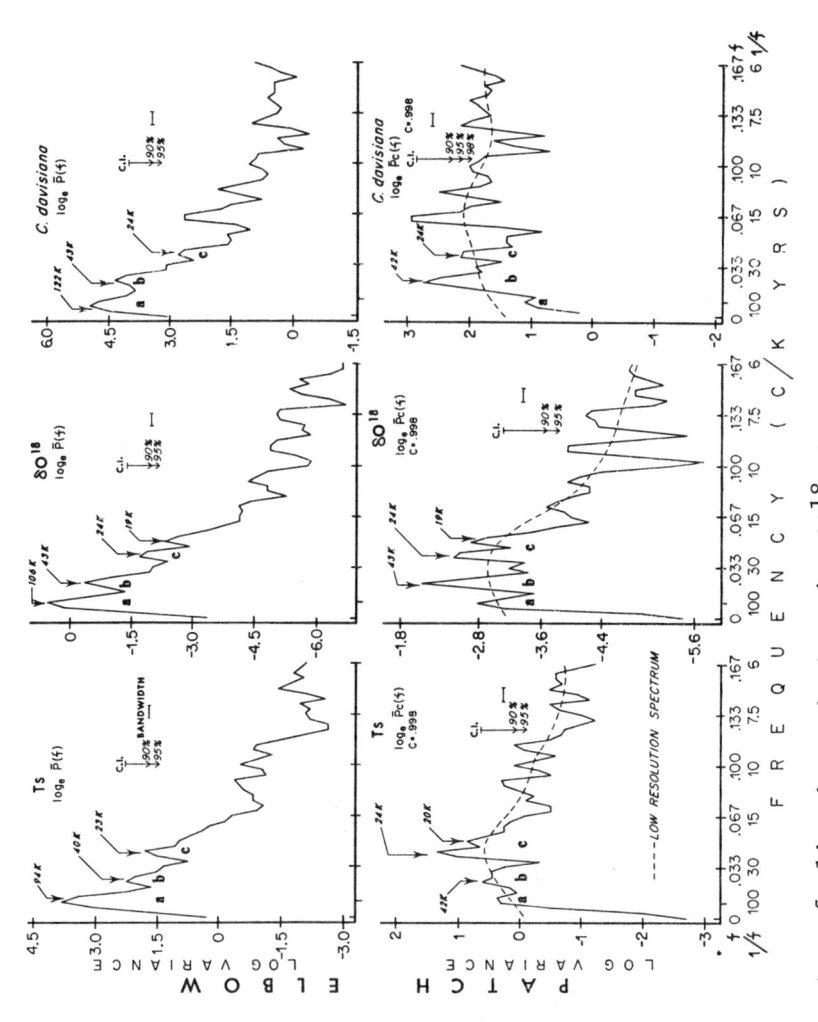

Figure 14. Spectra of climatic variations (Ts, δO^{18}, %C. davisiana) in combined subantarctic deep sea cores. Arrows without crossbars indicate weighted mean cycle length of spectral peak (in thousands of years). Arrows with crossbars show one-sided confidence intervals attached to estimates in the high-resolution spectrum. Prominent spectral peaks are labelled a, b, and c (top row). High-resolution spectra expressed as a natural log of the variance as a function of frequency, cycles per thousand years (bottom row). High-resolution spectra (solid line) and low-resolution spectra (dashed line) after pre-whitening with a first-difference filter (Hays et al., 1976).

year periods are close to those predicted by Milankovitch (1941) on the basis of an orbital hypothesis of climatic change where the obliquity (with a period of about 41,000 years) and the precession of the equinoxes (with a period of about 21,000 years) are the underlying control on the climate. The 100,000 year cycle is accounted for by relating it either to the variation in eccentricity or to a phasing of the obliquity with precession.

These results are very encouraging. If this work is confirmed then we will have a paleontological reference frame related only to the earth's orbital parameters. As these parameters are known with some certainty this may lead to a time scale accurate to tens of thousands of years which might extend over the whole oceanic record.

PALEOSEDIMENTOLOGY

In the prior discussion I concentrated on the record recovered in near surface cores. However there has been an equally major development in our understanding of the sediment in the cores recovered by the Deep Sea Drilling Project. To illustrate some of the advances I will summarize work presented earlier at this Congress by Kennett, Shackleton, Van Andel, and Thiede.

Figure 15 shows the δO^{18} record as a function of age in a core from the DSDP holes in the southern oceans. The variation of $\delta O^{18}/O^{16}$ ratio in forams is thought to be only a function of temperature and ice volume. Decreasing ice volume and increasing temperature will send the record to the right. As we know the total volume of ice in the world, we can calculate its effect. Simple calculations show that all of it must have gone by the middle Miocene and that everything prior to that was probably due to temperature variation. Thus the Antarctic ice sheet was not formed until 15 Ma and before that we have a relatively accurate record of the temperature of the waters in which the forams resided. This last point raises the interesting possibility of being able to predict past temperatures in the ocean as a function of time and depth by analyzing the δO^{18} composition of pre 15 Ma forams. However, at present, it is the first point which has received more attention. Kennett et al. (1974) have related the abrupt early Oligocene discontinuity in the sediment record in cores south of Australia to the separation of Australia from Antarctica and the opening just prior to this time of a major seaway around Antarctica in the early Oligocene (35 Ma). However by the late Oligocene (28 Ma) the seawater is free to flow through the passage and around Antarctica. These authors have related this flow to the continuing cooling of the ocean and then within about 10 million years the formation of the ice sheet in the middle Miocene, 15 Ma. This is an interesting but still speculative idea (Fig 16).

Another major advance summarized very well by Berger and

THE MARINE GEOSCIENCES

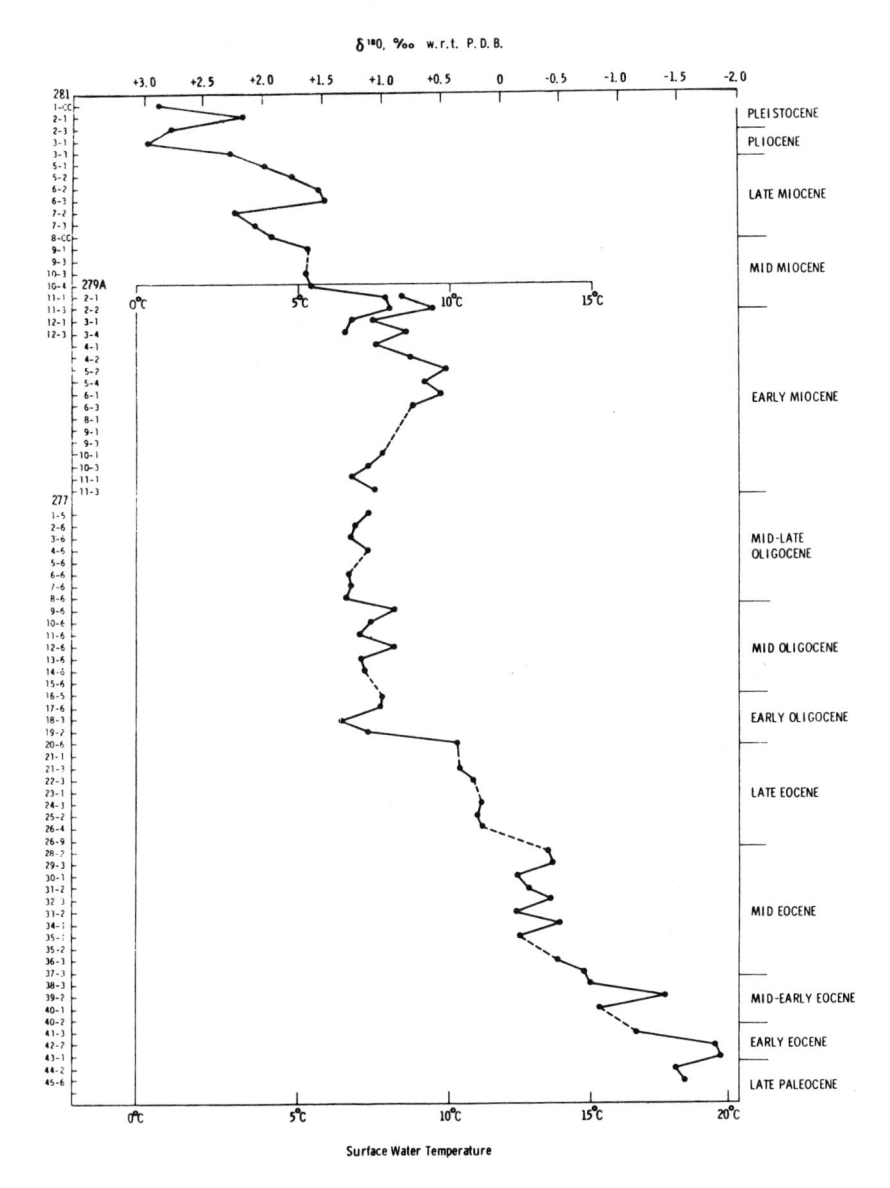

Figure 15. Oxygen isotope data for planktonic foraminifera at sites 277, 279 and 281. The stratigraphic plot is not based on thickness or age difference between samples, but samples are arbitrarily plotted at constant divisions. The surface temperature values are calculated from an equation given in Shackleton and Kennett (1975) from the base of the sequence to Miocene only. Above core 10, site 281 the surface temperature estimation is complicated by ocean isotopic changes (Shackleton and Kennett, 1975).

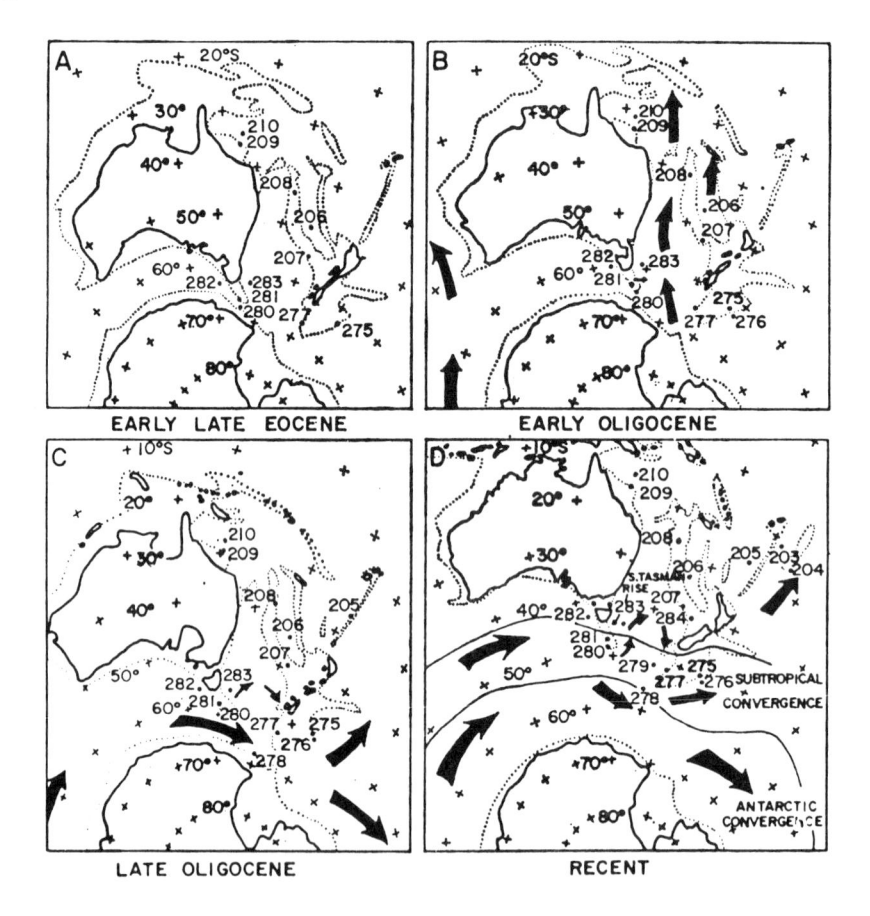

Figure 16. Successive Cenozoic reconstructions of Australia, Antarctica, New Zealand, and associated ridges during (A) early late Eocene (45 million years ago), (B) early Oligocene (37 million years ago), (C) late Oligocene (30 million years ago), and (D) recent. Directions of bottom-water circulation are shown by arrows. Land areas and shallow ridges and rises are marked. During the early Oligocene (B) extensive erosive bottom currents flowed northward through the Tasman and Coral Seas. The Circum-Antarctic Current did not develop until about the late Oligocene (C), and a strong northward-flowing western boundary current formed to the east of New Zealand. These directions have been retained to the present day. Site numbers indicate sites drilled during Legs 21 and 29 of the Deep Sea Drilling Project.

Winterer (1974) is that the oceans have a very definite pattern of sedimentation. At depths down to 4,000 to 4,500 m carbonate sediments predominate. However at depths greater than this clays start to predominate and at very deep depths clay radiolarian oozes and terriginous material become totally dominant. The boundary between the carbonate and clay sediments which is fairly sharp (about 300 m) is called the carbonate clay line or alternately the calcite compensation depth (CCD). It is the depth at which calcite is dissolved at the same rate as it is deposited. It is a feature of most DSDP holes in deep oceans that the sediments recovered immediately above basement are predominantly carbonates. Berger (1972) has shown that this is due to the initial sediments being deposited above the CCD when the ocean crust was young and considerably elevated with respect to its present depth. If one assumes that the present depth age relation has held through time then it is possible to backtrack all the sites to determine the exact depth at which all sediments of known age were deposited (Figure 17a). By backtracking sites in a specific basin and plotting the position of the boundary between the clay and the carbonate sediments it is possible to determine in any one basin the variation of CCD with time (Figure 17b).

Van Andel et al. (in press) have done exactly this for the South Atlantic. First they backtracked the individual sites to determine a history of the CCD in the South Atlantic as a function of time (Figure 18). To obtain the distribution of surface sediments for a given epoch in a basin it is then only necessary to match the CCD in a basin with the paleobathymetry. All depths above the CCD will have predominantly carbonate sediments deposited and all depths below the CCD will consist mainly of clays and radiolarian oozes. Van Andel et al. (in press) have compared their plot of the CCD through time with the paleobathymetry of the South Atlantic shown in Figure 6. Using this technique and other data gathered from the DSDP cores they have presented a sketch of the development of the sediments in the South Atlantic through time (Figure 19). Perhaps the most important points are the pelagic clays and black shales observed to the north and south during the early history and the remarkable change in the surface distribution of carbonate sediments between the late Eocene, Early Miocene, late Miocene and present.

The approach taken by Van Andel et al. (in press) is an attempt using the presently available data and some simple geological constraints to predict the past history of ocean sedimentation. At present this approach is only in its infancy. However, for the most recent sediments, those deposited before the Miocene, the basic proposed distribution of sediments can be compared with the actual sediment recovered in surface cores as well as the Deep Sea Drilling material. McCoy and Zimmerman (1977) have presented an analysis for the South Atlantic of DSDP cores and cores in the Lamont-Doherty

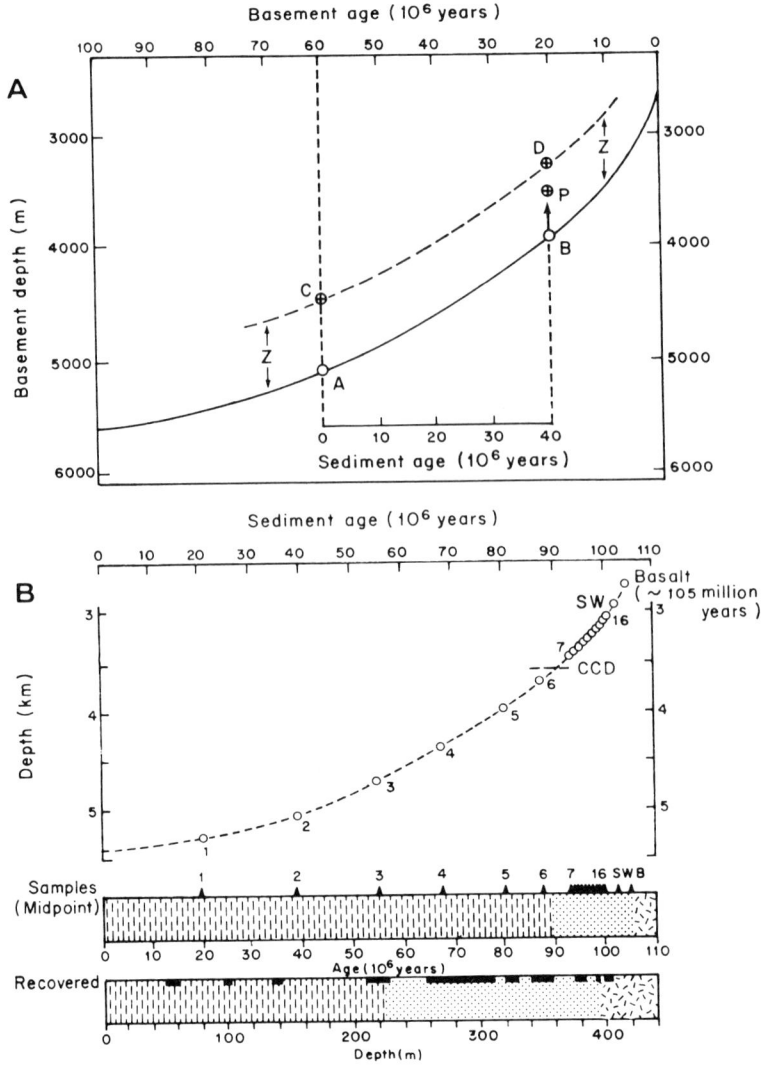

Figure 17. A. Paleodepth determination by vertical backtracking parallel to idealized subsidence track: A and B: present depth and paleodepth 40 Ma on idealized curve; C: actual site depth; D: analogue to B on parallel curve; Z: depth difference between A and C; P: final paleodepth after correction for isostatic loading (see Sclater et al., in press ; Berger and Winterer, 1974).

B. Backtracking of site 137, leg 14 (Hays et al., 1976). Symbols from left to right: clay, calcareous ooze, basalt. The sample numbers represent core numbers from site 137 (Berger and Winterer, 1974).

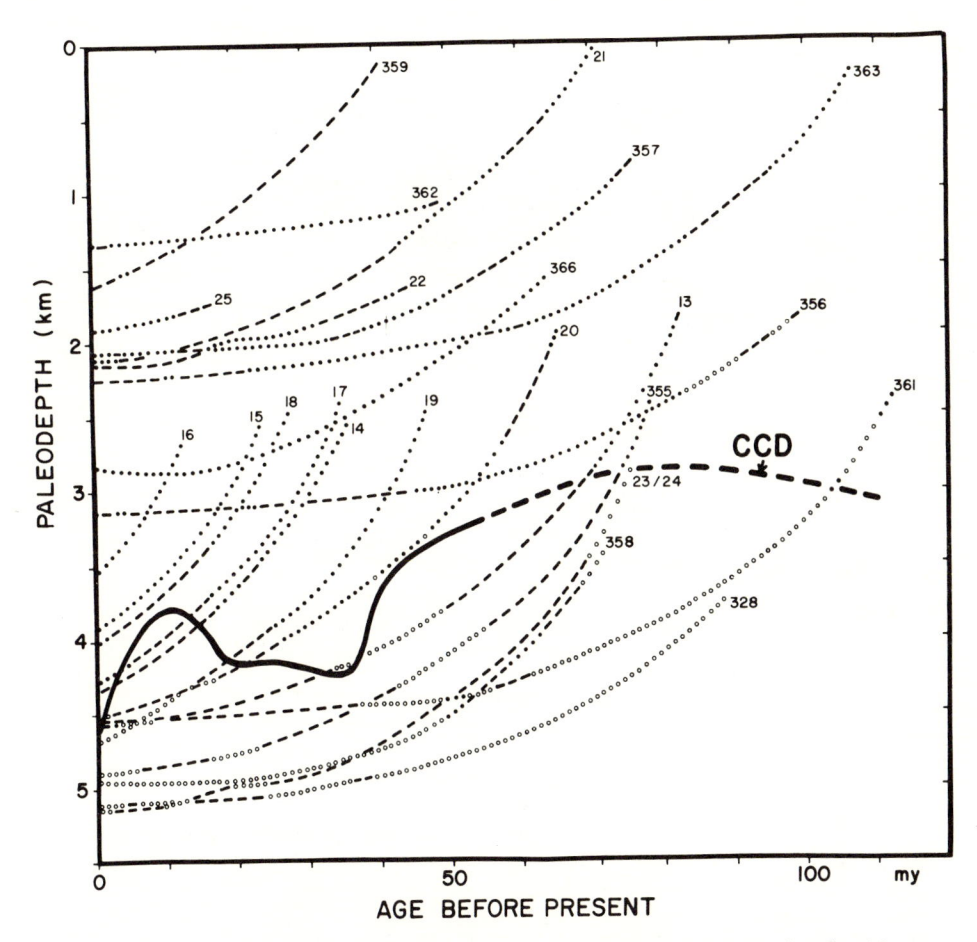

Figure 18. Variation of calcite compensation depth in South Atlantic based on the carbonate content of drilled sections. Dots: $CaCO_3$ more than 20 percent; Circles: $CaCO_3$ below 20 percent. Dashes indicate hiatus or coring gap. Numbers are drill sites. Solid CCD curve obtained by computation from $CaCO_3$ accumulation rates: dashed curve based only on lithology (Van Andel et al., in press).

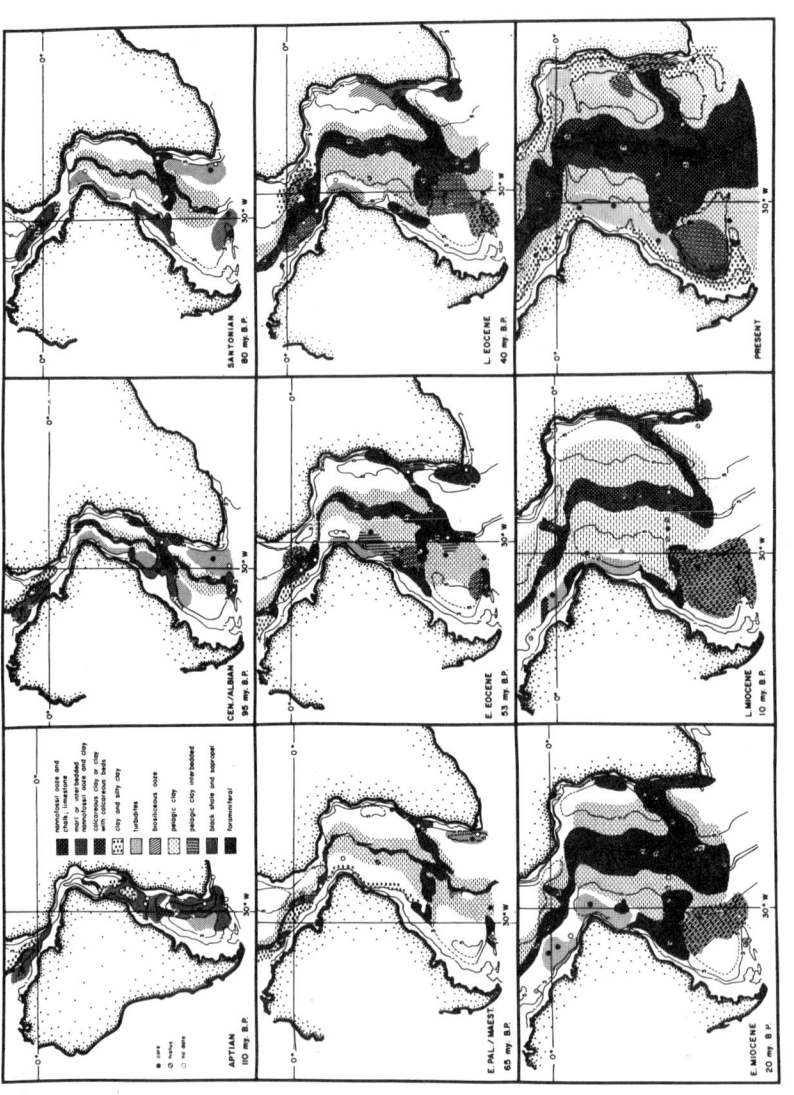

Figure 19. Lithofacies sketch maps for time slices between the early Cretaceous and present. Interval represented is age shown ± 2.5 million years. The bathymetry is from the paleobathymetric charts of the Atlantic of which a few have been shown in Figure 6. The recent sediment distribution is from Ellis and Moore (1973) and Lizitsin (1972). Lambert equal area projection (Van Andel et al., in press).

Geological Observatory collection. They show that between the present and the Miocene there was a considerable diminution in the distribution of calcareous sediments. Their presentation (Figure 20) of the mid-Miocene distribution of the calcareous sediments is remarkably similar to the mid-Miocene chart of Van Andel et al. (in press) (Figure 20, L. Miocene). Thus the work of McKay and Zimmerman (1977) appears to give considerable observational support for the interpretative approach of Van Andel et al. (in press).

CONCLUSION

These last few figures have been presented to demonstrate the overlapping nature of the field of Marine Geosciences. The paleobathymetry of the South Atlantic developed by purely tectonic arguments has given the framework within which the past history of surface sediment distribution has been interpreted. This coupled with the quantitative knowledge we are accumulating from surface sediments places us in the very exciting position of possibly seeing the quantitative interpretation of deep sea sediments through to the Eocene or Cretaceous within the next ten or twenty years.

However, in order to do this we will need to make at least two major advances, one in data coverage, the other in technique.

Figure 21 presents the world late Miocene data coverage. Note how little data has been gathered from the deep southern oceans. A major effort must be made to obtain a good data base from this area if our ability to make world-wide correlations between past oceanographic currents, sedimentation, and bathymetric changes is to advance. Another problem that we need to tackle is in coring technique. We need the ability to get long uninterrupted continuous cores from the deep ocean. For it is only then that we will be able to advance our quantitative understanding of the past 1 million years back into the Cretaceous.

To conclude I would like to thank you for staying to the end and listening to this review. I hope that in some way I have managed to transmit some of the excitement I felt when listening to the various talks in paleoceanography at this congress. I think that we are on the verge of a major breakthrough in ocean sedimentation and that ten years from now we may know as much about the ocean sediments as we do at present about the past and relative motions of the various major plates that cover the earth. For a geophysicist whose first introduction to the oceans was through the fluid dynamics lectures of Dr Ross and the meteorological lectures of Dr McIntosh it is interesting to speculate that the understanding we gather of the past environment of the oceans from surface sediments may be the key to our understanding the general circulation of the oceans.

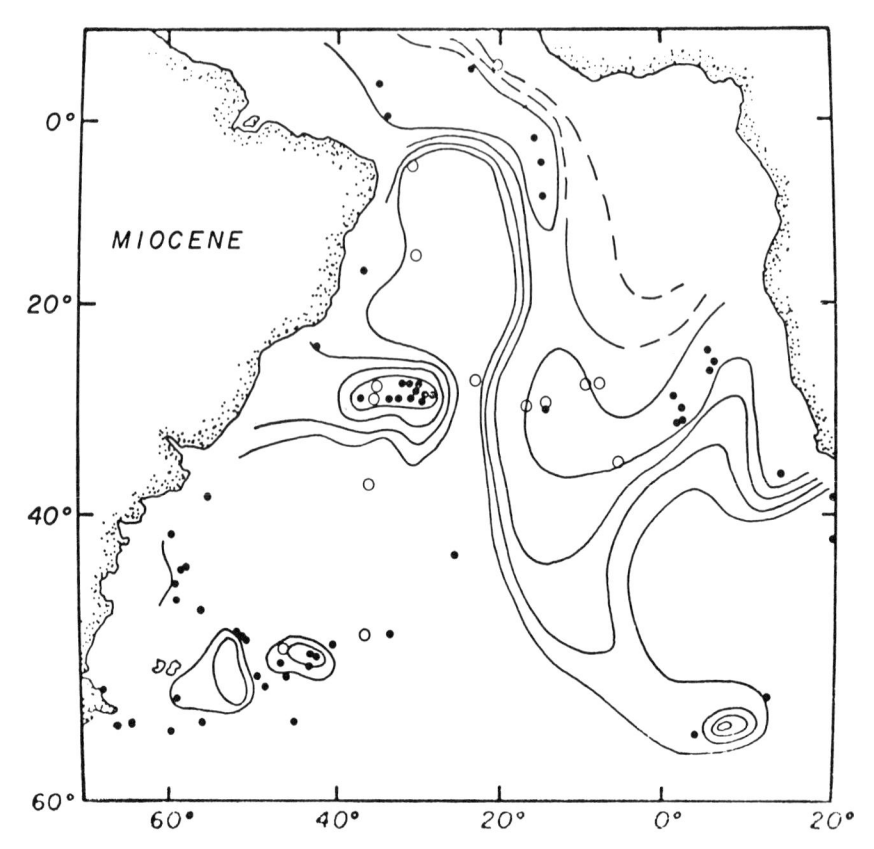

Figure 20. Contours of the Miocene calcium carbonate distribution in the South Atlantic. Note the striking similarity between this chart based on an extensive data coverage and the predicted distribution of calcium carbonate sediments predicted for the miocene shown in Figure 18 (McCoy and Zimmerman, 1977).

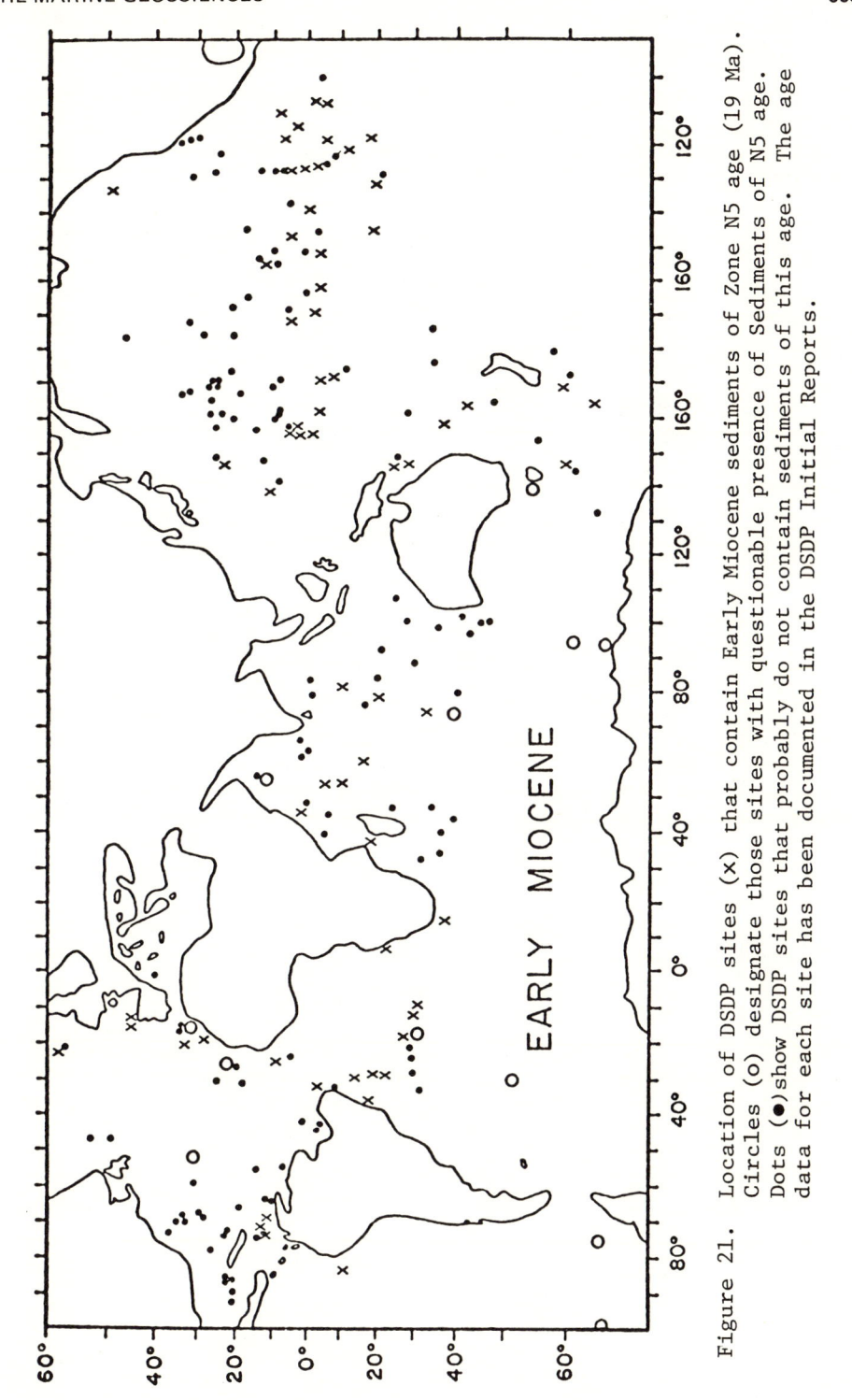

Figure 21. Location of DSDP sites (x) that contain Early Miocene sediments of Zone N5 age (19 Ma). Circles (o) designate those sites with questionable presence of Sediments of N5 age. Dots (●) show DSDP sites that probably do not contain sediments of this age. The age data for each site has been documented in the DSDP Initial Reports.

ACKNOWLEDGEMENTS

This review could not have been completed without the generous help of Nick Shackleton, Jim Kennett, Jim Hays, John Imbrie and Bill Berggren. I am also grateful to the Office of Naval Research, contract N00014-75-C-0291, and the National Academy of Sciences for support to give the paper in Edinburgh.

REFERENCES

Berger, W.H., 1972. Deep sea carbonates: dissolution facies and age-depth constancy, Nature, 236, 392-395.

Berger, W.H. and E.L. Winterer, 1974. Plate stratigraphy and the fluctuating carbonate line, International Association of Sedimentology, Spec. Publ., 1, 11-48.

Cline, R.M. and J.D. Hays, 1976. Investigation of late Quaternary paleoceanography and paleoclimatology, Geol. Soc. Amer. Memoir, 145, 464.

Cox, A., R.R. Doell, and G. Brent Dalrymple, 1964. Reversals of the Earth's magnetic field, Science, 144, 1537-1543.

Davis, E.E. and C.R.B. Lister, 1974. Fundamentals of ridge crest topography, Earth Planet. Sci. Lett., 21, 405-413.

Detrick, R., J.G. Sclater and J. Thiede, in press. Subsidence of aseismic ridges: geological and geophysical implications, Earth Planet. Sci. Lett.

Ellis, D.B. and T.C. Moore, 1973. Calcium carbonate, opal and quartz in Holocene pelagic sediments and the calcite compensation level in the South Atlantic Ocean, Jour. Mar. Res., 31, 210-227.

Emiliani, C., 1955. Pleistocene temperatures, J. Geology, 63, 538-578.

Hardenbol, J. and W.A. Berggren, in press. A new Paleogene numerical time scale.

Hays, J.D., J. Imbrie and N.J. Shackleton, 1976. Variations in the earth's orbit: pacemaker of the ice ages, Science, 194, 1121-1132.

Imbrie, J. and Kipp, N., 1971. A new micropaleontological method for quantitative paleoclimatology: application to a late Pleistocene Caribbean core. In Turekian, K.K., ed., Late Cenozoic glacial ages: New Haven, Conn., Yale Univ. Press, pp. 71-181.

Kennett, J.P., R.E. Houtz, P.B. Andrews, A.R. Edwards, V.A. Goston, M. Hajos, M.A. Hampton, D.G. Jenkins, S.V. Margolis, A.T. Ovenshine, and K. Perch-Nielsen, 1974, Development of the Circum-Antarctic Current, Science, 186, 144-147.

Kipp, N.G., 1976. New transfer function for estimating past sea-surface conditions from sea-bed distribution of planktonic foraminiferal assemblages in the North Atlantic, in Cline, R.M., and Hays, J.D., eds. Investigation of late Quaternary paleoceanography and paleoclimatology; Geol. Soc. Am. Memoir, 145, 3-42.

Kristoffersson, Y., S. Cande and M. Talwani, 1976. Sea-floor spreading and the early opening of the North Atlantic, Abstract, Trans. Amer. Geophys. Un., April, 38.

LePichon, X., 1968. Sea floor spreading and continental drift, J. Geophys, Res., 73, 3661-3697.

Lizitsin, A.P., 1972. Sedimentation in the world ocean, Soc. Econ. Pal. Min., Spec. pub., 17, 218 pages.

McCoy, F.W. and H.B. Zimmerman, 1977. A history of sediment lithofacies in the South Atlantic, in P. Supko and K. Perch-Nielson et al., Initial Reports of the Deep Sea Drilling Project, XXXIX, Washington (U.S. Govt. Printing Office).

McIntyre, A., N.G. Kipp, A.W.H. Be, T. Crowley, T. Kellog, Jr., Gardner, W. Prell and W.F. Ruddiman, 1976. Glacial North Atlantic 18,000 BP: A CLIMAP reconstruction, in Cline, R.M. and Hays, J.D. eds., Investigation of late Quaternary paleoceanography and paleoclimatology, Geol. Soc. Amer. Mem., 145, 43-76.

McKenzie, D. and R.L. Parker, 1967. The North Pacific, an example of tectonics on a sphere, Nature, 216, 1276-1280.

Milankovitch, M., 1941, K. Serb. Akad. Beogr. Spec. Pub. 132 (trans. by the Israel Program for Scientific Trans., Jerusalem, 1969).

Morgan, W.J., 1968. Rises, trenches and crustal blocks, J. Geophys. Res., 73, 1959.

Parker, R.L. and D.W. Oldenburg, 1973. Thermal model of ocean ridges, Nature, 242, 122, 137-139.

Parsons, B. and J.G. Sclater, in press. An analysis of the variation of ocean floor bathymetry and heat flow with age, J. Geophys. Res.

Phillips, J.D. and D. Forsyth, 1972. Plate tectonics, paleomagnetism and the opening of the Atlantic, Bull. Geol. Soc. Amer., 83, 6, 1579-1600.

Phillips, J., C. Tapscott, H. Fleming and R. Fedden, 1976. An aeromagnetic survey of the Arctic, Trans. Amer. Geophys. Un., Washington, D.C., April, 123.

Pitman, W.C., III, and Talwani, M., 1972. Sea floor spreading in the North Atlantic, Geol. Soc. Amer. Bull., 83, 619-646.

Pitman, W.C., III, et al., 1974. Magnetic lineations of the oceans, Geol. Soc. Amer., Spec. Paper.

Ryan, W.B.F., M.B. Cita, M. Dreyfus Rawson, L.H. Burkle and T. Saito, 1972. A paleomagnetic assignment of Neogene stage boundaries and the development of isochronous datum planes between the Mediterranean, the Pacific and Indian Oceans in order to investigate the response of the world ocean to the Mediterranean salinity crisis, Riv. Ital. Paleont., 80, 4, 631-688.

Sclater, J.G., S. Hellinger, and C. Tapscott, in press. Paleobathymetry of the Atlantic Ocean, J. Geology.

Shackleton, N.J. and J.P. Kennett, 1975. Paleotemperature history of the Cenozoic and the initiation of Antarctic glaciation: oxygen and carbon isotope analysis in DSDP sites 277, 279 and 281, in Kennett, J.P. and Houtz, R.E. et al., Initial Reports of the Deep Sea Drilling Project, Vol. XXIX; Washington, (U.S. Govt. Printing Office).

Shackleton, N.J. and Opdyke, N.D., 1973. Oxygen isotope and paleomagnetic stratigraphy of equatorial Pacific core V28-238: oxygen isotope temperatures and ice volumes on a 10^5 and 10^6 year scale, Quat. Res., 3, 39-55.

Shackleton, N.J. and N.D. Opdyke, 1976. Oxygen-isotope and paleomagnetic stratigraphy of Pacific core V28-239 late Pliocene to latest Pleistocene, in Cline, R.M., and Hays, J.D., eds., Investigation of late Quaternary paleoceanography and paleoclimatology, Geol. Soc. Amer. Mem., 145, 449-464.

Van Andel, T.H., J. Thiede, J.G. Sclater, and W.H. Hey, in press. Depositional history of the South Atlantic Ocean during the last 125 million years, J. Geol.

Vine, F., 1966. Spreading of the ocean floor: new evidence, Science, 154, 1405-1415.

Wilson, J.T., 1965. A new class of faults and their bearing on continental drift, Nature, 207, 343, 1965.

APPLIED ASPECTS OF OCEANOGRAPHY

T.D. PATTEN

INSTITUTE OF OFFSHORE ENGINEERING

HERIOT-WATT UNIVERSITY, EDINBURGH, SCOTLAND

INTRODUCTION

As the close of this Assembly quickly approaches and faced with this all-embracing topic, I find it impossible to resist the temptation to indulge in some philosophy. The papers and the discussion have been dominantly scientific, and collectively give a good account of the state of oceanography, the science of the oceans, the inhabitants and the interface. This study of nature continues to uncover facts of differing kind, good and bad, significant and otherwise, and new or confirmatory. It is when application is considered that thought must be given to differentiating between these facts and to endeavouring to discover those which can be classed as useful. This succeeding stage and the progress to fulfilment introduces judgement, uncertainties and even trial and error, quite unlike the initial research phase.

In the Assembly proceedings, infrequent reference has been made to application and what balance there is, was provided in the sessions entitled, "Man and the Sea", "Oceanography and Fisheries", "Ocean Engineering", and "Geoscience, Minerals and Petroleum". It is a risky business to speculate on future applications and anyone with a proven record of success in this area would surely be tempted to keep potential projects to himself.

Pressure to apply oceanographic knowledge follows from the growing appreciation of the resources to be found in or via the oceans. On the other hand, the overall importance of the seas in maintaining a suitable survival environment for man, demands a responsible attitude with adequate attention to protection.

In order to make any headway in the summary it was necessary to make some classification of oceanography and this ranged from the simplest system used in the summary lectures of

(i) physical oceanography

(ii) biological oceanography

and (iii) marine geoscience

to the twenty-five or so titles under which the many Assembly papers have been distributed. I have used the simplest system but not rigorously and the end result is that the non-allocable Ocean Engineering items reappear in the class "applied aspects".

For very many reasons it has not been possible to attend the presentation of all relevant papers, or even to read them, and therefore the insight into the contents of the scientific papers has been derived from the advance abstracts. An analysis of the results of this rudimentary appraisal has given some pointers to the direction of possible application as seen by the authors concerned. It might appear to be invidious to identify such authors because in the longest term something of use might develop from any one contribution. Nevertheless the object of this paper is to pinpoint some contributions which are capable of assisting in the solution of known needs and therefore selective reference is made.

PHYSICAL OCEANOGRAPHY

No justification is offered for turning first to physical oceanography and particularly to ocean circulation. Ocean circulation is a recurrent theme of the assembly and is a factor vital to marine life, to climate and to many other process loops of our complex universe system. Some papers 1,2,3,4,5,6 link together ocean circulation and climate and ocean circulation and water composition and my reason for referring to them at this time is primarily to stress that it is essential that we understand the nature and totality of the mutual interdependence between man and the oceans if we are to regulate a pattern of life which will ensure a healthy and lasting balance of interaction between one and the other. The direct application of available knowledge on ocean circulation is probably many years ahead, and the variability of ocean behaviour is a major complication, but quantitative modelling on a large-scale is in progress. Given success in this pursuit and the ability to estimate future trends, proper application should at least assist in lessening the impact of pollution on the natural environment.

At the local level a need frequently exists for more authentic local wave data for example to improve the design of offshore

APPLIED ASPECTS OF OCEANOGRAPHY

structures and this possibility is admitted in the study 7 to provide a data base for wave prediction.

It is a truism that scientific investigation will benefit from the application and the further development of existing techniques and experimental devices but we can appreciate the possible operational benefits of high-resolution remote-sensing devices 8,9 and their place in exploration. Underwater, acoustic waves provide a penetrating and sensitive probe whereas surface behaviour may be measured from orbiting platforms.

Of wide significance, if practicable, is the proposal 10 to extract energy available through differences in salt concentration at sea. Attractive possibilities 11 for application arise through harnessing dynamic behaviour of waves and flows to inhibit sediment disposal in estuarine channels and to design effective breakwaters.

BIOLOGICAL OCEANOGRAPHY

The interaction between ocean circulation and marine life has already been referred to and established knowledge 12,13,14 will be applicable in due time, for example through a better appreciation of seasonal variations. Studies may be on an ocean scale or aimed at a partially enclosed sea 15,16, such as the Baltic or Mediterranean. The eventual target for application is improved productivity of fish resources 17,18 and therefore it may be necessary to give particular attention to the species 19,20 concerned.

Much has yet to be clarified about the phenomenon of upwelling which is responsible for major feeding grounds at preferential ocean locations, but it is receiving intensive attention 21,22,23 and will bring further changes in fishing technology. The ability to control or induce this process is an aim worthy of the most serious deliberations 24 and of great potential value.

Unwelcome changes in species abundance due to climatic disturbances such as the "El Nino" merit study 25, 26 because of the destructive effect and the possible eventual consequences for other species higher in the food chain. We must also be able to apply the conclusions of studies 27,28 into the impact of man-made disturbances on the marine ecosystem. Potentially dangerous processes include offshore oil production, deep-sea mining, marine sand and gravel extraction, land reclamation and waste disposal.

Undoubtedly this assembly was the occasion for a very large number of contributions on biological oceanographic matter. From a reading of titles only it would appear that much of the information might be capable of application, but as the information was provided via poster sessions it has been impossible for this reviewer to give it even superficial assessment. However the need for continuing close collaboration between the scientist and fishery technologist was expressed clearly in the invited paper 29 on living resources of the ocean.

MARINE GEOSCIENCE (APPLIED ASPECTS)

Two review papers 30,31 dealt authoritatively with the state-of-the-art and future possibilities in mineral deposit exploitation. The political and legal complications of deep sea mining still defy solution and may be the principal existing barriers to progress. Nevertheless it would be premature to write off the technological difficulties. Despite this hiatus, new information 32 on seabed minerals and their origin will assist in assessments of the economic feasibility of recovering different metallic ores. While the manganese nodule is widely regarded as the most important mineral resource on the ocean floor the exact mechanism of formation is not known and therefore further relevant information 33,34 on growth and growth rates is welcome.

The metalliferous sediments which can be associated with volcanic ocean ridges evoke interest 35,36 although practical methods of mining and processing have yet to be developed.

The continuing importance of oil as a world-wide energy resource, the high consumption rate, and the dwindling proven reserves, guarantee interest in suggestions 37,38 of prospective oil-bearing structures beneath the continental slopes. For some years the major operating oil companies have been giving active consideration to the development of the various elements of advanced production systems proposed for these greater depths.

A SUBJECTIVE VIEW

It has been a most educative experience for me to have this birds eye view of so much oceanographic science and I welcome the opportunity of paying public tribute to the initiative and effort of all those involved. Progress continues to be made in improving our basic understanding. At first sight many of the emerging facts must be useful. Rightly or wrongly, I have gained the impression that oceanography is moving from a political and military course to a socio-economic one.

What sign is there of a strong drive to apply usefully this improved understanding? Who is concerned with trying to match the needs with the potential of your findings and if he or she is not here, why not? Could some of the work reported be diverted more usefully in a different direction? For example now that we know so much about the process and variation of natural upwelling can we "seed" the essential nutrient concentrations or plankton communities and can we optimise the location and timing of this interference.

The outcome of this survey is that there appears to be surprisingly little international effort and financial support for applications based on new technology. In the UK the main incentive for application is the exploitation of offshore hydrocarbons. Outsiders may be concerned at the apparent narrowness of the UK interest, but such reservations must not be allowed to denigrate the scale and difficulty of present offshore applications. A UK movement to have the Science Research Council adopt Marine Technology as a priority R and D topic, appears to have preliminary acceptance.

Finance is a major obstacle to technology R and D because the costs are approximately ten times greater than research to establish the basic knowledge. Hard and painful experience has taught me that the level of necessary funding is a major disincentive to achieving programmes directed more towards applications and I must assume that international programmes suffer the same restrictions.

APPLIED ASPECTS

Clearly the ocean engineering contributions fall in the class of applied aspects because the work referred to has been carried out with the specific intention of application.

The paper 39, on energy sources other than petroleum, reviews a wide range of possibilities in which the sea is central or secondary to the process. The content is largely futuristic and speculative.

Mention has already been made of the review of marine mining methods 31 with accent on shelf mining. It is important to note the explicit reference therein to the problem of ship and vessel motion at the sea surface, which is general to most ocean and marine resource exploitation activities. Designers of offshore operational systems can expect to be faced with problems of motion and force compensation throughout the foreseeable future. Presumably in the past the experimental marine scientist has dealt summarily with this inconvenience.

Design cannot be carried out in the abstract and in the case of floating vehicles or fixed structures it is necessary to have quantitative data on waves, winds and currents. This was the subject of a review paper 40, containing extensive references and in it the case was put for a wave data directory. Also included in the same symposium were descriptive presentations on offshore structures 41 and pollution prevention 42. How do we strike an acceptable balance in this matter of pollution? Any commercial scale operation to exploit marine resources carries the potential danger of pollution - responsible society seeks to limit the risk to a tolerable level - the scientist must assist in setting these levels but it will fall to the engineer/technologist to ensure that the operational equipment meets the specification.

Successful monitoring of offshore operation is dependent on devising new full-proof, on-line detectors and instrumentation. In other areas of endeavour, progress in technological development is often traceable to new effective instrumentation. Safety in complex industrial plant requires efficient instrumentation. It was Lord Kelvin, a famous Past President of the Royal Society of Edinburgh who wrote,

"I often say that when you can measure what you are speaking about and express it in numbers, you know something about it; but when you cannot measure it in numbers your knowledge is of a meagre and unsatisfactory kind; it may be the beginning of knowledge, but you have scarcely in your thoughts advanced to the stage of science, whatever the matter may be".

How fitting therefore to have a paper 43 on ocean instrumentation in which the author stresses the need to improve the level and quality of marine instrument application, and the in situ accuracy and reliability. For many of the audience or future readers the issue may be summarised to the need for improved engineering of equipment for scientific exploration purposes. For me the problem of the future is to satisfy the needs for instrumentation to meet engineering and application needs offshore and underwater. In addition to accurate and reliable equipment we must develop the technology and techniques to ensure that the whole operation of installing the equipment, transmitting and recording the data can be carried out with a predictable and acceptable probability of success. Engineering information may differ substantially from scientific data, for example the acceptable frequency limits for wave date for oceanographic purposes would probably exclude the frequencies capable of leading to fatigue of conductor pipe.

It is clear to me that ocean instrumentation, for application purposes in particular, will remain an important and continuing future area of study and the paper referred to provides additional justification.

Arising from this thought are a number of other identifiable future concerns.

In considering possibilities of offshore oil production under different and varying conditions there emerge many problem areas in which oceanographers and related scientists may be able to contribute, that is areas of scientific research which arise out of the needs of the application. The different conditions for oil production include:-

> (1) water depths as at present, but a significantly higher degree of safety to apply to all offshore personnel,
>
> (2) greater water depths, and
>
> (3) more severe environmental loading, including ice.

The additional and relevant identificable future concerns include:

> (a) underwater visibility
>
> (b) corrosion
>
> (c) marine fouling and its inhibition
>
> (d) remote-controlled systems
>
> (e) autonomous power generation at the seabed
>
> (f) location of sophisticated plant at the seabed
>
> (g) relevant ice properties and ice movement.

Conversely, in addition to making demands for new information from scientists the new applications in operation at sea are capable of providing new scientific opportunities, for example, offshore structures can assist in furnishing new environmental data.

CLOSURE

After life itself and faith, the oceans may yet be shown to be man?s greatest resource. We cannot excape using them and we should strive to do so wisely and well.

REFERENCES (to papers presented at the Joint Oceanographic Assembly)

1. FIADEIRO, M. Tracer models and deep ocean circulation.
2. MUNK, W.H. The variable ocean structure.
3. STEWART, R.W. Ocean variability - influences of atmospheric processes.
4. BRYAN, K. A model of water mass formation in the world ocean.
5. GOLDBERG, E.D. Climate and the composition of surface ocean waters.
6. DUCE, R.A. The sea/air interface and its effects on marine aerosol composition.
7. HASSELMANN, K. and JONES, L. Joint north sea wave project.
8. BREKHOVSKIKH, L.M. Acoustical sensing of oceanic processes.

9. APEL, J.R. Ocean science from space.

10. LOEB, S. and BLOCH, M.R. Salinity power, potential and processes, especially membrane processes.

11. ISAACS, J.D., SEYMOUR, R.J. and COSTA, S.L. Exploiting highly non-linear interactions in ocean processes.

12. REID, J.L., McGOWAN, J.A., BRINTON, E., FLEMINGER, A and VENRICK, E.L. Ocean circulation and marine life.

13. LASKER, R. Ocean variability and its biological effects: regional review-northeast Pacific.

14. HARDEN JONES, F.R. Tactics of fish movement in relation to migration strategy and water circulation.

15. JANSSON, B. The Baltic - a system's analysis of a semi-enclosed sea.

16. GASCARD, J.C. and NIVAL, P. Mediterranean dynamics and productivity.

17. SAETERSDAL, G. Fishery oceanography, concepts and problem areas.

18. LASKER, R. Variations in reproductive success of marine fish stocks as influenced by environmental factors.

19. YAMANAKA, I. Oceanography in tuna research.

20. HERMANN, F. and HORSTED, S.A. Fluctuations in the cod stock and the ocean climate at West Greenland.

21. SMITH, R.L. Physical aspects of coastal upwelling.

22. VINOGRADOV, M.E. Equatorial upwelling - its physical and biological peculiarities.

23. BARBER, R.T. Biological aspects of coastal upwelling.

24. HOOD, D.W. Upwelled impoundments as means of enhancing primary productivity.

25. KESTEVEN, G.L. and GONZALES, J.V. The anchovy and the "El Nino".

26. WALSH, J.J. The biological consequences of interaction of the event and climatic scales of variability in the eastern tropical Pacific.

27. THIEL, H. Deep-sea bottom community components, their relative importance and suggestions for future investigations.

28. LEE, A. Effects of man on the fish resources of the North Sea.

29. PARRISH, B.B. Living resources of the ocean.

30. NEUWEILER, H. Non-renewable resources.

31. VELZEBOER, P. Th.   Marine mining.
32. RYAN, W.B.F. Changing patterns of productivity and the origin of mineral resources.
33. GUNDLACH, H., MA CHIG, V., HEYE, D., SCHNIER, C. and SCHUETT, C. New results from the research on manganese nodules.
34. LALOU, C., BRICHET, E. and BONTE, P. Does manganese nodules represent the slowest accretion phenomenon ever known or a very fast episodic reaction?
35. ROSS, D.A. Mid-ocean ridge mineralization and hot brines.
36. LALOU, C., BRICHET, E., KU, T.L. and JEHANNO, C. The Famous hydrothermal deposit, radiochemical and SEM/EDAX study.
37. MONTADERT, L. and TISSOT, B. Oil prospects in continental slopes and deep seas.
38. EMERY, K.O. Types of continental margins and their oil potential.
39. HOOG, H. Ocean engineering - energy other than petroleum.
40. HOGBEN, N. Environmental parameters.
41. GOODFELLOW, R. Offshore structures
    CRESSWELL, N.J.
42. RANKEN, M.B.F. The prevention of pollution of the sea.
43. NICHOLSON, W.M. Ocean instrumentation.

INDEX

Abra alba, 160
Acartia danae, 226
Acartia negligens, 226
Age-depth
  formulae, 17
  curve, 16
Agulhas basin, 7, 9
Algae blooms, 150
  carbon content of, 150
  nitrogen content of, 150
Angola basin, 3, 7, 11, 14,
    18, 28, 34, 40
Angola continental margin,
    11, 18, 28, 29, 34
Antarctic Bottom Current, 5, 7,
    9, 11, 14, 27, 28, 29
Antarctic Bottom Water, 5, 7, 9,
    11, 14, 27, 39, 40
Antarctic glaciation, 27
Anthropogenic fluxes, 51
Argentina continental
    margin, 11, 14, 28
Argentine basin, 2-11, 14, 18,
    23-28, 34, 39, 40, 314
Astarte borealis, 160
Atmospheric influence
  on the ocean, 213-219
  cloudiness, 213, 218
  heat and water vapour
    transfer, 213, 218
  indeterminacy, 219
  momentum transfer, 213
  pressure effects, 213, 217
  response of upper layer, 218
  wind stirring, 213, 215, 218
Aurelia aurita, 140, 169

Baltic
  aerobic bacteria in, 153
  annual primary production of,
    152
  as a detritus based system, 156
  benthic ecosystem, 160
  biogeochemical cycles in, 132
  biological systems in, 138-160
  biomass, 147
  birds, 169, 170, 171
  bottom sediments, 134
  bottom water, stagnant, 165
  carbon budget, 165
  DDT in, 169
  desulphovibrio bacteria, 158
  downwelling in, 138
  ecosystem, recent changes in,
    164
  eutrophication, 164-169
  fish, 146, 152, 160, 162, 164,
    169, 171
  heavy metals, 171
  Kelvin waves in, 138
  macrofauna, 158
  meiofauna, 158
  morphology, 132-134
  nutrient transport in, 137
  oceanization of, 164, 169
  oil spills in, 171
  outlook for, 172
  particulate organic matter in,
    153
  photoautotrophic purple
    bacteria, 156, 158
  pollution of, 169-172
  pycnocline of, 135-136

Baltic (cont'd)
  redox-cline of, 156
  residence time, 137
  salinity, 135
  seals, 169, 171
  species, 141
  subsystem, pelagic, 132, 148-153
  subsystem, phytal, 132, 142-143
  subsystem, soft bottom, 132, 153-160
  systems analysis of, 131-183
  upwelling in, 138
  water balance of, 134
  zooplankton production in, 152
Barents Sea, 66
Beggiatoa, 15
Benguela current, 11
Biological oceanography, trends in, 297-306
  abyssal food webs, 299
  climatic effects, 300
  coastal upwelling, 301
  food web structures, 305
  marine resource management, 302
  mathematical models, 300
  mesoscale features, 301
  microbial degradation, 299
  organic material, recycling of, 299
  paleo-oceanography, 300
  patchiness, 302
  pollution threat, 302, 303
  reproduction strategy, 305
  SCUBA diving, 299
  submersibles, 299
  taxonomy, 303
  zoogeography, 301
Biomass distribution, upper layer, 81, 84
Black shales, 23, 329
Brazil basin, 2, 3, 5, 7, 9, 18, 23, 24, 26, 28, 34, 40

Calanoides acutus, 232, 234, 236
Calanoides glacialis, 89, 93
Calanus jobei, 226
Calanus propinquus, 8, 9, 93, 116, 232, 234, 236
Calanus finmarchicus, 89, 92, 119

Calanus simillimus, 89
Calcareous ooze, 14
Calothrix scopulorum, 142
Cape basin, 2, 3, 7, 9, 11, 18, 21, 23, 24, 25, 28, 34, 39, 40
Carbonate compensation depth, 7, 9, 11, 13, 14, 18, 21, 28, 34, 329
  paleobathymetry and, 329
Carbon fallout, 57
Carinaria japonica, 107
Carolinia inflexa, 95
Carolinia uncinata, 99
Ceramium, 142
Ceratium furca, 120
Chalk, 14
Circum-Antarctic current, 328
Cladophora glomerata, 142
Clauso calanus farrani, 226
Clauso calanus furcatus, 226
Clauso calanus lividus, 95, 98
Clauso calanus mastigophorus, 226
Clay facies, pelagic, 14, 39
CLIMAP program, 316
Climate and surface sea water 49-63
Climatic change
  Milankovich theory, 326
  oxygen isotope ratios, 326
Coastal lagoons
  birds in, 250
  conservation of, 252
  definition of, 245, 246
  energy from, 249
  fish in, 250
  food from, 249, 250
  geographic importance of, 248
  multiple use conflicts in, 250, 252
  resources of, 245-253
  salt from, 249
  shorelines of, 248
  types of, 246, 247
Columbus, Christopher, 253
Coccoliths, 9, 14, 23
Coral Sea, 328
Crustal zone volatility, 61

INDEX

Currents, ocean,
  Agulhas, 99
  Atlantic, 86
  Aries observations, 209
  atlases, 292
  atmospheric influence on, 213-219
  baroclinic instability of, 219
  Benguela, 99
  California, 107
  casual, 209
  climate and, 294
  deep water, 293
  Equatorial, 67
  fish distribution and, 294
  fluctuations in, 293
  GEOSECS project and, 293
  kinetic energy of, 209
  mean circulation of, 292
  mesoscale features, 210
  MODE project and, 210, 291
  North equatorial, 67
  Pacific, 68, 69
  paleo, 34
  POLYGON expedition and, 210, 291
  POLYMODE project and, 210, 291
  sub-equatorial, 81
  synoptic-scale eddies in, 291
  time and space scales of, 214
  variability in, 209-211
  vertical displacement of, 215, 216
Crustacean herbivores, 116
Cyanea capillata, 169
Cyprina islandica, 158

DDT, 54
  in Baltic, 169, 170
Deep water circulation,
  South Atlantic 5, 27
Denticula, 121
Diatoms, antarctic, 14
  california current, 120
Deep Sea Drilling Project, 1, 15, 316, 326, 328, 329
  classification used, 13
  sites, South Atlantic, 4

Ecosystems
  continuous primary production 226-231
  interrupted production, 231-239
  variability in, 221-243
Emiliana huxleyi, 120
Engraulis mordax, 302
Equatorial zones, 85, 110, 113
Equatorial fracture zone barrier, 29
Eucalanus attenuatus, 226
Eucalanus bungii, 110
Eucalanus californicus, 107, 110
Eucalanus elongatus, 110
Eucalanus hyalinus, 110
Eucalanus inermis, 110
Eucalanus subtenuis, 226
Euchaeta longicornis, 228
Euchaeta marina, 226
Euphausia brevis, 95, 96, 116
Euphausia crystallorophias, 89
Euphausia diomedeae, 99, 102, 116
Euphausia distinguenda, 99, 106, 110
Euphausia frigida, 89
Euphausia gibba, 95
Euphausia hemigibba, 95, 97
Euphausia nana, 90
Euphausia pacifica, 89, 90, 116
Euphausia paragibba, 99, 101
Euphausia pseudogibba, 99
Euphausia superba, 95, 116, 225
Evaporites, 29

Falkland-Agulhas barrier, 29
Falkland-Agulhas fracture zone, 3, 5, 34
Falkland Channel, 39
Falkland plateau, 3, 11, 14, 18, 34, 314
Fish
  Baltic, 146
  migration, 185-207
  tagging experiments, 189-192
  tracking experiments, 193-196
Fish migration,
  climate and, 188

Fish migration (cont'd)
  distribution and, 187
  energy problems in, 184
  homing and, 188
  regularity of, 187
  selective tidal stream
      transport in, 197-203
  strategy of, 185-207
  straying in relation to, 187
  tactics in, 189
  water circulation and, 185-207
Fish movement
  migration strategy and, 185-207
  water circulation and, 185-207
Fly ash, 58
Foraminifera, 14
  assemblages, 320, 321
  planktonic species, 320
Forest and plant fires, 50
Forest fire particulates, 57
  latitudinal transport of, 57
  organics in, 58
Fragilariopsis, 120

Fucus, Baltic, 146
Fucus vesiculosus, 142
Fumaroles, 54
  minerals in, 54
Furcellaria fastigiata, 142

Gammarus oceanicus, 172
Geological stages
  Albian, 15
  Aptian, 26
  Aptian Albian, 23, 26, 29
  Cenozoic, 5
    early, 27
  Campanian, 23, 34
  Campanian Maastrichtian, 23
  Cenozoic, 5
  Coniacian, 27, 28
  Coniacian-Santonian, 15, 23
  Cretacean, 333
    lower, 28
  Cretacean-Cenozoic, 27
  Eocene, 18, 21, 24, 329, 333
    lower middle, 28
    middle, 25
  Eocene Oligocene, 27

Geological stages (cont'd)
  Maastrichtian, 15, 23, 27, 34, 36, 37, 38
  Maastrichtian Paleocene, 23
  Miocene 21, 27, 326, 329, 333, 334
    middle, 25
  Oligocene, 18, 128, 39, 326, 328
    lower, 15, 26, 39, 40
    upper, 7, 27, 39
  Paleocene, 23, 24, 34, 38
    upper, 23
  Paleocene Eocene, 27
  Pleistocene, 27
  Pliocene, 27
  Santonian, 27, 28, 34
  Santonian Campanian, 23, 34, 35
  Valanginian, 5, 28
Glomar Challenger, 1, 2
Gonyaulax catenata, 148
Guinea basin, 11
Gyres
  anticyclonic, 65, 67, 70, 79, 81, 85
  convergent, 70
  cyclonic, 65, 70, 75, 79, 81, 110
  habitats of, 81
  North Atlantic tropical, 66
  pelagic communities in, 221, 223, 239
  subantarctic, 85, 89, 160
  subarctic, 75, 85, 88, 89, 110, 113, 120
  subtropical, 85, 88, 95, 110, 113
  windspun, 209

Heavy metals
  Baltic, 171
  in plants, 50
  from cement production, 59
  from sea surface, 61
Holmes, Sherlock, 185, 203

Isopach maps, 5

INDEX

Janthina, 116
JOIDES, 17, 18, 314

Limacina helicina, 89, 91
Limacina inflata, 99
Limacina lesueri, 95
Limacina trochiformis, 107
Luciosudis normani, 302
Lithosphere
  continental, 17
  oceanic, 17
Lysocline, 7, 9, 11, 18, 21, 23, 26

Macoma baltica, 140, 160
Macoma calcarea, 160
Manganese nodules, 255, 258, 259, 264
  effect on global economy, 264
Marine resources
  non-renewable, 255-266
Marine geosciences, advances in, 307-347
Marls, 14, 18
  dolomitic, 29
Metals
  antimony, 58
  arsenic, 54, 58, 59
  barium, 58
  bismuth, 58
  boron, 59
  cadmium, 58
  copper, 49, 52, 54, 55
  lead, 49, 52, 58, 59
  manganese, 49
  mercury, 49, 50, 56, 58
    in Baltic, 171
  nickel, 49, 52
  selenium, 59
  silver, 58
  thallium, 58
  tin, 54
  zinc, 49, 52, 54, 56, 58, 59
Methylation processes, biological, 57
Micro-oceanography, 29
Microfossils, calcareous, 9
Mixing, interhemispheric, 52
Mytilus edulis, 140, 146

Nannocalanus minor, 226
Nematobrachion boopis, 116
Nematoscelis atalantica, 119
Nematoscelis difficilis, 107, 108
Nematoscelis gracilis, 99, 119
Nematoscelis megalops, 107, 108
Nematoscelis microps, 99, 105, 116
Neocalanus gracilis, 226
Nitzchia, 120
Nodularia spumigena, 131, 150, 165
North Atlantic
  core coverage in, 316
  Deep Current, 28
  Deep Water, 7
  reconstruction, 308
  tropical gyre of, 66

Ocean atmosphere comparisons, 214
Ocean circulation and marine life, 65-130
Ocean composition, 50
Ocean crust
  morphology, 1
  productivity, 79
  sedimentary cover, 1
Ocean evolution, methodology for reconstruction, 1, 14-27
Onocottus quadricornis, 140
Oxygen, 75
  Pacific Ocean, 78, 79, 80
Oxygen isotope ratios in sediments, 321, 326

Paleo-
  bathymetry, 1, 14, 15, 28, 314-316, 333
  carbonate compensation depth, 18
  current maps, 28
  facies distribution maps, 28
  isopach maps, 15
  sedimentology, 320-333
  sedimentary facies maps, 1
  water circulation maps, 1
Paracalanus parvus, 226

PCB in Baltic, 169, 170
Pelagic clay, 14
  South Atlantic, 329
Pelagic communities, 221
Phosphate
  nutrient, 70, 85
  Pacific Ocean, 74, 75, 76
  world ocean, 77
Photoreduction, 158
Physical oceanography
  time-scale of interest, 289
  trends in, 289-295
Phytoplankton distribution
  Antarctic, 231
  bipolar species, 120
  subarctic, 121
  warm water, 121
Planktonic foraminifera, 9
Plant exudates, 49, 56
  metals from, 57
Plant volatiles, 56
  α and β pinenes, 56
  myrcene, 56
  isoprene, 56
Plate tectonics, 1, 308-314
Pontellina, 99, 100
Pontellina morii, 99, 100
Pontellina platychela, 99
Pontellina sobrina, 99
Pontoporeia affinis, 160
Pontoporeia femorata, 160
Provinces, biological, 88, 113
Pseudocalanus minutus
          elongatus, 150
Pterosagitta draco, 226

Rennell, James, 209
Rio Grande Rise, 3, 7, 13, 17, 27, 34, 314, 316
Rio Grande-Walvis Bay barrier, 29, 34
Rhinocalanus gigas, 232, 234, 236
Romanche fracture zone, 3, 7, 28

Saduria (Mesidothea) entomon, 160
Sagitta elegans, 89
Sagitta enflata, 226
Sagitta ferox, 99

Sagitta regulatis, 226
Sagitta scrippsae, 107
Salinity, 70
  Pacific Ocean, 72, 73
Salinity power, 267-288
  dialytic battery, 267, 278-281
  pressure-retarded osmosis, 267, 270-278
  reverse electrodialysis, 267, 278-281
  vapour pressure difference engines, 281-283
Salinity tolerance, Baltic species, 140, 142
Salpa fusiformis, 119
Salpa thompsoni, 89
Sao Paulo plateau, 27, 29, 34
Sapropels, 17, 29, 34
Seabed minerals, 255-266
  concentration processes, 256-259
  law and, 265, 266
  manganese nodules, 255, 258, 259, 264
  metalliferous sediments, 255, 255, 258, 259
  phosphorites, 255, 262
  placer deposits, 255, 257, 261, 262
  politics and, 265, 266
  sand and gravel, 255, 260
  vulcanism and, 258
Sea spray, 52
Sea water
  heavy metal content, 61
  major constituents, 50
  surface, composition of, 49
Sedimentation
  ocean patterns of, 329
  recent, 316-326
  time scales of, 316
Sedimentary facies
  calcareous ooze, 12, 13, 14
  chalk, 12, 13, 14
  clay, 12
  coccolith bearing mud, 12, 13, 14
  diatomaceous bearing mud, 12, 13, 14

INDEX

Sedimentary facies (cont'd)
  dolomitic limestone, 12
  evaporites, 12
  foraminifer bearing mud marl, 12, 13, 14
  mud, 12, 13, 14
  pelagic clay, 12, 13, 14
  radiolarian bearing mud, 12, 13, 14
  sapropel, 17, 29, 34
  shale and sandstone, 12
  zeolitic mud, 12
Sediment thickness, South Atlantic, 5
Sediments, paleodistribution, 18
Sediments, terrigeneous, 11
Siliceous mud, 14
Skeletonema costatum, 148
South Atlantic
  bathymetry, 7
  boundaries of, 3
  evolution of, 1-48, 314
  fertility of surface waters of, 11
  reconstruction, 1-48, 314
  rivers
    Amazon, 11
    Berg, 11
    Congo, 11
    Kunene, 11
    Niger, 11
    Olifants, 11
    Orange, 11
    Parana, 11
  surface currents, 7
  surface sediments, 7, 11
    calcium carbonate content, 10
Speciation, 122
  asexual reproduction, 122
  sexual reproduction, 122
Species
  neritic, 122
  oceanic, 122
  transitional, 107
Stictyosiphon, 142
Stylocheiron carinatum, 99, 103, 119
Stylocheiron elongatum, 116, 117
Stylocheiron maximum, 99, 104, 116, 117

Stylocheiron suhmii, 95
Stylocheiron subula, 95
Syracosphaera, 120

Tasman Sea, 99, 328
Tectonics - recent advances, 308-314
Temora longicornis, 150
Temperature
  Pacific ocean, 71
  World ocean, 87
Tethys, 34
Thalassionema nitzschiodes, 120
Thalassiosira, 120
Thalassiosira baltica, 148
Theodoxus fluviatilis, 142
Thysanoessa gregaria, 107, 109
Thysanoessa inermis, 89, 94
Thysanoessa longpipes, 89
Trace metals
  atmospheric, 51
  enrichment factors of, 51, 52
  concentrations, 53
Transparency, 81, 83
  Secchi disc readings, 85, 116

Undinula darwini, 226
Upper layer communities, 223
Upwelling, coastal, 216

Vaporisation, biological, 52
Vema Channel, 7, 13, 39
Vertical migration, 196
  patterns of, 226-227
  seasonal, 110
Volatilization, low temperature, 53
Volcanoes, 54
  Hawaiian, 54

Walvis Ridge, 3, 7, 11, 27, 29, 34
Walvis Rise, 17, 314, 316
Water color, 85, 116
Waves, 294
  IWEX, 295
  JONSWAP, 295

Zeolites, 13

Zooplankton
  size, 116
  upper level, open ocean, 67-130
  volume, 81, 82, 84
Zosteria marina, 147

COLUMBIA UNIVERSITY LIBRARIES

This book is due on the date indicated below, or at the expiration of a definite period after the date of borrowing, as provided by the library rules or by special arrangement with the Librarian in charge.

DATE BORROWED	DATE DUE	DATE BORROWED	DATE DUE
	NEW BOOK		
DOES NOT	CIRCULATE		
UNTIL	2-13-79		
	3-6-79		
	4-3-79		
	6-5-79		
	1-4-83		
	6-21-83		

C28 (871) 50M